APOLLONIUS O

TREATISE ON CONIC SECTIONS

BY

T. L. HEATH, M.A.

SOMETIME FELLOW OF TRINITY COLLEGE, CAMBRIDGE.

ζηλοῦντες τοὺς Πυθαγορείους, οἷς πρόχειρον ἦν καὶ τοῦτο σύμβολον σχᾶμα καὶ βᾶμα, ἀλλ' οὐ σχᾶμα καὶ τριώβολον. PROCLUS.

1896

British Library Cataloguing-in-Publication Data
A catalogue record for this book is available from the
British Library

Thomas Little Heath

Thomas Little Heath was born on 5 October 1861, in Barnetby-le-Wold, Lincolnshire, England. He was the son of a farmer, Samuel Heath, and was educated at the local Caistor Grammar School and later Clifton College, Bristol. Heath rapidly displayed a keen intellectual prowess and took his university education at *Trinity College, Cambridge,* where he was awarded a Doctorate of Science and became a fellow of the college. In 1884, at the age of twenty-three, Heath passed his Civil Service examination and became an Assistant Secretary to the Treasury, becoming Permanent Secretary in 1913. He spent six years in this position, but subsequently moved to the *National Debt Office*. Here, Heath utilised his considerable financial and mathematical skills, and retained a post there until his retirement in 1926. Although hard to fathom, during this period Heath found the spare time to author several books on Greek mathematicians. His tracts included *Archimedes: Works* (1897) *The Thirteen Books of Euclid's Elements* (1908) *Aristarchus of Samos; The Ancient Copernicus* (1913) and *A History of Greek Mathematics* (1921). It is primarily through Heath's translations that modern English-speaking readers are aware of the astounding range of Archimedes' achievements. Heath has been honoured for his work in the Civil Service by being appointed C.B. in 1903, K.C.B. in 1909, and K.C.V.O. in 1916. He was elected a Fellow of the Royal Society in May 1912. After his retirement from the civil service, Heath continued to write on the Greeks, publishing a study of *Greek Astronomy* in 1932 and a well-respected treatise on *Mathematics in Aristotle* (posthumously published in 1949). He died on 16 March, 1940, at the age of seventy-nine.

PREFACE.

IT is not too much to say that, to the great majority of mathematicians at the present time, Apollonius is nothing more than a name and his *Conics*, for all practical purposes, a book unknown. Yet this book, written some twenty-one centuries ago, contains, in the words of Chasles, "the most interesting properties of the conics," to say nothing of such brilliant investigations as those in which, by purely geometrical means, the author arrives at what amounts to the complete determination of the evolute of any conic. The general neglect of the "great geometer," as he was called by his contemporaries on account of this very work, is all the more remarkable from the contrast which it affords to the fate of his predecessor Euclid; for, whereas in this country at least the *Elements* of Euclid are still, both as regards their contents and their order, the accepted basis of elementary geometry, the influence of Apollonius upon modern text-books on conic sections is, so far as form and method are concerned, practically *nil*.

Nor is it hard to find probable reasons for the prevailing absence of knowledge on the subject. In the first place, it could hardly be considered surprising if the average mathematician were apt to show a certain faintheartedness when confronted with seven Books in Greek or Latin which contain 387 propositions in all; and doubtless the apparently portentous bulk of the treatise has deterred many from attempting to make its acquaintance. Again, the form of the propositions is an additional difficulty, because the reader finds in them none of the ordinary aids towards the comprehension of somewhat complicated geometrical work, such as the conventional appropriation, in modern text-books, of definite letters to denote particular points on the various conic sections. On the contrary, the enunciations of propositions which, by the aid of a notation once agreed upon, can now be stated in a few lines, were by Apollonius invariably given in words like the enunciations of Euclid. These latter are often sufficiently unwieldy; but the inconvenience is greatly intensified in Apollonius, where the greater complexity of the conceptions entering into the investigation of conics, as compared with the more elementary notions relating to the line and circle, necessitates in many instances an enunciation extending over a space equal to (say) half a page of this book. Hence it is often a matter of considerable labour even to grasp the enunciation of a proposition. Further, the propositions are, with the exception that separate paragraphs mark the formal divisions, printed continuously; there are no breaks for the purpose of enabling the eye to take in readily the successive steps in the demonstration and so facilitating the comprehension of the argument as a whole. There is no uniformity of notation, but in almost every fresh proposition a different letter is employed to denote the same point: what wonder then if there are the most serious obstacles in the way of even remembering the results of certain propositions? Nevertheless these propositions, though unfamiliar to mathematicians of the present day, are of the very essence of Apollonius' system, are being constantly used, and must therefore necessarily be borne in mind.

The foregoing remarks refer to the editions where Apollonius can be read in the Greek or in a Latin translation, i.e. to those of Halley and Heiberg; but the only attempt which has been made to give a complete view of the substance of Apollonius in a form more accessible to the modern reader is open to much the same objections. This reproduction of the *Conics* in German by H. Balsam (Berlin, 1861) is a work deserving great praise both for its accuracy and the usefulness of the occasional explanatory notes, but perhaps most of all for an admirable set of figures to the number of 400 at the end of the book; the enunciations of the propositions are, however, still in words, there are few breaks in the continuity of the printing, and the notation is not sufficiently modernised to make the book of any more real service to the ordinary reader than the original editions.

An edition is therefore still wanted which shall, while in some places adhering even more closely than Balsam to the original text, at the same time be so entirely remodelled by the aid of accepted modern notation as to be thoroughly readable by any competent mathematician; and this want it is the object of the present work to supply.

In setting myself this task, I made up my mind that any satisfactory reproduction of the *Conics* must fulfil certain essential conditions: (1) it should be Apollonius and nothing but Apollonius, and nothing should be altered either in the substance or in the order of his thought, (2) it should be complete, leaving out nothing of any significance or importance, (3) it should exhibit under different headings the successive divisions of the subject, so that the definite scheme followed by the author may be seen as a whole.

Accordingly I considered it to be the first essential that I should make myself thoroughly familiar with the whole work at first hand. With this object I first wrote out a perfectly literal translation of the whole of the extant seven Books. This was a laborious task, but it was not in other respects difficult, owing to the excellence of the standard editions. Of these editions, Halley's is a monumental work, beyond praise alike in respect of its design and execution; and for Books V—VII it is still the only complete edition. For Books I—IV I used for the most part the new Greek text of Heiberg, a scholar who has earned the undying gratitude of all who are interested in the history of Greek mathematics by successively bringing out a critical text (with Latin translation) of Archimedes, of Euclid's *Elements*, and of all the writings of Apollonius still extant in Greek. The only drawback to Heiberg's Apollonius is the figures, which are poor and not seldom even misleading, so that I found it a great advantage to have Halley's edition, with its admirably executed diagrams, before me even while engaged on Books I—IV.

The real difficulty began with the constructive work of re-writing the book, involving as it did the substitution of a new and uniform notation, the condensation of some propositions, the combination of two or more into one, some slight re-arrangements of order for the purpose of bringing together kindred propositions in cases where their separation was rather a matter of accident than indicative of design, and so on. The result has been (without leaving out anything essential or important) to diminish the bulk of the work by considerably more than one-half and to reduce to a corresponding extent the number of separate propositions.

When the re-editing of the *Conics* was finished, it seemed necessary for completeness to prefix an Introduction for the purposes (1) of showing the relation of Apollonius to his predecessors in the same field both as regards matter and method, (2) of explaining more fully than was possible in the few notes inserted in square brackets in the body of the book the mathematical significance of certain portions of the *Conics* and the probable connexion between this and other smaller treatises of Apollonius about which we have information, (3) of describing and illustrating fully the form and language of the propositions as they stand in the original Greek text. The first of these purposes required that I should give a sketch of the history of conic sections up to the time of Apollonius; and I have accordingly considered it worth while to make this part of the

Introduction as far as possible complete. Thus e.g. in the case of Archimedes I have collected practically all the propositions in conics to be found in his numerous works with the substance of the proofs where given; and I hope that the historical sketch as a whole will be found not only more exhaustive, for the period covered, than any that has yet appeared in English, but also not less interesting than the rest of the book.

For the purposes of the earlier history of conics, and the chapters on the mathematical significance of certain portions of the *Conics* and of the other smaller treatises of Apollonius, I have been constantly indebted to an admirable work by H. G. Zeuthen, *Die Lehre von den Kegelschnitten im Altertum* (German edition, Copenhagen, 1886), which to a large extent covers the same ground, though a great portion of his work, consisting of a mathematical analysis rather than a reproduction of Apollonius, is of course here replaced by the re-edited treatise itself. I have also made constant use of Heiberg's *Litterargeschichtliche Studien über Euklid* (Leipzig, 1882), the original Greek of Euclid's *Elements*, the works of Archimedes, the *συναγωγή* of Pappus and the important *Commentary on Eucl. Book* I. by Proclus (ed. Friedlein, Leipzig, 1873).

The frontispiece to this volume is a reproduction of a quaint picture and attached legend which appeared at the beginning of Halley's edition. The story is also told elsewhere than in Vitruvius, but with less point (cf. Claudii Galeni Pergameni Προτρεπτικὸς ἐπὶ τέχνας c. v. § 8, p. 108, 3–8 ed. I. Marquardt, Leipzig, 1884). The quotation on the title page is from a vigorous and inspiring passage in Proclus' *Commentary on Eucl. Book* I. (p. 84, ed. Friedlein) in which he is describing the scientific purpose of his work and contrasting it with the useless investigations of paltry lemmas, distinctions of cases, and the like, which formed the stock-in-trade of the ordinary Greek commentator. One merit claimed by Proclus for his work I think I may fairly claim for my own, that it at least contains ὅσα πραγματειωδεστέραν ἔχει θεωρίαν; and I should indeed be proud if, in the judgment of competent critics, it should be found possible to apply to it the succeeding phrase, *συντελεῖ πρὸς τὴν ὅλην φιλοσοφίαν*.

Lastly, I wish to express my thanks to my brother, Dr R. S. Heath, Principal of Mason College, Birmingham, for his kindness in reading over most of the proof sheets and for the constant interest which he has taken in the progress of the work.

T. L. HEATH.

March, 1896.

LIST OF PRINCIPAL AUTHORITIES.

EDMUND HALLEY, *Apollonii Pergaei Conicorum libri octo et Sereni Antissensis de sectione cylindri et coni libri duo.* (Oxford, 1710.)

EDMUND HALLEY, *Apollonii Pergaei de Sectione Rationis libri duo, ex Arabico versi.* (Oxford, 1706.)

J. L. HEIBERG, *Apollonii Pergaei quae Graece exstant cum commentariis antiquis.* (Leipzig, 1891–3.)

H. BALSAM, *Des Apollonius von Perga sieben Bücher über Kegelschnitte nebst dem durch* Halley *wieder hergestellten achten Buche deutsch bearbeitet.* (Berlin, 1861.)

J. L. HEIBERG, *Litterargeschichtliche Studien über Euklid.* (Leipzig, 1882.)

J. L. HEIBERG, *Euclidis elementa.* (Leipzig, 1883–8.)

G. FRIEDLEIN, *Procli Diadochi in primum Euclidis elementorum librum commentarii.* (Leipzig, 1873.)

J. L. HEIBERG, *Quaestiones Archimedeae.* (Copenhagen, 1879.)

J. L. HEIBERG, *Archimedis opera omnia cum commentariis Eutocii.* (Leipzig, 1880–1.)

F. HULTSCH, *Pappi Alexandrini collectionis quae supersunt.* (Berlin, 1876–8.)

C. A. BRETSCHNEIDER, *Die Geometrie und die Geometer vor Euklides.* (Leipzig, 1870.)

M. CANTOR, *Vorlesungen über Geschichte der Mathematik.* (Leipzig, 1880.)

H. G. ZEUTHEN, *Die Lehre von den Kegelschnitten im Altertum. Deutsche Ausgabe.* (Copenhagen, 1886.)

CONTENTS.

INTRODUCTION.

PART I. THE EARLIER HISTORY OF CONIC SECTIONS AMONG THE GREEKS.

		PAGE
CHAPTER I.	THE DISCOVERY OF CONIC SECTIONS: MENAECHMUS	xvii
CHAPTER II.	ARISTAEUS AND EUCLID	xxxi
CHAPTER III.	ARCHIMEDES	xli

PART II. INTRODUCTION TO THE *CONICS* OF APOLLONIUS.

CHAPTER I.	THE AUTHOR AND HIS OWN ACCOUNT OF THE *CONICS*	lxviii
CHAPTER II.	GENERAL CHARACTERISTICS	lxxxvii
§ 1.	Adherence to Euclidean form, conceptions and language	lxxxvii
§ 2.	Planimetric character of the treatise	xcvii
§ 3.	Definite order and aim	xcviii
CHAPTER III.	THE METHODS OF APOLLONIUS	ci
§ 1.	Geometrical algebra	ci
	(1) The theory of proportions	ci
	(2) The application of areas	cii
	(3) Graphic representation of areas by means of auxiliary lines	cxi
	(4) Special use of auxiliary points in Book VII.	cxiii
§ 2.	The use of coordinates	cxv
§ 3.	Transformation of coordinates	cxviii
§ 4.	Method of finding two mean proportionals	cxxv
§ 5.	Method of constructing normals passing through a given point	cxxvii
CHAPTER IV.	THE CONSTRUCTION OF A CONIC BY MEANS OF TANGENTS	cxxx

PAGE

CHAPTER V. THE THREE-LINE AND FOUR-LINE LOCUS . . cxxxviii

CHAPTER VI. THE CONSTRUCTION OF A CONIC THROUGH FIVE POINTS cli

APPENDIX. NOTES ON THE TERMINOLOGY OF GREEK GEOMETRY clvii

THE *CONICS* OF APOLLONIUS.

THE CONE 1

THE DIAMETER AND ITS CONJUGATE 15

TANGENTS 22

PROPOSITIONS LEADING TO THE REFERENCE OF A CONIC TO ANY NEW DIAMETER AND THE TANGENT AT ITS EXTREMITY 31

CONSTRUCTION OF CONICS FROM CERTAIN DATA . . 42

ASYMPTOTES 53

TANGENTS, CONJUGATE DIAMETERS AND AXES. . . 64

EXTENSIONS OF PROPOSITIONS 17—19 84

RECTANGLES UNDER SEGMENTS OF INTERSECTING CHORDS 95

HARMONIC PROPERTIES OF POLES AND POLARS . . 102

INTERCEPTS MADE ON TWO TANGENTS BY A THIRD . 109

FOCAL PROPERTIES OF CENTRAL CONICS 113

THE LOCUS WITH RESPECT TO THREE LINES ETC. . 119

INTERSECTING CONICS. 126

NORMALS AS MAXIMA AND MINIMA 139

PROPOSITIONS LEADING IMMEDIATELY TO THE DETERMINATION OF THE *EVOLUTE* 168

CONSTRUCTION OF NORMALS 180

OTHER PROPOSITIONS RESPECTING MAXIMA AND MINIMA 187

EQUAL AND SIMILAR CONICS 197

PROBLEMS 209

VALUES OF CERTAIN FUNCTIONS OF THE LENGTHS OF CONJUGATE DIAMETERS 221

INTRODUCTION.

PART I.

THE EARLIER HISTORY OF CONIC SECTIONS AMONG THE GREEKS.

CHAPTER I.

THE DISCOVERY OF CONIC SECTIONS: MENAECHMUS.

There is perhaps no question that occupies, comparatively, a larger space in the history of Greek geometry than the problem of the Doubling of the Cube. The tradition concerning its origin is given in a letter from Eratosthenes of Cyrene to King Ptolemy Euergetes quoted by Eutocius in his commentary on the second Book of Archimedes' treatise *On the Sphere and Cylinder**; and the following is a translation of the letter as far as the point where we find mention of **Menaechmus**, with whom the present subject begins.

"Eratosthenes to King Ptolemy greeting.

"There is a story that one of the old tragedians represented Minos as wishing to erect a tomb for Glaucus and as saying, when he heard that it was a hundred feet every way,

> Too small thy plan to bound a royal tomb.
> Let it be double; yet of its fair form
> Fail not, but haste to double every side †.

* In quotations from Archimedes or the commentaries of Eutocius on his works the references are throughout to Heiberg's edition (*Archimedis opera omnia cum commentariis Eutocii.* 3 vols. Leipzig, 1880-1). The reference here is III. p. 102.

† μικρόν γ' ἔλεξας βασιλικοῦ σηκὸν τάφου·
διπλάσιος ἔστω· τοῦ καλοῦ δὲ μὴ σφαλεὶς
δίπλαζ' ἕκαστον κῶλον ἐν τάχει τάφου.

Valckenaer (*Diatribe de fragm. Eurip.*) suggests that the verses are from the

But he was clearly in error; for, when the sides are doubled, the area becomes four times as great, and the solid content eight times as great. Geometers also continued to investigate the question in what manner one might double a given solid while it remained in the same form. And a problem of this kind was called the doubling of the cube; for they started from a cube and sought to double it. While then for a long time everyone was at a loss, Hippocrates of Chios was the first to observe that, if between two straight lines of which the greater is double of the less it were discovered how to find two mean proportionals in continued proportion, the cube would be doubled; and thus he turned the difficulty in the original problem* into another difficulty no less than the former. Afterwards, they say, some Delians attempting, in accordance with an oracle, to double one of the altars fell into the same difficulty. And they sent and begged the geometers who were with Plato in the Academy to find for them the required solution. And while they set themselves energetically to work and sought to find two means between two given straight lines, Archytas of Tarentum is said to have discovered them by means of half-cylinders, and Eudoxus by means of the so-called curved lines. It is, however, characteristic of them all that they indeed gave demonstrations, but were unable to make the actual construction or to reach the point of practical application, except to a small extent Menaechmus and that with difficulty."

Some verses at the end of the letter, in commending Eratosthenes' own solution, suggest that there need be no resort to Archytas' unwieldy contrivances of cylinders or to "cutting the cone in the triads of Menaechmus†." This last phrase of Eratosthenes appears

Polyidus of Euripides, but that the words after *σφαλεὶς* (or *σφαλῇς*) are Eratosthenes' own, and that the verses from the tragedy are simply

μικρόν γ' ἔλεξας βασιλικοῦ σηκὸν τάφου·
διπλάσιος ἔστω· τοῦ κύβου δὲ μὴ σφαλῇς.

It would, however, be strange if Eratosthenes had added words merely for the purpose of correcting them again: and Nauck (*Tragicorum Graecorum Fragmenta*, Leipzig, 1889, p. 874) gives the three verses as above, but holds that they do not belong to the *Polyidus*, adding that they are no doubt from an earlier poet than Euripides, perhaps Aeschylus.

* *τὸ ἀπόρημα αὐτοῦ* is translated by Heiberg "haesitatio eius," which no doubt means "*his* difficulty." I think it is better to regard *αὐτοῦ* as neuter, and as referring to the problem of doubling the cube.

† *μηδὲ Μενεχμείους κωνοτομεῖν τριάδας.*

again, by way of confirmatory evidence, in a passage of Proclus*, where, quoting Geminus, he says that the conic sections were discovered by Menaechmus.

Thus the evidence so far shows (1) that Menaechmus (a pupil of Eudoxus and a contemporary of Plato) was the discoverer of the conic sections, and (2) that he used them as a means of solving the problem of the doubling of the cube. We learn further from Eutocius† that Menaechmus gave two solutions of the problem of the two mean proportionals, to which Hippocrates had reduced the original problem, obtaining the two means first by the intersection of a certain parabola and a certain rectangular hyperbola, and secondly by the intersection of two parabolas‡. Assuming that a, b are the two given unequal straight lines and x, y the two required mean proportionals, the discovery of Hippocrates amounted to the discovery of the fact that from the relation

$$\frac{a}{x} = \frac{x}{y} = \frac{y}{b} \quad \text{.............................(1)}$$

it follows that $$\left(\frac{a}{x}\right)^3 = \frac{a}{b},$$

and, if $a = 2b$, $$a^3 = 2x^3.$$

The equations (1) are equivalent to the three equations

$$x^2 = ay, \quad y^2 = bx, \quad xy = ab \quad \text{..................(2)},$$

and the solutions of Menaechmus described by Eutocius amount to the determination of a point as the intersection of the curves represented in a rectangular system of Cartesian coordinates by any two of the equations (2).

Let AO, BO be straight lines placed so as to form a right angle at O, and of length a, b respectively§. Produce BO to x and AO to y.

* *Comm. on Eucl.* I., p. 111 (ed. Friedlein). The passage is quoted, with the context, in the work of Bretschneider, *Die Geometrie und die Geometer vor Euklides*, p. 177.

† *Commentary on Archimedes* (ed. Heiberg, III. p. 92—98).

‡ It must be borne in mind that the words *parabola* and *hyperbola* could not have been used by Menaechmus, as will be seen later on; but the phraseology is that of Eutocius himself.

§ One figure has been substituted for the two given by Eutocius, so as to make it serve for both solutions. The figure is identical with that attached to the *second* solution, with the sole addition of the portion of the rectangular hyperbola used in the first solution.

It is a curious circumstance that in Eutocius' second figure the straight line

The *first* solution now consists in drawing a parabola, with vertex O and axis Ox, such that its parameter is equal to BO or b, and a hyperbola with Ox, Oy as asymptotes such that the rectangle under the distances of any point on the curve from Ox, Oy respectively is equal to the rectangle under AO, BO, i.e. to ab. If P be

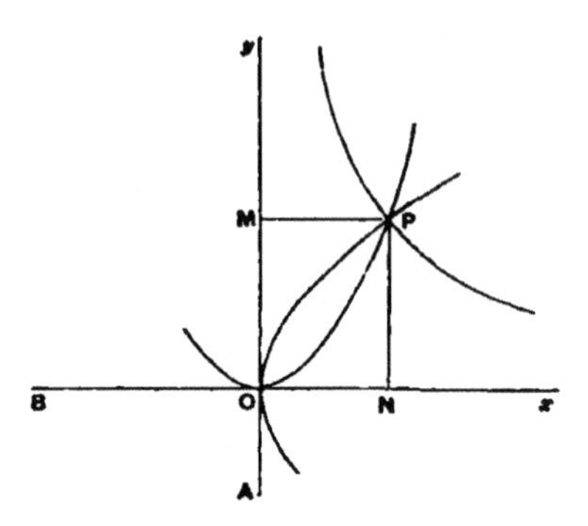

the point of intersection of the parabola and hyperbola, and PN, PM be drawn perpendicular to Ox, Oy, i.e. if PN, PM be denoted by y, x, the coordinates of the point P, we shall have

$$\left.\begin{array}{l} y^2 = b\,.\,ON = b\,.\,PM = bx \\ xy = PM\,.\,PN = ab \end{array}\right\},$$

and

whence

$$\frac{a}{x} = \frac{x}{y} = \frac{y}{b}.$$

In the *second* solution of Menaechmus we are to draw the parabola described in the first solution and also the parabola whose

representing the length of the parameter of each parabola is drawn in the same straight line with the axis of the parabola, whereas Apollonius always draws the parameter as a line starting from the vertex (or the end of a diameter) and perpendicular to the axis (or diameter). It is possible that we may have here an additional indication that the idea of the parameter as ὀρθία or the *latus rectum* originated with Apollonius; though it is also possible that the selection of the directions of AO, BO was due to nothing more than accident, or may have been made in order that the successive terms in the continued proportion might appear in the figure in cyclic order, which corresponds moreover to their relative positions in the mechanical solution attributed to Plato. For this solution see the same passage of Eutocius (*Archimedes*, ed. Heiberg, III. p. 66—70).

vertex is O, axis Oy and parameter equal to a. The point P where the two parabolas intersect is given by

$$\left.\begin{array}{l} y^2 = bx \\ x^2 = ay \end{array}\right\},$$

whence, as before,
$$\frac{a}{x} = \frac{x}{y} = \frac{y}{b}.$$

We have therefore, in these two solutions, the parabola and the rectangular hyperbola in the aspect of *loci* any points of which respectively fulfil the conditions expressed by the equations in (2); and it is more than probable that the discovery of Menaechmus was due to efforts to determine *loci* possessing these characteristic properties rather than to any idea of a systematic investigation of the sections of a cone as such. This supposition is confirmed by the very special way in which, as will be seen presently, the conic sections were originally produced from the right circular cone; indeed the special method is difficult to explain on any other assumption. It is moreover natural to suppose that, after the discovery of the convertibility of the cube-problem into that of finding two mean proportionals, the two forms of the resulting equations would be made the subject of the most minute and searching investigation. The form (1) expressing the equality of three ratios led naturally to the solution attributed to Plato, in which the four lines representing the successive terms of the continued proportion are placed mutually at right angles and in cyclic order round a fixed point, and the extremities of the lines are found by means of a rectangular frame, three sides of which are fixed, while the fourth side can move freely parallel to itself. The investigation of the form (2) of the equations led to the attempt of Menaechmus to determine the loci corresponding thereto. It was known that the locus represented by $y^2 = x_1 x_2$, where y is the perpendicular from any point on a fixed straight line of given length, and x_1, x_2 are the segments into which the line is divided by the perpendicular, was a circle; and it would be natural to assume that the equation $y^2 = bx$, differing from the other only in the fact that a constant is substituted for one of the variable magnitudes, would be capable of representation as a locus or a continuous curve. The only difficulty would be to discover its form, and it was here that the cone was introduced.

If an explanation is needed of the circumstance that Menaechmus should have had recourse to any solid figure, and to a cone in particular, for the purpose of producing a plane locus, we find it in the fact that solid geometry had already reached a high state of development, as is shown by the solution of the problem of the two mean proportionals by Archytas of Tarentum (born about 430 B.C.). This solution, in itself perhaps more remarkable than any other, determines a certain point as the intersection of three surfaces of revolution, (1) a right cone, (2) a right cylinder whose base is a circle on the axis of the cone as diameter and passing through the apex of the cone, (3) the surface formed by causing a semicircle, whose diameter is the same as that of the circular base of the cylinder and whose plane is perpendicular to that of the circle, to revolve about the apex of the cone as a fixed point so that the diameter of the semicircle moves always in the plane of the circle, in other words, the surface consisting of half a *split ring* whose centre is the apex of the cone and whose inner diameter is indefinitely small. We find that in the course of the solution (*a*) the intersection of the surfaces (2) and (3) is said to be a certain curve (γραμμήν τινα), being in fact a curve of double curvature, (*b*) a circular section of the right cone is used in the proof, and (*c*), as the penultimate step, two mean proportionals are found in one and the same plane (triangular) section of the cone*.

* The solution of Archytas is, like the others, given by Eutocius (p. 98—102) and is so instructive that I cannot forbear to quote it. Suppose that *AC*, *AB* are the straight lines between which two mean proportionals are to be found. *AC* is then made the diameter of a circle, and *AB* is placed as a chord in the circle.

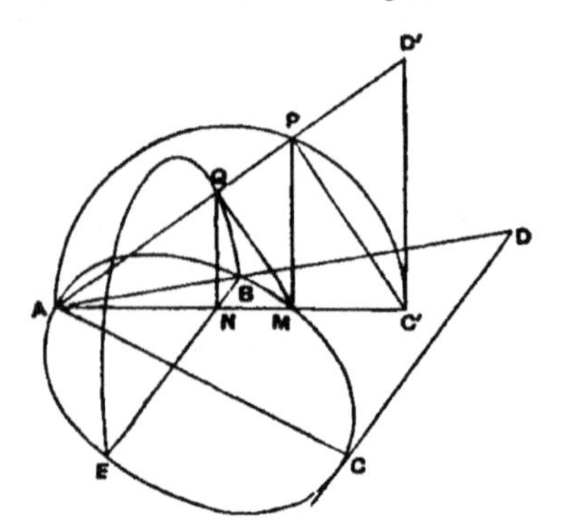

A semicircle is drawn with diameter *AC* but in a plane perpendicular to that of *ABC*, and revolves about an axis through *A* perpendicular to the plane of *ABC*.

Thus the introduction of cones by Menaechmus should not in itself be a matter for surprise.

Concerning Menaechmus' actual method of deducing the properties of the conic sections from the cone we have no definite information; but we may form some idea of his probable procedure

A half-cylinder (right) is now erected with ABC as base: this will cut the surface described by the moving semicircle APC in a certain curve.

Lastly let CD, the tangent to the circle ABC at the point C, meet AB produced in D; and suppose the triangle ACD to revolve about AC as axis. This will generate the surface of a right circular cone, and the point B will describe a semicircle BQE perpendicular to the plane of ABC and having its diameter BE at right angles to AC. The surface of the cone will meet in some point P the curve described on the cylinder. Let APC' be the corresponding position of the revolving semicircle, and let AC' meet the circle ABC in M.

Drawing PM perpendicular to the plane of ABC, we see that it must meet the circumference of the circle ABC because P is on the cylinder which stands on ABC as base. Let AP meet the circumference of the semicircle BQE in Q, and let AC' meet its diameter BE in N. Join PC', QM, QN.

Then, since both semicircles are perpendicular to the plane ABC, so is their line of intersection QN. Therefore QN is perpendicular to BE.

Hence $$QN^2 = BN \cdot NE = AN \cdot NM.$$

Therefore the angle AQM is a right angle.

But the angle $C'PA$ is also right: therefore MQ is parallel to $C'P$.

It follows, by similar triangles, that

$$C'A : AP = AP : AM = AM : AQ.$$

That is, $$AC : AP = AP : AM = AM : AB,$$

and AB, AM, AP, AC are in continued proportion.

In the language of analytical geometry, if AC is the axis of x, a line through A perpendicular to AC in the plane of ABC the axis of y, and a line through A parallel to PM the axis of z, then P is determined as the intersection of the surfaces

$$x^2 + y^2 + z^2 = \frac{a^2}{b^2} x^2 \quad \text{.................................... (1)},$$

$$x^2 + y^2 = ax \quad \text{.................................... (2)},$$

$$x^2 + y^2 + z^2 = a\sqrt{x^2 + y^2} \quad \text{.................................... (3)},$$

where $$AC = a, \; AB = b.$$

From the first two equations

$$x^2 + y^2 + z^2 = \frac{(x^2 + y^2)^2}{b^2},$$

and from this equation and (3) we have

$$\frac{a}{\sqrt{x^2 + y^2 + z^2}} = \frac{\sqrt{x^2 + y^2 + z^2}}{\sqrt{x^2 + y^2}} = \frac{\sqrt{x^2 + y^2}}{b},$$

or $$AC : AP = AP : AM = AM : AB.$$

if we bear in mind (1) what we are told of the manner in which the earlier writers on conics produced the three curves from particular kinds of right circular cones, and (2) the course followed by Apollonius (and Archimedes) in dealing with sections of *any* circular cone, whether right or oblique.

Eutocius, in his commentary on the *Conics* of Apollonius, quotes with approval a statement of Geminus to the effect that the ancients defined a cone as the surface described by the revolution of a right-angled triangle about one of the sides containing the right angle, and that they knew no other cones than right cones. Of these they distinguished three kinds according as the vertical angle of the cone was less than, equal to, or greater than, a right angle. Further they produced only one of the three sections from each kind of cone, always cutting it by a plane perpendicular to one of the generating lines, and calling the respective curves by names corresponding to the particular kind of cone; thus the "section of a right-angled cone" was their name for a parabola, the "section of an acute-angled cone" for an ellipse, and the "section of an obtuse-angled cone" for a hyperbola. The sections are so described by Archimedes.

Now clearly the parabola is the one of the three sections for the production of which the use of a right-angled cone and a section at right angles to a generator gave the readiest means. If N be a point on the diameter BC of any circular section in such a cone, and if NP be a straight line drawn in the plane of the section and perpendicular to BC, meeting the circumference of the circle (and therefore the surface of the cone) in P,

$$PN^2 = BN . NC.$$

Draw AN in the plane of the axial triangle OBC meeting the generator OB at right angles in A, and draw AD parallel to BC meeting OC in D; let DEF, perpendicular to AD or BC, meet BC in E and AN produced in F.

Then AD is bisected by the axis of the cone, and therefore AF is likewise bisected by it. Draw CG perpendicular to BC meeting AF produced in G.

Now the angles BAN, BCG are right; therefore B, A, C, G are concyclic, and

$$BN . NC = AN . NG.$$

But $$AN = CD = FG;$$

therefore, if AF meets the axis of the cone in L,

$$NG = AF = 2AL.$$

Hence
$$PN^2 = BN . NC$$
$$= 2AL . AN,$$

and, if A is fixed, $2AL$ is constant.

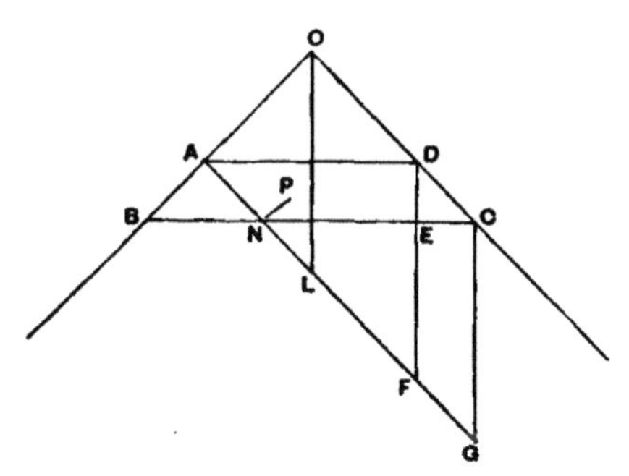

Thus P satisfies the equation

$$y^2 = 2AL . x,$$

where $y = PN$, $x = AN$.

Therefore we have only to select A as a point on OB such that AL (or AO) $= \frac{b}{2}$, and the curve corresponding to the equation $y^2 = bx$ is found.

The 'parameter' of the parabola is equal to twice the distance between A and the point where AN meets the axis of the cone, or ἁ διπλασία τᾶς μέχρι τοῦ ἄξονος, as Archimedes calls it*.

The discovery that the hyperbola represented by the equation $xy = ab$, where the asymptotes are the coordinate axes, could be obtained by cutting an obtuse-angled cone by a plane perpendicular to a generator was not so easy, and it has been questioned whether Menaechmus was aware of the fact. The property, $xy =$ (const.), for a hyperbola referred to its asymptotes does not appear in Apollonius until the second Book, after the diameter-properties have been proved. It depends on the propositions (1) that every series of parallel chords is bisected by one and the same diameter, and (2) that the parts of any chord intercepted between the curve and the asymptotes are equal. But it is not necessary to assume that

* Cf. *On Conoids and Spheroids*, 3, p. 304.

Menaechmus was aware of these general propositions. It is more probable that he obtained the equation referred to the asymptotes from the equation referred to the *axes*; and in the particular case which he uses (that of the rectangular hyperbola) this is not difficult.

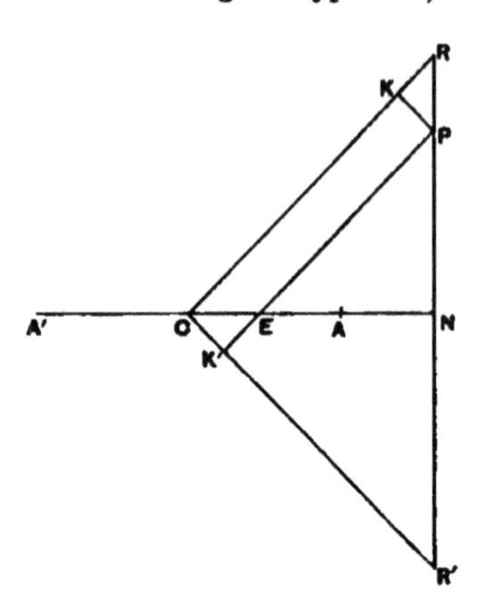

Thus, if P be a point on the curve and PK, PK' be perpendicular to the asymptotes CR, CR' of a rectangular hyperbola, and if $RPNR'$ be perpendicular to the bisector of the angle between the asymptotes,

$$PK \cdot PK' = \text{the rect. } CKPK'$$
$$= \text{the quadrilateral } CRPE,$$

since $\triangle CEK' = \triangle PRK$.

Hence
$$PK \cdot PK' = \triangle RCN - \triangle PEN$$
$$= \tfrac{1}{2}(CN^2 - PN^2)$$
$$= \frac{x^2 - y^2}{2},$$

where x, y are the coordinates of P referred to the axes of the hyperbola.

We have then to show how Menaechmus could obtain from an obtuse-angled cone, by a section perpendicular to a generator, the rectangular hyperbola

$$x^2 - y^2 = (\text{const.}) = \frac{a^2}{4}, \text{ say,}$$

or
$$y^2 = x_1 x_2,$$

where x_1, x_2 are the distances of the foot of the ordinate y from the points A, A' respectively, and $AA' = a$.

Take an obtuse-angled cone, and let BC be the diameter of any circular section of it. Let A be any point on the generator OB, and through A draw AN at right angles to OB meeting CO produced in A' and BC in N.

Let y be the length of the straight line drawn from N perpendicular to the plane of the axial triangle OBC and meeting the surface of the cone. Then y will be determined by the equation

$$y^2 = BN \,.\, NC.$$

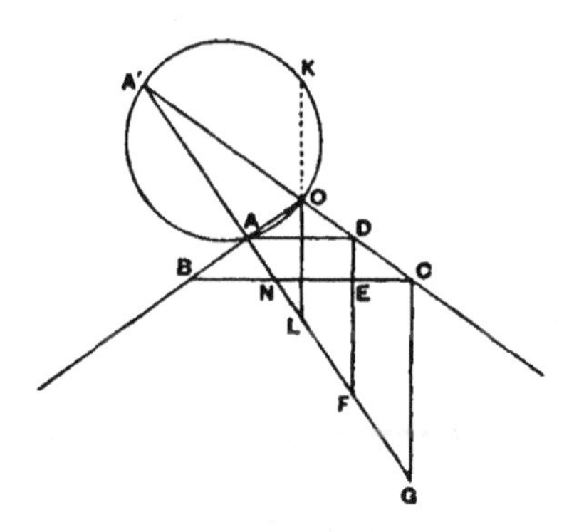

Let AD be drawn, as before, parallel to BC and meeting OC in D, and let OL, DF, CG be drawn perpendicular to BC meeting AN produced in L, F, G respectively.

Then, since the angles BAG, BCG are right, the points B, A, C, G are concyclic;

$$\therefore\ y^2 = BN \,.\, NC = AN \,.\, NG.$$

But $\qquad NG : AF = CN : AD$, by similar triangles,

$$= A'N : AA'.$$

Hence
$$y^2 = AN \,.\, \frac{AF}{AA'} \,.\, A'N$$

$$= \frac{2AL}{AA'} \,.\, x_1 x_2\,;$$

and the locus of the extremity of y for different positions of the circular section, or (in other words) the section of the cone by a plane through AN perpendicular to the plane of the axial triangle, satisfies the desired condition *provided that* $\dfrac{2AL}{AA'} = 1$.

This relation, together with the fact that the angle AOL is equal to half the supplement of the angle $A'OA$, enables us to determine the position of the apex O, and therefore the vertical angle, of the desired cone which is to contain the rectangular hyperbola.

For suppose O determined, and draw the circle circumscribing AOA'; this will meet LO produced in some point K, and OA' will be its diameter. Thus the angle $A'KO$ is right;

$\therefore \angle AA'K =$ complement of $\angle ALK = \angle AOL = \angle LOC = \angle A'OK$, whence it follows that the segments AK, $A'K$ are equal, and therefore K lies on the line bisecting AA' at right angles.

But, since the angle $A'KL$ is right, K also lies on the semicircle with $A'L$ as diameter.

K is therefore determined by drawing that semicircle and then drawing a line bisecting AA' at right angles and meeting the semicircle. Thus, K being found and KL joined, O is determined.

The foregoing construction for a rectangular hyperbola can be equally well applied to the case of the hyperbola generally or of an ellipse; only the value of the constant $\frac{2AL}{AA'}$ will be different from unity. In every case $2AL$ is equal to the parameter of the ordinates to AA', or the parameter is equal to twice the distance between the vertex of the section and the axis of the cone, ἁ διπλασία τᾶς μέχρι τοῦ ἄξονος (as Archimedes called the principal parameter of the parabola).

The assumption that Menaechmus discovered all three sections in the manner above set forth agrees with the reference of Eratosthenes to the "Menaechmean triads," though it is not improbable that the ellipse was known earlier as a section of a right cylinder. Thus a passage of Euclid's *Phaenomena* says, "if a cone or cylinder be cut by a plane not parallel to the base, the resulting section is a section of an acute-angled cone which is similar to a θυρεός," showing that Euclid distinguished the two ways of producing an ellipse. Heiberg (*Litterargeschichtliche Studien über Euklid*, p. 88) thinks it probable that θυρεός was the name by which Menaechmus called the curve*.

It is a question whether Menaechmus used mechanical contriv-

* The expression ἡ τοῦ θυρεοῦ for the ellipse occurs several times in Proclus and particularly in a passage in which Geminus is quoted (p. 111); and it would seem as though this name for the curve was more common in Geminus' time than the name "ellipse." [Bretschneider, p. 176.]

ances for effecting the construction of his curves. The idea that he did so rests (1) upon the passage in the letter of Eratosthenes* to the effect that all who had solved the problem of the two mean proportionals had written theoretically but had not been able to effect the actual construction and reduce the theory to practice except, to a certain extent, Menaechmus and that only with difficulty, (2) upon two well known passages in Plutarch. One of these latter states that Plato blamed Eudoxus, Archytas and Menaechmus for trying to reduce the doubling of the cube to instrumental and mechanical constructions (as though such methods of finding two mean proportionals were not legitimate), arguing that the good of geometry was thus lost and destroyed, as it was brought back again to the world of sense instead of soaring upwards and laying hold of those eternal and incorporeal images amid which God is and thus is ever God†; the other passage (*Vita Marcelli* 14, § 5) states that, in consequence of this attitude of Plato, mechanics was completely divorced from geometry and, after being neglected by philosophers for a long time, became merely a part of the science of war. I do not think it follows from these passages that Menaechmus and Archytas made machines for effecting their constructions; such a supposition would in fact seem to be inconsistent with the direct statement of Eratosthenes that, with the partial exception of Menaechmus, the three geometers referred to gave theoretical solutions only. The words of Eratosthenes imply that Archytas did not use any mechanical contrivance, and, as regards Menaechmus, they rather suggest such a method as the finding of a large number of points on the curve‡. It seems likely therefore that Plato's criticism referred, not to the

* See the passage from Eratosthenes, translated above, p. xviii. The Greek of the sentence in question is: συμβέβηκε δὲ πᾶσιν αὐτοῖς ἀποδεικτικῶς γεγραφέναι, χειρουργῆσαι δὲ καὶ εἰς χρείαν πεσεῖν μὴ δύνασθαι πλὴν ἐπὶ βραχύ τι τοῦ Μενέχμου καὶ ταῦτα δυσχερῶς.

† Διὸ καὶ Πλάτων αὐτὸς ἐμέμψατο τοὺς περὶ Εὔδοξον καὶ Ἀρχύταν καὶ Μέναιχμον εἰς ὀργανικὰς καὶ μηχανικὰς κατασκευὰς τὸν τοῦ στερεοῦ διπλασιασμὸν ἀπάγειν ἐπιχειροῦντας (ὥσπερ πειρωμένους διὰ λόγου [scr. δι' ἀλόγου] δύο μέσας ἀνάλογον μὴ [scr. ᾗ] παρείκοι λαβεῖν). ἀπόλλυσθαι γὰρ οὕτω καὶ διαφθείρεσθαι τὸ γεωμετρίας ἀγαθόν, αὖθις ἐπὶ τὰ αἰσθητὰ παλινδρομούσης καὶ μὴ φερομένης ἄνω, μηδ' ἀντιλαμβανομένης τῶν ἀϊδίων καὶ ἀσωμάτων εἰκόνων, πρὸς αἷσπερ ὢν ὁ θεὸς ἀεὶ θεός ἐστι. (*Quaest. conviv.* VIII. 2. 1.)

‡ This is partly suggested by Eutocius' commentary on Apollonius I. 20, 21, where it is remarked that it was often necessary for want of instruments to describe a conic by a continuous series of points. This passage is quoted by Dr Taylor, *Ancient and Modern Geometry of Conics*, p. xxxiii.

use of machines, but simply to the introduction of mechanical *considerations* in each of the three solutions of Archytas, Eudoxus, and Menaechmus.

Much has been written on the difficulty of reconciling the censure on Archytas and the rest with the fact that a mechanical solution is attributed by Eutocius to Plato himself. The most probable explanation is to suppose that Eutocius was mistaken in giving the solution as Plato's; indeed, had the solution been Plato's, it is scarcely possible that Eratosthenes should not have mentioned it along with the others, seeing that he mentions Plato as having been consulted by the Delians on the duplication problem.

Zeuthen has suggested that Plato's objection may have referred, in the case of Menaechmus, to the fact that he was not satisfied to regard a curve as completely defined by a fundamental plane property such as we express by the equation, but must needs give it a geometrical definition as a curve arrived at by cutting a cone, in order to make its form realisable by the senses, though this presentation of it was not made use of in the subsequent investigations of its properties; but this explanation is not so comprehensible if applied to the objection to *Archytas'* solution, where the curve in which the revolving semicircle and the fixed half-cylinder intersect is a curve of double curvature and not a plane curve easily represented by an equation.

CHAPTER II.

ARISTAEUS AND EUCLID.

WE come next to the treatises which **Aristaeus** 'the elder' and **Euclid** are said to have written; and it will be convenient to deal with these together, in view of the manner in which the two names are associated in the description of Pappus, who is our authority upon the contents of the works, both of which are lost. The passage of Pappus is in some places obscure and some sentences are put in brackets by Hultsch, but the following represents substantially its effect*. "The four books of Euclid's *conics* were completed by Apollonius, who added four more and produced eight books of *conics*. Aristaeus, who wrote the still extant five books of *solid loci* connected with the conics, called one of the conic sections the section of an acute-angled cone, another the section of a right-angled cone and the third the section of an obtuse-angled cone.... Apollonius says in his third book that the 'locus with respect to three or four lines' had not been completely investigated by Euclid, and in fact neither Apollonius himself nor any one else could have added in the least to what was written by Euclid with the help of those properties of conics only which had been proved up to Euclid's time; Apollonius himself is evidence for this fact when he says that the theory of that locus could not be completed without the propositions which he had been obliged to work out for himself. Now Euclid—regarding Aristaeus as deserving credit for the discoveries he had already made in conics, and without anticipating him or wishing to construct anew the same system (such was his scrupulous fairness and his exemplary kindliness towards all who could advance mathematical science to however small an extent), being moreover in no wise contentious and, though exact, yet no braggart like the other—wrote so much about the locus as was possible by means of the conics of Aristaeus, without claiming completeness for his demonstrations.

* See Pappus (ed. Hultsch), pp. 672—678.

Had he done so he would certainly have deserved censure, but, as matters stand, he does not by any means deserve it, seeing that neither is Apollonius called to account, though he left the most part of his *conics* incomplete. Apollonius, too, has been enabled to add the lacking portion of the theory of the locus through having become familiar beforehand with what had already been written about it by Euclid and having spent a long time with the pupils of Euclid in Alexandria, to which training he owed his scientific habit of mind. Now this 'locus with respect to three and four lines,' the theory of which he is so proud of having added to (though he should rather acknowledge his obligations to the original author of it), is arrived at in this way. If three straight lines be given in position and from one and the same point straight lines be drawn to meet the three straight lines at given angles, and if the ratio of the rectangle contained by two of the straight lines so drawn to the square of the remaining one be given, then the point will lie on a solid locus given in position, that is on one of the three conic sections. And, if straight lines be drawn to meet, at given angles, four straight lines given in position, and the ratio of the rectangle under two of the lines so drawn to the rectangle under the remaining two be given, then in the same way the point will lie on a conic section given in position."

It is necessary at this point to say a word about the *solid locus* (στερεὸς τόπος). Proclus defines a *locus* (τόπος) as "a position of a line or a surface involving one and the same property" (γραμμῆς ἢ ἐπιφανείας θέσις ποιοῦσα ἓν καὶ ταὐτὸν σύμπτωμα), and proceeds to say that loci are divided into two classes, *line-loci* (τόποι πρὸς γραμμαῖς) and *surface-loci* (τόποι πρὸς ἐπιφανείαις). The former, or loci which are lines, are again divided by Proclus into *plane loci* and *solid loci* (τόποι ἐπίπεδοι and τόποι στερεοί), the former being simply generated in a plane, like the straight line, the latter from some section of a solid figure, like the cylindrical helix and the conic sections. Similarly Eutocius, after giving as examples of the plane locus (1) the circle which is the locus of all points the perpendiculars from which on a finite straight line are mean proportionals between the segments into which the line is divided by the foot of the perpendicular, (2) the circle which is the locus of a point whose distances from two fixed points are in a given ratio (a locus investigated by Apollonius in the τόπος ἀναλυόμενος), proceeds to say that the so-called solid loci have derived their name from the fact that they arise from the cutting of solid figures, as for instance the sections of the cone and several others*. Pappus makes a further division of those line-loci which are not *plane loci*, i.e. of the class which Proclus and Eutocius call by the one name of *solid loci*, into *solid loci* (στερεοὶ τόποι) and *linear loci* (τόποι γραμμικοί). Thus, he says, *plane loci* may be generally described as those which are straight lines or circles, *solid loci* as those which are sections of cones, i.e. parabolas or ellipses or hyperbolas, while *linear loci* are lines such as are not straight lines, nor circles, nor any of the said three conic sections†. For example, the curve described on the cylinder in Archytas' solution of the problem of the two mean proportionals is a *linear locus* (being in fact a curve of double curvature), and such a locus arises out of, or is traced upon, a locus which is a surface (τόπος πρὸς ἐπιφανείᾳ). Thus *linear loci* are those which have a more complicated and unnatural origin than straight lines, circles and conics, "being generated from more irregular surfaces and intricate movements‡."

It is now possible to draw certain conclusions from the passage of Pappus above reproduced.

1. The work of Aristaeus on *solid loci* was concerned with those loci which are parabolas, ellipses, or hyperbolas; in other words, it was a treatise on conics regarded as loci.

2. This book on *solid loci* preceded that of Euclid on conics and was, at least in point of originality, more important. Though both treatises dealt with the same subject-matter, the object and the point of view were different; had they been the same, Euclid could scarcely have refrained, as Pappus says he did, from an attempt to improve upon the earlier treatise. Pappus' meaning must therefore be that, while Euclid wrote on the general theory of conics as Apollonius did, he yet confined himself to those properties which were necessary for the analysis of the *solid loci* of Aristaeus.

3. Aristaeus used the names "section of a right-angled, acute-angled, and obtuse-angled cone," by which up to the time of Apollonius the three conic sections were known.

4. The *three-line* and *four-line locus* must have been, albeit imperfectly, discussed in the treatise of Aristaeus; and Euclid, in

* Apollonius, Vol. II. p. 184. † Pappus, p. 662.

‡ Pappus, p. 270: γραμμαὶ γὰρ ἕτεραι παρὰ τὰς εἰρημένας εἰς τὴν κατασκευὴν λαμβάνονται ποικιλωτέραν ἔχουσαι τὴν γένεσιν καὶ βεβιασμένην μᾶλλον, ἐξ ἀτακτοτέρων ἐπιφανειῶν καὶ κινήσεων ἐπιπεπλεγμένων γεννώμεναι.

dealing synthetically with the same locus, was unable to work out the theory completely because he only used the conics of Aristaeus and did not add fresh discoveries of his own.

5. The *Conics* of Euclid was superseded by the treatise of Apollonius, and, though the *Solid Loci* of Aristaeus was still extant in Pappus' time, it is doubtful whether Euclid's work was so.

The subject of the *three-line* and *four-line locus* will be discussed in some detail in connexion with Apollonius; but it may be convenient to mention here that Zeuthen, who devotes some brilliant chapters to it, conjectures that the imperfection of the investigations of Aristaeus and Euclid arose from the absence of any conception of the hyperbola with two branches as forming one curve (which was the discovery of Apollonius, as may be inferred even from the fulness with which he treats of the double-hyperbola). Thus the proposition that the rectangles under the segments of intersecting chords in fixed directions are in a constant ratio independent of the position of the point of intersection is proved by Apollonius for the double-hyperbola as well as for the single branch and for the ellipse and parabola. So far therefore as the theorem was not proved for the double-hyperbola before Apollonius, it was incomplete. On the other hand, had Euclid been in possession of the proof of the theorem in its most general form, then, assuming e.g. that the three-line or four-line locus was reduced by Aristaeus' analysis to this particular property, Euclid would have had the means (which we are told that he had not) of completing the synthesis of the locus also. Apollonius probably mentions Euclid rather than Aristaeus as having failed to complete the theory for the reason that it was Euclid's treatise which was on the same lines as his own; and, as Euclid was somewhat later in time than Aristaeus, it would in any case be natural for Apollonius to regard Euclid as the representative of the older and defective investigations which he himself brought to completion.

With regard to the contents of the *Conics* of Euclid we have the following indications.

1. The scope must have been generally the same as that of the first three Books of Apollonius, though the development of the subject was more systematic and complete in the later treatise. This we infer from Apollonius' own preface as well as from the statement of Pappus quoted above.

2. A more important source of information, in the sense of

giving more details, is at hand in the works of Archimedes, who frequently refers to propositions in conics as well known and not requiring proof. Thus

(*a*) The fundamental property of the ellipse,

$$PN^2 : AN . NA' = P'N'^2 : AN' . N'A' = BC^2 : AC^2,$$

that of the hyperbola,

$$PN^2 : AN . NA' = P'N'^2 : AN' . N'A',$$

and that of the parabola,

$$PN^2 = p_a . AN,$$

are assumed, and must therefore presumably have been contained in Euclid's work.

(*b*) At the beginning of the treatise on the area of a parabolic segment the following theorems are simply cited.

(1) If PV be a diameter of a segment of a parabola and QVq a chord parallel to the tangent at P, $QV = Vq$.

(2) If the tangent at Q meet VP produced in T, $PV = PT$.

(3) If QVq, $Q'V'q'$ be two chords parallel to the tangent at P and bisected in V, V',

$$PV : PV' = QV^2 : Q'V'^2.$$

"*And these propositions are proved in the elements of conics*" (i.e. in Euclid and Aristaeus).

(*c*) The third proposition of the treatise *On Conoids and Spheroids* begins by enunciating the following theorem: If straight lines drawn from the same point touch any conic section whatever, and if there be also other straight lines drawn in the conic section parallel to the tangents and cutting one another, the rectangles contained by the segments (of the chords) will have to one another the same ratio as the squares of the (parallel) tangents. "*And this is proved in the elements of conics.*"

(*d*) In the same proposition we find the following property of the parabola: If p_a be the parameter of the ordinates to the axis, and QQ' be any chord not perpendicular to the axis such that the diameter PV bisects it in V, and if QD be drawn perpendicular to PV, then (says Archimedes), supposing p to be such a length that

$$QV^2 : QD^2 = p : p_a,$$

the squares of the ordinates to PV (which are parallel to QQ') are equal to the rectangles applied to a straight line equal to p and of width equal to the respective intercepts on PV towards P. "*For this has been proved in the conics.*"

In other words, if p_a, p are the parameters corresponding respectively to the axis and the diameter bisecting QQ',

$$p : p_a = QV^2 : QD^2.$$

(For a figure and a proof of this property the reader is referred to the chapter on Archimedes p. liii.)

Euclid still used the old names for the three conic sections, but he was aware that an ellipse could be obtained by cutting a cone in any manner by a plane not parallel to the base (assuming the section to lie wholly between the apex of the cone and its base), and also by cutting a cylinder. This is expressly stated in the passage quoted above (p. xxviii) from the *Phaenomena*. But it is scarcely possible that Euclid had in mind any other than a right cone; for, had the cone been oblique, the statement would not have been true without a qualification excluding the circular sections subcontrary to the base of the cone.

Of the contents of Euclid's *Surface-loci*, or τόποι πρὸς ἐπιφανείᾳ, we know nothing, though it is reasonable to suppose that the treatise dealt with such loci as the surfaces of cones, spheres and cylinders, and perhaps other surfaces of the second degree. But Pappus gives two lemmas to the *Surface-loci*, one of which (the second) is of the highest importance*. This lemma states, and gives a complete proof of, the proposition that *the locus of a point whose distance from a given point is in a given ratio to its distance from a fixed line is a conic section, and is an ellipse, a parabola, or a hyperbola according as the given ratio is less than, equal to, or greater than, unity.*

The proof in the case where the given ratio is different from unity is shortly as follows.

Let S be the fixed point, and let SX be the perpendicular from S on the fixed line. Let P be any point on the locus and PN perpendicular to SX, so that SP is to NX in the given ratio. Let e be this ratio, so that

$$e^2 = \frac{PN^2 + SN^2}{NX^2}.$$

Now let K be a point on the line SX such that

$$e^2 = \frac{SN^2}{NK^2};$$

* Pappus (ed. Hultsch) p. 1006 seqq.

then, if K' be another point so taken that $NK = NK'$, we shall have

$$e^2 = \frac{PN^2 + SN^2}{NX^2} = \frac{SN^2}{NK^2} = \frac{PN^2}{NX^2 - NK^2} = \frac{PN^2}{XK \, . \, XK'}.$$

The position of the points N, K, K' changes with the position of P. If we suppose A to be the point on which N falls when K coincides with X, we have

$$\frac{SA}{AX} = e = \frac{SN}{NK}.$$

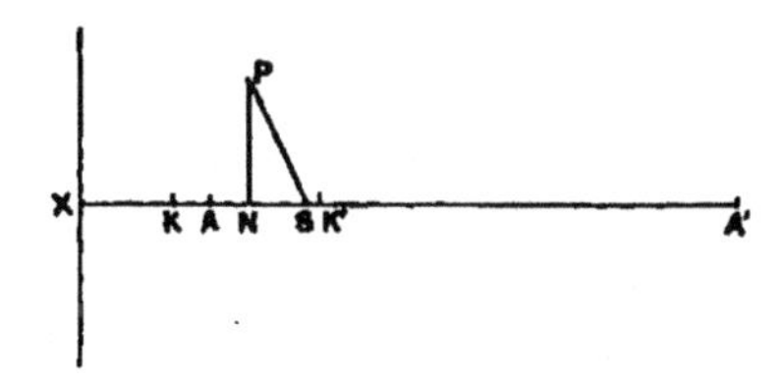

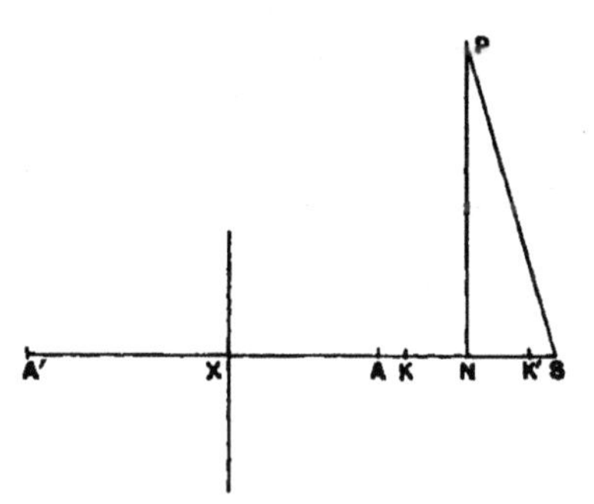

It follows that $\frac{AX}{SA}$, $\frac{NK}{SN}$ are both known and equal, and therefore $\frac{SX}{SA}$, $\frac{SK}{SN}$ are both known and equal. Hence either of the latter expressions is equal to

$$\frac{SX - SK}{SA - SN}, \text{ or } \frac{XK}{AN},$$

which is therefore known

$$\left[= \frac{SX}{AS} = 1 + \frac{1}{e} \right].$$

In like manner, if A' be the point on which N falls when K' coincides with X, we have $\frac{SA'}{A'X} = e$; and in the same way we shall find that the ratio $\frac{XK'}{A'N}$ is known and is equal to

$$\frac{SX}{A'S}\left[=1 \sim \frac{1}{e}\right].$$

Hence, by multiplication, the ratio $\frac{XK \cdot XK'}{AN \cdot A'N}$ has a known value.

And, since $\frac{PN^2}{XK \cdot XK'} = e^2$, from above,

we have $$\frac{PN^2}{AN \cdot A'N} = (\text{const.})\left[=e^2\left(1 \sim \frac{1}{e^2}\right) = 1 \sim e^2\right].$$

This is the property of a central conic, and the conic will be an ellipse or a hyperbola according as e is less or greater than 1; for in the former case the points A, A' will lie on the same side of X and in the latter case on opposite sides of X, while in the former case N will lie on AA' and in the latter N will lie on AA' produced.

The case where $e = 1$ is easy, and the proof need not be given here.

We can scarcely avoid the conclusion that Euclid must have used this proposition in the treatise on *surface-loci* to which Pappus' lemma refers, and that the necessity for the lemma arose out of the fact that Euclid did not prove it. It must therefore have been assumed by him as evident or quoted as well known. It may therefore well be that it was taken from some known work*, not impossibly that of Aristaeus on *solid loci*.

That Euclid should have been acquainted with the property of conics referred to the focus and directrix cannot but excite surprise

* It is interesting to note in this connexion another passage in Pappus where he is discussing the various methods of trisecting an angle or circular arc. He gives (p. 284) a method which "some" had used and which involves the construction of a hyperbola whose eccentricity is 2.

Suppose it is a segment of a circle which has to be divided into three equal

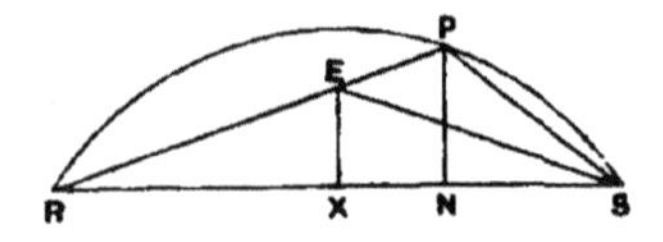

parts. Suppose it done, and let SP be one-third of the arc SPR. Join RP, SP. Then the angle RSP is equal to twice the angle SRP.

seeing that this property does not appear at all in Apollonius, and the focus of a parabola is not even mentioned by him. The explanation may be that, as we gather from the preface of Apollonius, he does not profess to give all the properties of conics known to him, and his third Book is intended to give the means for the synthesis of solid loci, not the actual determination of them. The focal property may therefore have been held to be a more suitable subject for a treatise on solid loci than for a work on conics proper. We must not assume that the focal properties had not, up to the time of Apollonius, received much attention. The contrary is indeed more probable, and this supposition is supported by a remarkable coincidence between Apollonius' method of determining the foci of a central conic and the theorem contained in Pappus' 31st lemma to Euclid's *Porisms*.

This theorem is as follows: Let $A'A$ be the diameter of a semicircle, and from A', A let two straight lines be drawn at right angles to $A'A$. Let any straight line RR' meet the two perpendiculars in R, R' respectively and the semicircle in Y. Further let YS be drawn perpendicular to RR', meeting $A'A$ produced in S.

It is to be proved that

$$AS \cdot SA' = AR \cdot A'R',$$

i.e. that $$SA : AR = A'R' : A'S.$$

Now, since R', A', Y, S are concyclic, the angle $A'SR'$ is equal to the angle $A'YR'$ in the same segment.

Let SE bisect the angle RSP, meeting RP in E and draw EX, PN perpendicular to RS.

Then the angle ERS is equal to the angle ESR, so that $RE = ES$;

$$\therefore RX = XS, \text{ and } X \text{ is given.}$$

Also $$RS : SP = RE : EP = RX : XN;$$

$$\therefore RS : RX = SP : NX.$$

But $$RS = 2RX;$$

$$\therefore SP = 2NX,$$

whence $$SP^2 = 4NX^2,$$

or $$PN^2 + SN^2 = 4NX^2.$$

"Since then the two points S, X are given, and PN is perpendicular to SX, while the ratio of NX^2 to $PN^2 + SN^2$ is given, P lies on a hyperbola."

This is obviously a particular case of the lemma to the τόποι πρὸς ἐπιφανείᾳ, and the ratio $\frac{NX^2}{PN^2 + SN^2}$ is stated in the same form in both cases.

Similarly, the angle ARS is equal to the angle AYS.

But, since $A'YA$, $R'YS$ are both right angles,

$$\angle A'YR' = \angle AYS;$$

$$\therefore \angle A'SR' = \angle ARS;$$

hence, by similar triangles,

$$A'R' : A'S = SA : AR,$$

or $$AS . SA' = AR . A'R'.$$

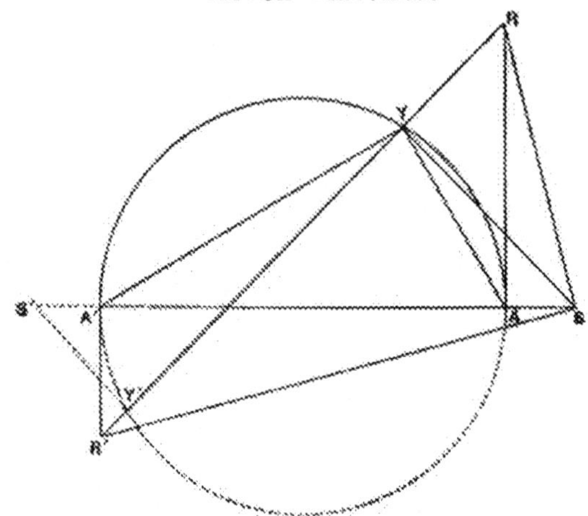

It follows of course from this that, if the rectangle $AR . A'R'$ is constant, $AS . SA$ is also constant and S is a fixed point.

It will be observed that in Apollonius, III. 45 [Prop. 69], the complete circle is used, AR, $A'R'$ are tangents at the extremities of the axis AA' of a conic, and RR' is any other tangent to the conic. He has already proved, III. 42 [Prop. 66], that in this case $AR . A'R' = BC^2$, and he now takes two points S, S' on the axis or the axis produced such that

$$AS . SA' = AS' . S'A' = BC^2.$$

He then proves that RR' subtends a right angle at each of the points S, S', and proceeds to deduce other focal properties.

Thus Apollonius' procedure is exactly similar to that in the lemma to Euclid's *Porisms*, except that the latter does not bring in the conic. This fact goes far to support the view of Zeuthen as to the origin and aim of Euclid's *Porisms*, namely, that they were partly a sort of by-product in the investigation of conic sections and partly a means devised for the further development of the subject.

CHAPTER III.

ARCHIMEDES.

No survey of the history of conic sections could be complete without a tolerably exhaustive account of everything bearing on the subject which can be found in the extant works of **Archimedes.**

There is no trustworthy evidence that Archimedes wrote a separate work on conics. The idea that he did so rests upon no more substantial basis than the references to κωνικὰ στοιχεῖα (without any mention of the name of the author) in the passages quoted above, which have by some been assumed to refer to a treatise by Archimedes himself. But the assumption is easily seen to be unsafe when the references are compared with a similar reference in another passage* where by the words ἐν τῇ στοιχειώσει the *Elements* of *Euclid* are undoubtedly meant. Similarly the words "this is proved in the elements of conics" simply mean that it is found in the text-books on the elementary principles of conics. A positive proof that this is so may be drawn from a passage in Eutocius' commentary on Apollonius. Heracleides†, the biographer of Archimedes, is there quoted as saying that Archimedes was the first to invent theorems in conics, and that Apollonius, having found that they had not been published by Archimedes, appropriated them‡;

* *On the Sphere and Cylinder*, I. p. 24. The proposition quoted is Eucl. XII. 2.

† The name appears in the passage referred to as Ἡράκλειος. Apollonius (ed. Heiberg) Vol. II. p. 168.

‡ Heracleides' statement that Archimedes was the first to "invent" (ἐπινοῆσαι) theorems in conics is not easy to explain. Bretschneider (p. 156) puts it, as well as the charge of plagiarism levelled at Apollonius, down to the malice with which small minds would probably seek to avenge themselves for the contempt in which they would be held by an intellectual giant like

and Eutocius subjoins the remark that the allegation is in his opinion not true, "for on the one hand Archimedes appears in many passages to have referred to the elements of conics as an older treatise (ὡς παλαιοτέρας), and on the other hand Apollonius does not profess to be giving his own discoveries." Thus Eutocius regarded the reference as being to earlier expositions of the elementary theory of conics by other geometers: otherwise, i.e. if he had thought that Archimedes referred to an earlier work of his own, he would not have used the word παλαιοτέρας but rather some expression like πρότερον ἐκδεδομένης.

In searching for the various propositions in conics to be found in Archimedes, it is natural to look, in the first instance, for indications to show how far Archimedes was aware of the possibility of producing the three conic sections from cones other than right cones and by plane sections other than those perpendicular to a generator of the cone. We observe, first, that he always uses the old names "section of a right-angled cone" &c. employed by Aristaeus, and there is no doubt that in the three places where the word ἔλλειψις appears in the MSS. it has no business there. But, secondly, at the very beginning of the treatise *On Conoids and Spheroids* we find the following: "If a cone be cut by a plane meeting all the sides of the cone, the section will be either a circle or a section of an acute-angled cone" [i.e. an ellipse]. The way in which this proposition was proved in the case where the plane of section is at right angles to the plane of symmetry can be inferred from propositions 7 and 8 of the same treatise, where it is shown that it is possible to *find* a cone of which a given ellipse is a section and whose apex is on a straight line drawn from the centre of the ellipse (1) perpendicular to the plane of the ellipse, (2) not perpendicular to its plane, but lying in a plane at right angles to it and passing through one of the axes of the ellipse. The problem evidently amounts to determining the

Apollonius. Heiberg, on the other hand, thinks that this is unfair to Heracleides, who was probably misled into making the charge of plagiarism by finding many of the propositions of Apollonius already quoted by Archimedes as known. Heiberg holds also that Heracleides did not intend to ascribe the actual *invention* of conics to Archimedes, but only meant that the elementary theory of conic sections as formulated by Apollonius was due to Archimedes; otherwise Eutocius' contradiction would have taken a different form and he would not have omitted to point to the well-known fact that Menaechmus was the discoverer of the conic sections.

circular sections of the cone, and this is what Archimedes proceeds to do.

(1) Conceive an ellipse with BB' as its minor axis and lying in a plane perpendicular to the plane of the paper: suppose the line CO drawn perpendicular to the plane of the ellipse, and

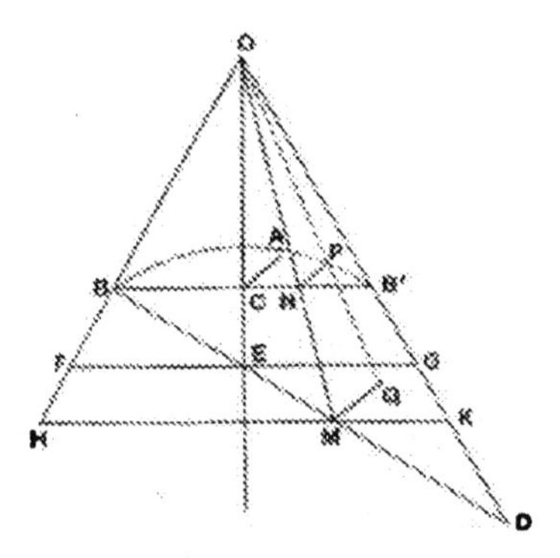

let O be the apex of the required cone. Produce OB, OC, OB', and in the same plane with them draw BED meeting OC, OB' produced in E, D respectively, and in such a direction that

$$BE \,.\, ED : EO^2 = CA^2 : CO^2$$

(where CA is half the major axis of the ellipse).

And this is possible, since

$$BE \,.\, ED : EO^2 > BC \,.\, CB' : CO^2.$$

[Both the construction and this last proposition are assumed as known.]

Now conceive a circle with BD as diameter drawn in a plane perpendicular to that of the paper, and describe a cone passing through this circle and having O for its apex.

We have then to prove that the given ellipse is a section of this cone, or, if P is any point on the ellipse, that P lies on the surface of the cone.

Draw PN perpendicular to BB'. Join ON, and produce it to meet BD in M, and let MQ be drawn in the plane of the circle on BD as diameter and perpendicular to BD, meeting the circumference of the circle in Q. Also draw FG, HK through E, M respectively each parallel to BB'.

Now
$$QM^2 : HM . MK = BM . MD : HM . MK$$
$$= BE . ED : FE . EG$$
$$= (BE . ED : EO^2) . (EO^2 : FE . EG)$$
$$= (CA^2 : CO^2) . (CO^2 : BC . CB')$$
$$= CA^2 : BC . CB'$$
$$= PN^2 : BN . NB'.$$
$$\therefore\ QM^2 : PN^2 = HM . MK : BN . NB'$$
$$= OM^2 : ON^2,$$

whence, since PN, QM are parallel, OPQ is a straight line.

But Q is on the circumference of the circle on BD as diameter; therefore OQ is a generator of the cone, and therefore P lies on the cone.

Thus the cone passes through all points of the given ellipse.

(2) Let OC not be perpendicular to AA', one of the axes of the given ellipse, and let the plane of the paper be that containing AA' and OC, so that the plane of the ellipse is perpendicular to that plane. Let BB' be the other axis of the ellipse.

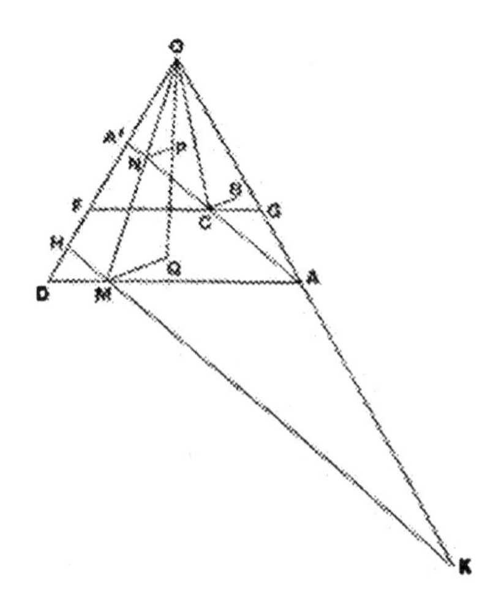

Now OA, OA' are unequal. Produce OA' to D so that $OA = OD$. Join AD, and draw FG through C parallel to it.

Conceive a plane through AD perpendicular to the plane of the paper, and in it describe

either (a), if $CB^2 = FC \cdot CG$, a circle with diameter AD,

or (b), if not, an ellipse on AD as axis such that if d be the other axis

$$d^2 : AD^2 = CB^2 : FC \cdot CG.$$

Take a cone with apex O and passing through the circle or ellipse just drawn. This is possible even when the curve is an ellipse, because the line from O to the middle point of AD is perpendicular to the plane of the ellipse, and the construction follows that in the preceding case (1).

Let P be any point on the given ellipse, and we have only to prove that P lies on the surface of the cone so described.

Draw PN perpendicular to AA'. Join ON, and produce it to meet AD in M. Through M draw HK parallel to $A'A$. Lastly, draw MQ perpendicular to the plane of the paper (and therefore perpendicular to both HK and AD) meeting the ellipse or circle about AD (and therefore the surface of the cone) in Q.

Then

$$\begin{aligned} QM^2 : HM \cdot MK &= (QM^2 : DM \cdot MA) \cdot (DM \cdot MA : HM \cdot MK) \\ &= (d^2 : AD^2) \cdot (FC \cdot CG : A'C \cdot CA) \\ &= (CB^2 : FC \cdot CG) \cdot (FC \cdot CG : A'C \cdot CA) \\ &= CB^2 : A'C \cdot CA \\ &= PN^2 : A'N \cdot NA. \end{aligned}$$

$$\begin{aligned} \therefore\ QM^2 : PN^2 &= HM \cdot MK : A'N \cdot NA \\ &= OM^2 : ON^2. \end{aligned}$$

Hence OPQ is a straight line, and, Q being on the surface of the cone, it follows that P is also on the surface of the cone.

The proof that the *three* conics can be produced by means of sections of any circular cone, whether right or oblique, which are made by planes perpendicular to the plane of symmetry, but not necessarily perpendicular to a generating line of the cone, is of course essentially the same as the proof for the ellipse. It is therefore to be inferred that Archimedes was equally aware of the fact that the parabola and the hyperbola could be found otherwise than by th old method. The continued use of the old names of the curves is of no importance in this connexion because the ellipse was still called the "section of an acute-angled cone" after it was discovered that it could be produced by means of a plane cutting all the generating lines of any cone, whatever its vertical angle. Heiberg concludes that Archimedes only obtained the parabola in the old way because he describes the parameter as double of the line between the vertex of the parabola and the axis of the cone, which is only correct in the case of the right-angled cone; but this is no more an objection to the continued use of the term as a well-known description of the parameter than it is an objection to the continued use by Archimedes of the term "section of an acute-angled cone" that the ellipse had been found to be obtainable in a different manner. Zeuthen points out, as further evidence, the fact that we have the following propositions enunciated by Archimedes without proof (*On Conoids and Spheroids*, 11):

(1) "If a right-angled conoid [a paraboloid of revolution] be cut by a plane through the axis or parallel to the axis, the section will be a section of a right-angled cone the same as that comprehending the figure (*ἁ αὐτὰ τᾷ περιλαμβανούσᾳ τὸ σχῆμα*). And its diameter [axis] will be the common section of the plane which cuts the figure and of that which is drawn through the axis perpendicular to the cutting plane.

(2) "If an obtuse-angled conoid [a hyperboloid of revolution] be cut by a plane through the axis or parallel to the axis or through the apex of the cone enveloping (*περιέχοντος*) the conoid, the section will be a section of an obtuse-angled cone: if [the cutting plane passes] through the axis, the same as that comprehending the figure: if parallel to the axis, similar to it: and if through the apex of the cone enveloping the conoid, not similar. And the diameter [axis] of the section will be the common section of the plane which cuts the figure and of that drawn through the axis at right angles to the cutting plane.

(3) "If any one of the spheroidal figures be cut by a plane through the axis or parallel to the axis, the section will be a section of an acute-angled cone: if through the axis, the actual section which comprehends the figure: if parallel to the axis, similar to it."

Archimedes adds that the proofs of all these propositions are obvious. It is therefore tolerably certain that they were based on the same essential principles as his earlier proofs relating to the sections of conical surfaces and the proofs given in his later investigations of the elliptic sections of the various surfaces of revolution. These depend, as will be seen, on the proposition that, if two chords drawn in fixed directions intersect in a point, the ratio of the rectangles under the segments is independent of the position of the point. This corresponds exactly to the use, in the above proofs with

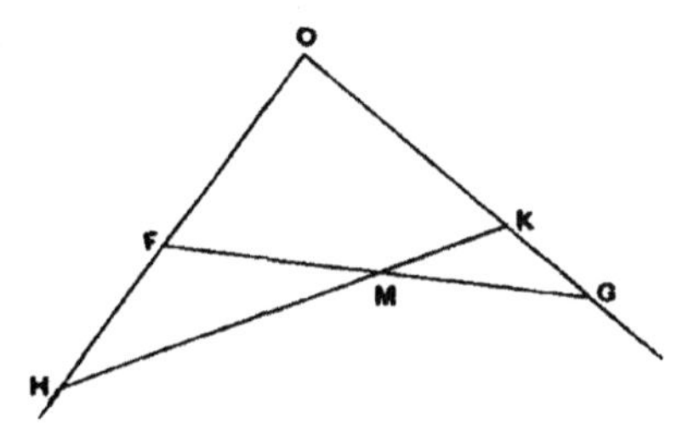

regard to the cone, of the proposition that, if straight lines *FG*, *HK* are drawn in fixed directions between two lines forming an angle, and if *FG*, *HK* meet in any point *M*, the ratio $FM . MG : HM . MK$ is constant; the latter property being in fact the particular case of the former where the conic reduces to two straight lines.

The following is a reproduction, given by way of example, of the proposition (13) of the treatise *On Conoids and Spheroids* which proves that the section of an obtuse-angled conoid [a hyperboloid of revolution] by any plane which meets all the generators of the enveloping cone, and is not perpendicular to the axis, is an ellipse whose major axis is the part intercepted within the hyperboloid of the line of intersection of the cutting plane and the plane through the axis perpendicular to it.

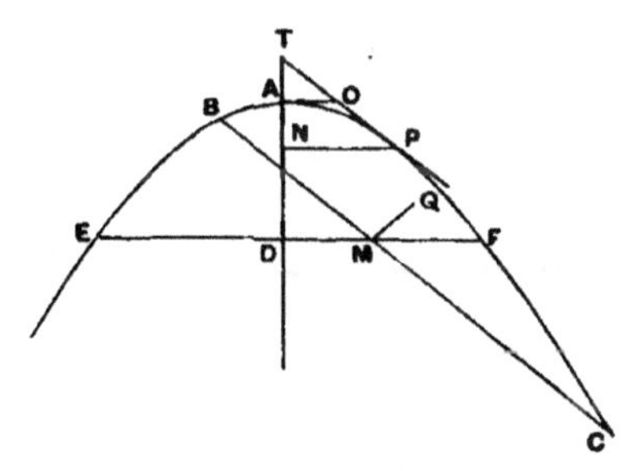

Suppose the plane of the paper to be this latter plane, and the line *BC* to be its intersection with the plane of section which is perpendicular to the plane of the paper. Let *Q* be any point on the section of the hyperboloid, and draw *QM* perpendicular to *BC*.

Let EAF be the hyperbolic section of the hyperboloid made by the plane of the paper and AD its axis. Through M in this plane draw EDF at right angles to AD meeting the hyperbola in E, F.

Then the section of the hyperboloid by the plane through EF perpendicular to AD is a circle, QM lies in its plane, and Q is a point on it.

Therefore $$QM^2 = EM . MF.$$

Now let PT be that tangent to the hyperbola which is parallel to BC, and let it meet the axis in T and the tangent at A in O. Draw PN perpendicular to AD.

Then $$QM^2 : BM . MC = EM . MF : BM . MC$$
$$= OA^2 : OP^2,$$

which is constant for all positions of Q on the section through BC.

Also $OA < OP$, *because it is a property of hyperbolas that*

$$AT < AN, \text{ and therefore } OT < OP,$$

whence *a fortiori* $OA < OP$.

Therefore Q lies on an ellipse whose major axis is BC.

It is also at once evident that all parallel elliptic sections are similar.

Archimedes, it will be seen, here assumes two propositions

(a) that the ratio of the rectangles under the segments of intersecting chords in fixed directions is equal to the constant ratio of the squares on the parallel tangents to the conic, and

(b) that in a hyperbola $AN > AT$.

The first of these two propositions has already been referred to as having been known before Archimedes' time [p. xxxv]; the second assumption is also interesting. It is not easy to see how the latter could be readily proved except by means of the general property that, if PP' be a diameter of a hyperbola and from any point Q on the curve the ordinate QV be drawn to the diameter, while the tangent QT meets the diameter in T, then

$$TP : TP' = PV : P'V,$$

so that we may probably assume that Archimedes was aware of this property of the hyperbola, or at least of the particular case of it where the diameter is the axis.

It is certain that the corresponding general proposition for the parabola, $PV = PT$, was familiar to him; for he makes frequent use of it.

As a preliminary to collecting and arranging in order the other properties of conics either assumed or proved by Archimedes, it may be useful to note some peculiarities in his nomenclature as compared with that of Apollonius. The term *diameter*, when used with reference to the complete conic as distinguished from a segment, is only applied to what was afterwards called the axis. In an *ellipse* the major axis is ἁ μείζων διάμετρος and the minor axis ἁ ἐλάσσων διάμετρος. For the *hyperbola*, by the 'diameter' is only understood that part of it which is within the (single-branch) hyperbola. This we infer from the fact that the 'diameter' of a hyperbola is identified with the axis of the figure described by its revolution about the diameter, while the axis of the hyperboloid does not extend outside it, as it meets (ἅπτεται) the surface in the vertex (κορυφά), and the distance between the vertex and the apex of the enveloping cone [the centre of the revolving hyperbola] is 'the line *adjacent* to the axis' (ἁ ποτεοῦσα τῷ ἄξονι). In the *parabola* diameters other than the axis are called 'the lines *parallel to the diameter*'; but in a *segment* of a parabola that one which bisects the base of the segment is called the *diameter of the segment* (τοῦ τμάματος). In the ellipse diameters other than the axes have no special name, but are simply 'lines drawn through the centre.'

The term *axis* is only used with reference to the solids of revolution. For the complete figure it is the axis of revolution; for a segment cut off by a plane it is the portion intercepted within the segment of the line, (1) in the paraboloid, drawn through the vertex of the segment parallel to the axis of revolution, (2) in the hyperboloid, joining the vertex of the segment and the apex of the enveloping cone, (3) in the spheroid, joining the vertices of the *two* segments into which the figure is divided, the *vertex* of any segment being the point of contact of the tangent plane parallel to the base. In a spheroid the 'diameter' has a special signification, meaning the straight line drawn through the centre (defined as the middle point of the axis) at right angles to the axis. Thus we are told that "those spheroidal figures are called similar whose axes have the same ratio to the diameters*."

The two *diameters* (axes) of an ellipse are called *conjugate* (συζυγεῖς).

The asymptotes of a hyperbola are in Archimedes *the straight lines nearest to the section of the obtuse-angled cone* (αἱ ἔγγιστα

* *On Conoids and Spheroids*, p. 282.

εὐθεῖαι τᾶς τοῦ ἀμβλυγωνίου κώνου τομᾶς), while what we call the centre of a hyperbola is for Archimedes *the point in which the nearest lines meet* (τὸ σαμεῖον, καθ' ὃ αἱ ἔγγιστα συμπίπτοντι). Archimedes never speaks of the 'centre' of a hyperbola: indeed the use of it implies the conception of the two branches of a hyperbola as forming one curve, which does not appear earlier than in Apollonius.

When the asymptotes of a hyperbola revolve with the curve round the axis they generate the cone *enveloping* or *comprehending* the hyperboloid, (τὸν δὲ κῶνον τὸν περιλαφθέντα ὑπὸ τᾶν ἔγγιστα τᾶς τοῦ ἀμβλυγωνίου κώνου τομᾶς περιέχοντα τὸ κωνοειδὲς καλεῖσθαι).

The following enumeration* gives the principal properties of conics mentioned or proved in Archimedes. It will be convenient to divide them into classes, taking first those propositions which are either quoted as having been proved by earlier writers, or assumed as known. They fall naturally under four heads.

I. General.

1. The proposition about the rectangles under the segments of intersecting chords has been already mentioned (p. xxxv and xlviii).

2. *Similar conics.* The criteria of similarity in the case of central conics and of segments of conics are practically the same as those given by Apollonius.

The proposition that *all parabolas are similar* was evidently familiar to Archimedes, and is in fact involved in his statement that all paraboloids of revolution are similar (τὰ μὲν οὖν ὀρθογώνια κωνοειδέα πάντα ὁμοιά ἐντι).

3. Tangents at the extremities of a 'diameter' (axis) are perpendicular to it.

II. The Ellipse.

1. The relations

$$PN^2 : AN \,.\, A'N = P'N'^2 : AN' \,.\, A'N'$$
$$= BB'^2 : AA'^2 \text{ or } CB^2 : CA^2$$

* A word of acknowledgement is due here to Heiberg for the valuable summary of "Die Kenntnisse des Archimedes über die Kegelschnitte," contained in the *Zeitschrift für Mathematik und Physik* (*Historisch-literarische Abtheilung*) 1880, pp. 41—67. This article is a complete guide to the relevant passages in Archimedes, though I have of course not considered myself excused in any instance from referring to the original.

are constantly used as expressing the fundamental property and the criterion by which it is established that a curve is an ellipse.

2. The more general proposition

$$QV^2 : PV.P'V = Q'V'^2 : PV'.P'V'$$

also occurs.

3. If a circle be described on the major axis as diameter, and an ordinate PN to the axis of the ellipse be produced to meet the circle in p, then

$$pN : PN = (\text{const.}).$$

4. The straight line drawn from the centre to the point of contact of a tangent bisects all chords parallel to the tangent.

5. The straight line joining the points of contact of parallel tangents passes through the centre; and, if a line be drawn through the centre parallel to either tangent and meeting the ellipse in two points, the parallels through those points to the chord of contact of the original parallel tangents will touch the ellipse.

6. If a cone be cut by a plane meeting all the generators, the section is either a circle or an ellipse.

Also, if a cylinder be cut by two parallel planes each meeting all the generators, the sections will be either circles or ellipses equal and similar to one another.

III. The Hyperbola.

1. We find, as fundamental properties, the following,

$$PN^2 : P'N'^2 = AN.A'N : AN'.A'N',$$

$$QV^2 : Q'V'^2 = PV.P'V : PV'.P'V';$$

but Archimedes does not give any expression for the constant ratios $PN^2 : AN.A'N$ and $QV^2 : PV.P'V$, from which we may infer that he had no conception of diameters or radii of a hyperbola not meeting the curve.

If C be the point of concourse of the asymptotes, A' is arrived at by producing AC and measuring CA' along it equal to CA; and the same procedure is used for finding P', the other extremity of the diameter through P: the lengths AA', PP' are then in each case *double of the line adjacent to the axis* [in one case of the whole surface, and in the other of a segment of which P is the 'vertex']. This term for AA', PP' was, no doubt, only used in order to avoid mention of the cone of which the hyperbola is a section, as the introduction of this cone might have complicated matters (seeing that the enveloping cone also appears); for it is obvious that AA' appeared first as the distance along the principal diameter of the hyperbola intercepted between the vertex and the point where it meets the surface of the opposite half of the double cone, and the notion of the asymptotes came later in the order of things.

2. If from a point on a hyperbola two straight lines are drawn in any directions to meet the asymptotes, and from another point two other straight lines are similarly drawn parallel respectively to the former, the rectangles contained by each pair will be equal*.

3. A line through the point of concourse of the asymptotes and the point of contact of any tangent bisects all chords parallel to the tangent.

4. If PN, the principal ordinate from P, and PT, the tangent at P, meet the axis in N, T respectively, then

$$AN > AT.$$

5. If a line between the asymptotes meets a hyperbola and is bisected at the point of concourse, it will touch the hyperbola†.

IV. THE PARABOLA.

1.
$$\left.\begin{array}{l} PN^2 : P'N'^2 = AN : AN' \\ QV^2 : Q'V'^2 = PV : PV' \end{array}\right\}.$$
and

We find also the forms

$$\left.\begin{array}{l} PN^2 = p_a \cdot AN \\ QV^2 = p \cdot PV \end{array}\right\}.$$

p_a (the principal parameter) is called by Archimedes the *parameter* of the ordinates (parallel to the tangent at the vertex), παρ' ἂν δύνανται αἱ ἀπὸ τᾶς τομᾶς, and is also described as *the double of the line extending* [from the vertex] *to the axis* [of the cone] ἁ διπλασία τᾶς μέχρι τοῦ ἄξονος.

The term 'parameter' is not applied by Archimedes to p, the constant in the last of the four equations just given. p is simply described as the line to which the rectangle equal to QV^2 and of width equal to PV is applied.

2. Parallel chords are bisected by one line parallel to the axis;

* This proposition and its converse appear in a fragment given by Eutocius in his note on the 4th proposition of Book II. *On the Sphere and Cylinder*.

† This is also used in the fragment quoted by Eutocius.

and a line parallel to the axis bisects chords parallel to the tangent at the point where the said line cuts the parabola.

3. If QD be drawn perpendicular to the diameter PV bisecting the chord QVQ', and if p be the parameter of the ordinates parallel to QQ', while p_a is the principal parameter,

$$p : p_a = QV^2 : QD^2.$$

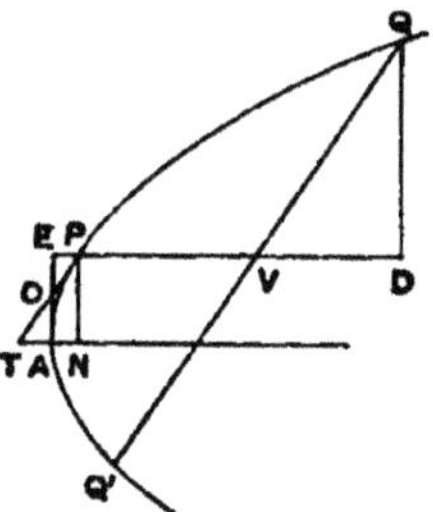

[This proposition has already been mentioned above (p. xxxv, xxxvi). It is easily derived from Apollonius' proposition I. 49 [Prop. 22]. If PV meet the tangent at A in E, and PT, AE intersect in O, the proposition in question proves that

$$OP : PE = p : 2PT,$$

and $$OP = \tfrac{1}{2}PT;$$

$$\therefore\ PT^2 = p \,.\, PE$$

$$= p \,.\, AN.$$

Thus $$QV^2 : QD^2 = PT^2 : PN^2, \text{ by similar triangles,}$$

$$= p \,.\, AN : p_a \,.\, AN$$

$$= p : p_a.]$$

4. If the tangent at Q meet the diameter PV in T, and QV be an ordinate to the diameter,

$$PV = PT.$$

5. By the aid of the preceding, tangents can be drawn to a parabola (*a*) from a point on it, (*b*) parallel to a given chord.

6. In the treatise *On floating bodies* (περὶ τῶν ὀχουμένων), II. 5, we have this proposition: If K be a point on the axis, and KF be measured along the axis away from the vertex and equal to half the principal parameter, while KH is drawn perpendicular to the diameter through any point P, then FH is perpendicular to the tangent at P. (See the next figure.)

It is obvious that this is equivalent to the proposition that *the subnormal at any point P is constant and equal to half the principal parameter.*

7. If QAQ' be a segment of a parabola such that QQ' is perpendicular to the axis, while QVq, parallel to the tangent at P, meets the diameter through P in V, and if R be any other point on the curve the ordinate from which RHK meets PV in H and the axis in K, then (M being the middle point of QQ')

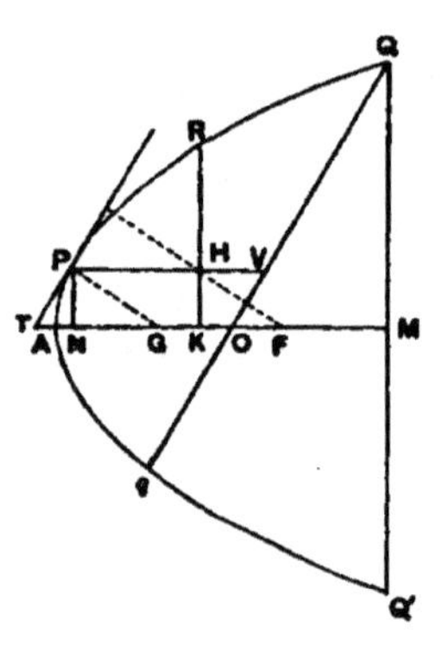

$$PV : PH \underset{\text{or} =}{>} MK : KA,$$

"*for this is proved.*" (*On floating bodies*, II. 6.)

[There is nothing to show where or by whom the proposition was demonstrated, but the proof can be supplied as follows:

We have to prove that $\dfrac{PV}{PH} - \dfrac{MK}{KA}$ is *positive* or *zero*.

Let Qq meet AM in O.

Now
$$\frac{PV}{PH} - \frac{MK}{KA} = \frac{PV \,.\, AK - PH \,.\, MK}{PH \,.\, KA}$$
$$= \frac{AK \,.\, PV - (AK - AN)(AM - AK)}{AK \,.\, PH}$$
$$= \frac{AK^2 - AK(AM + AN - PV) + AM \,.\, AN}{AK \,.\, PH}$$
$$= \frac{AK^2 - AK \,.\, OM + AM \,.\, AN}{AK \,.\, PH},$$

(since $AN = AT$).

But
$$\frac{OM}{QM} = \frac{NT}{PN};$$

$$\therefore \frac{OM^2}{p_a \,.\, AM} = \frac{4AN^2}{p_a \,.\, AN},$$

whence
$$OM^2 = 4AM \,.\, AN,$$

or
$$AM \,.\, AN = \frac{OM^2}{4}.$$

It follows that

$$AK^2 - AK \,.\, OM + AM \,.\, AN = AK^2 - AK \,.\, OM + \frac{OM^2}{4},$$

which is a complete square, and therefore cannot be negative;

$$\therefore \left(\frac{PV}{PH} - \frac{MK}{KA}\right) \underset{\text{or}=}{>} 0,$$

whence the proposition follows.]

8. If any three similar and similarly situated parabolic segments have one extremity (B) of their bases common and their bases BQ_1, BQ_2, BQ_3 lying along the same straight line, and if EO

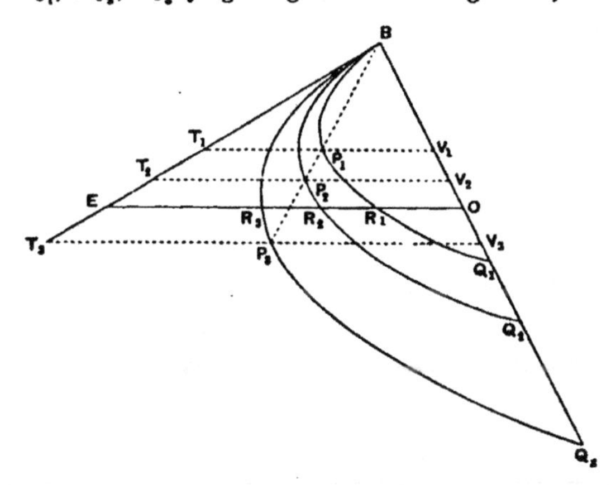

be drawn parallel to the axis of any of the segments meeting the tangent at B to one of them in E, the common base in O, and each of the three segments in R_1, R_2, R_3, then

$$\frac{R_3R_2}{R_2R_1} = \frac{Q_1Q_2}{BQ_3} \cdot \frac{BQ_1}{Q_1Q_2}.$$

[This proposition is given in this place because it is *assumed* without proof (*On floating bodies*, II. 10). But it may well be that it is assumed, not because it was too well known to need proof, but as being an easy deduction from another proposition proved in the *Quadrature of a parabola* which the reader could work out for himself. The latter proposition is given below (No. 1 of the next group) and demonstrates that, if EB be the tangent at B to the segment BR_1Q_1,

$$ER_1 : R_1O = BO : OQ_1.$$

To deduce from this the property enunciated above, we observe *first* that, if V_1, V_2, V_3 be the middle points of the bases of the three segments and the (parallel) diameters through V_1, V_2, V_3 meet the respective segments in P_1, P_2, P_3, then, since the segments are similar,

$$BV_1 : BV_2 : BV_3 = P_1V_1 : P_2V_2 : P_3V_3.$$

It follows that B, P_1, P_2, P_3 are in one straight line.

But, since BE is the tangent at B to the segment BR_1Q_1, $T_1P_1 = P_1V_1$ (where V_1P_1 meets BE in T_1).

Therefore, if V_2P_2, V_3P_3 meet BE in T_2, T_3,

$$T_2P_2 = P_2V_2,$$

and

$$T_3P_3 = P_3V_3,$$

and BE is therefore a tangent to all three segments.

Next, since $\quad ER_1 : R_1O = BO : OQ_1,$

$$\left.\begin{aligned} ER_1 : EO &= BO : BQ_1. \\ ER_2 : EO &= BO : BQ_2, \\ ER_3 : EO &= BO : BQ_3. \end{aligned}\right\}$$

Similarly
and

From the first two relations we derive

$$\frac{R_2R_1}{EO} = BO\left(\frac{1}{BQ_1} - \frac{1}{BQ_2}\right)$$

$$= \frac{BO \cdot Q_1Q_2}{BQ_1 \cdot BQ_2}.$$

Similarly $\quad \dfrac{R_3R_2}{EO} = \dfrac{BO \cdot Q_2Q_3}{BQ_2 \cdot BQ_3}.$

From the last two results it follows that

$$\left.\frac{R_3R_2}{R_2R_1} = \frac{Q_2Q_3}{BQ_3} \cdot \frac{BQ_1}{Q_1Q_2}\right].$$

9. If two similar parabolic segments with bases BQ_1, BQ_2 be placed as described in the preceding proposition, and if BR_1R_2 be any

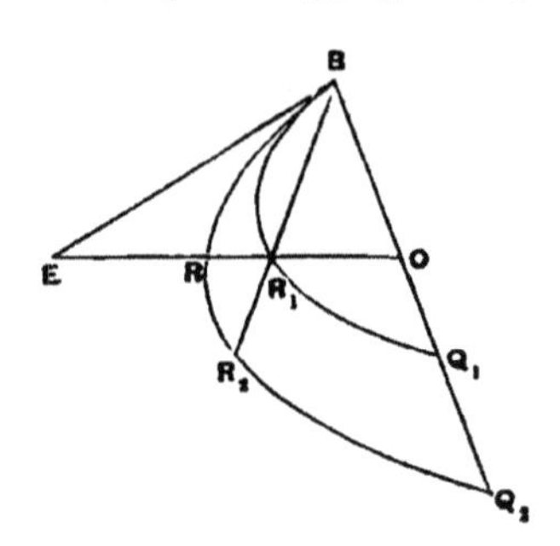

straight line through B cutting the segments in R_1, R_2 respectively, then

$$BQ_1 : BQ_2 = BR_1 : BR_2.$$

[Let the diameter through R_1 meet the tangent at B in E, the other segment in R, and the common base in O.

Then, as in the last proposition,

$$ER_1 : EO = BO : BQ_1,$$

and

$$ER : EO = BO : BQ_2;$$

$$\therefore\ ER : ER_1 = BQ_1 : BQ_2.$$

But, since R_1 is a point within the segment BRQ_2, and ERR_1 is the diameter through R_1, we have in like manner

$$ER : ER_1 = BR_1 : BR_2.$$

Hence

$$BQ_1 : BQ_2 = BR_1 : BR_2.]$$

10. Archimedes assumes the solution of the problem of placing, between two parabolic segments, similar and similarly situated as in the last case, a straight line of a given length and in a direction parallel to the diameters of either parabola.

[Let the given length be l, and assume the problem solved, RR_1 being equal to l.

Using the last figure, we have

$$\frac{BO}{BQ_1} = \frac{ER_1}{EO},$$

and

$$\frac{BO}{BQ_2} = \frac{ER}{EO}.$$

Subtracting, we obtain

$$\frac{BO \cdot Q_1Q_2}{BQ_1 \cdot BQ_2} = \frac{RR_1}{EO}:$$

whence

$$BO \cdot OE = l \cdot \frac{BQ_1 \cdot BQ_2}{Q_1Q_2},$$

which is known.

And the ratio $BO : OE$ is given.

Therefore BO^2, or OE^2, can be found, and therefore O.

Lastly, the diameter through O determines RR_1.]

It remains to describe the investigations in which it is either expressed or implied that they represent new developments of the theory of conics due to Archimedes himself. With the exception of certain propositions relating to the areas of ellipses, his discoveries mostly have reference to the parabola and, in particular, to the determination of the area of any parabolic segment.

The preface to the treatise on that subject (which was called by Archimedes, not *τετραγωνισμὸς παραβολῆς*, but *περὶ τῆς τοῦ ὀρθογωνίου κώνου τομῆς*) is interesting. After alluding to the attempts of the earlier geometers to square the circle and a segment of a circle, he proceeds: "And after that they endeavoured to square the area bounded by the section of the whole cone* and a straight line, assuming lemmas not easily conceded, so that it was recognised by most people that the problem was not solved. But I am not aware that any one of my predecessors has attempted to square the segment bounded by a straight line and a section of a right-angled cone, of which problem I have now discovered the solution. For it is here shown that every segment bounded by a straight line and a section of a right-angled cone is four-thirds of the triangle which has the same base and an equal altitude with the segment, and for the demonstration of this fact the following lemma is assumed†: that the excess by which the greater of (two) unequal areas exceeds the less can, by being added to itself, be made to exceed any given finite area. The earlier geometers have also used this lemma; for it is by the use of this same lemma that they have shown that circles are to one another in the duplicate ratio of their diameters, and that spheres are to one another in the triplicate ratio of their diameters, and further that every pyramid is one third part of the prism having the same base with the pyramid and equal altitude: also, that every cone is one third part of the cylinder having the same base as the cone and equal altitude they proved by assuming a certain lemma similar to that aforesaid. And, in the result, each of the aforesaid theorems has been accepted‡ no less than those proved

* There seems to be some corruption here: the expression in the text is *τᾶς ὅλου τοῦ κώνου τομᾶς*, and it is not easy to give a natural and intelligible meaning to it. The section of 'the whole cone' might perhaps mean a section cutting right through it, i.e. an ellipse, and the 'straight line' might be an axis or a diameter. But Heiberg objects to the suggestion to read *τᾶς ὀξυγωνίου κώνου τομᾶς*, in view of the addition of *καὶ εὐθείας*, on the ground that the former expression always signifies the whole of an ellipse, never a segment of it (*Quaestiones Archimedeae*, p. 149).

† The lemma is used in the mechanical proof only (Prop. 16 of the treatise) and not in the geometrical proof, which depends on Eucl. x. 1 (see p. lxi, lxiii).

‡ The Greek of this passage is: *συμβαίνει δὲ τῶν προειρημένων θεωρημάτων*

without the lemma. As therefore my work now published has satisfied the same test as the propositions referred to, I have written out the proof of it and send it to you, first as investigated by means of mechanics and next also as demonstrated by geometry. Prefixed are, also, the elementary propositions in conics which are of service in the proof" (στοιχεῖα κωνικὰ χρείαν ἔχοντα ἐς τὰν ἀπόδειξιν).

The first three propositions are simple ones merely stated without proof. The remainder, which are given below, were apparently not considered as forming part of the elementary theory of conics; and this fact, together with the circumstance that they appear only as subsidiary to the determination of the areas of parabolic segments, no doubt accounts for what might at first seem strange, viz. that they do not appear in the *Conics* of Apollonius.

1. *If Qq be the base of any segment of a parabola, and P the vertex* of the segment, and if the diameter through any other point R on the curve meet Qq in O, QP in F, and the tangent at Q in E, then*

$$(1) \quad QV : VO = OF : FR,$$

$$(2) \quad QO : Oq = ER : RO†.$$

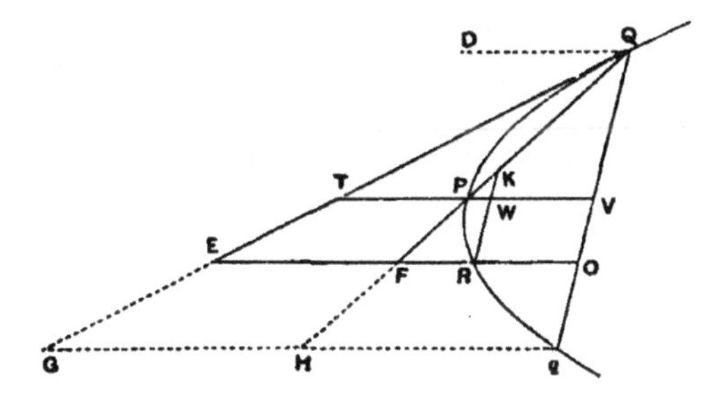

ἕκαστον μηδὲν ἧσσον τῶν ἄνευ τούτου τοῦ λήμματος ἀποδεδειγμένων πεπιστευκέναι. Here it would seem that πεπιστευκέναι must be wrong and that the Passive should have been used.

* According to Archimedes' definition the *height* (ὕψος) of the segment is "the greatest perpendicular from the curve upon the base," and the *vertex* (κορυφά) "the point (on the curve) from which the greatest perpendicular is drawn." The vertex is therefore *P*, the extremity of the diameter bisecting *Qq*.

† These results are used in the *mechanical* investigation of the area of a parabolic segment. The mechanical proof is here omitted both because it is more lengthy and because for the present purpose the geometrical proof given below is more germane.

To prove (1), we draw the ordinate RW to PV, meeting QP in K.

Now $$PV : PW = QV^2 : RW^2;$$
therefore, by parallels,
$$PQ : PK = PQ^2 : PF^2.$$

In other words, PQ, PF, PK are in continued proportion;
$$\therefore\ PQ : PF = PF : PK$$
$$= PF + PQ : PK + PF$$
$$= QF : KF;$$
therefore, by parallels,
$$QV : VO = OF : FR.$$

To prove (2), we obtain from the relation just proved
$$QV : qO = OF : OR.$$

Also, since $TP = PV$, $EF = OF$.

Accordingly, doubling the antecedents in the proportion,
$$Qq : qO = OE : OR,$$
or
$$QO : Oq = ER : RO.$$

It is clear that the equation (1) above is equivalent to a change of axes of coordinates from the tangent and diameter to the chord Qq (as axis of x, say) and the diameter through Q (as the axis of y).

For, if $$QV = a,\ PV = \frac{a^2}{p},$$
and if $$QO = x,\ RO = y,$$
we have at once from (1)
$$\frac{a}{x - a} = \frac{OF}{OF - y};$$
$$\therefore\ \frac{a}{2a - x} = \frac{OF}{y} = \frac{x \cdot \frac{a}{p}}{y},$$
whence $$py = x(2a - x).$$

Zeuthen points out (p. 61) that the results (1) and (2) above can be put in the forms
$$RO \cdot OV = FR \cdot qO \ldots\ldots\ldots\ldots\ldots\ldots\ldots\ldots\ldots(1)$$
and
$$RO \cdot OQ = ER \cdot qO \ldots\ldots\ldots\ldots\ldots\ldots\ldots\ldots\ldots(2)$$

and either of these equations represents a particular case of the parabola as a "locus with respect to four lines." Thus the first represents the equality of the rectangles formed, two and two, from the distances of the movable point R taken in fixed directions from the fixed lines Qq, PV, PQ and Gq (where Gq is the diameter through q); while the second represents the same property with respect to the lines Qq, QD (the diameter through Q), QT and Gq.

2. *If RM be a diameter bisecting QV in M, and RW be the ordinate to PV from R, then*

$$PV = \tfrac{4}{3} RM.$$

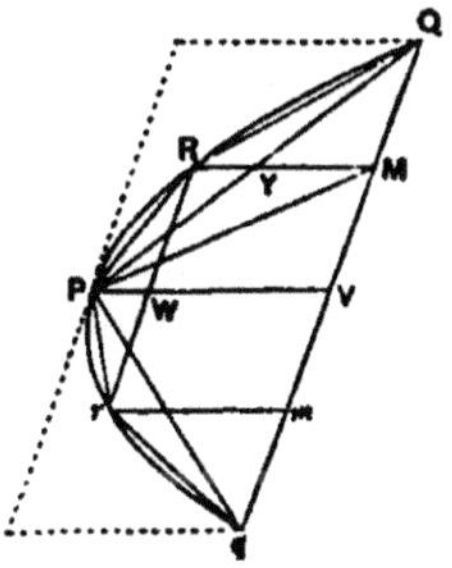

For $\quad PV : PW = QV^2 : RW^2$

$$= 4RW^2 : RW^2;$$

$$\therefore\ PV = 4PW,$$

and $\quad PV = \tfrac{4}{3} RM.$

3. *The triangle PQq is greater than half the segment PQq.*

For the triangle PQq is equal to half the parallelogram contained by Qq, the tangent at P, and the diameters through Q, q. It is therefore greater than half the segment.

COR. It follows that *a polygon can be inscribed in the segment such that the remaining segments are together less than any assignable area.*

For, if we continually take away an area greater than the half, we can clearly, by continually diminishing the remainders, make them, at some time, together less than any given area (Eucl. x. 1).

4. *With the same assumptions as in No. 2 above, the triangle PQq is equal to eight times the triangle RPQ.*

RM bisects QV, and therefore it bisects PQ (in Y, say).

Therefore the tangent at R is parallel to PQ.

Now $\quad PV = \tfrac{4}{3} RM,$

and $\quad PV = 2YM;$

$$\therefore\ YM = 2RY,$$

and $\quad \triangle PQM = 2 \triangle PRQ.$

Hence $\quad \triangle PQV = 4 \triangle PRQ,$

so that $\quad \triangle PQq = 8 \triangle PRQ.$

Also, if RW produced meet the curve again in r,

$$\triangle PQq = 8 \triangle Prq, \text{ similarly.}$$

5. *If there be a series of areas A, B, C, D... each of which is four times the next in order, and if the largest, A, is equal to the triangle PQq, then the sum of all the areas A, B, C, D... will be less than the area of the parabolic segment PQq.*

For, since $\triangle PQq = 8 \triangle PQR = 8 \triangle Pqr$,

$$\triangle PQq = 4(\triangle PQR + \triangle Pqr);$$

therefore, since $\triangle PQq = A$,

$$\triangle PQR + \triangle Pqr = B.$$

In like manner we can prove that the triangles similarly inscribed in the remaining segments are together equal to the area C, and so on.

Therefore $A + B + C + D + \ldots$

is equal to the area of a certain inscribed polygon, and therefore less than the area of the segment.

6. *Given the series A, B, C, D... just described, if Z be the last of the series, then*

$$A + B + C + \ldots + Z + \tfrac{1}{3}Z = \tfrac{4}{3}A.$$

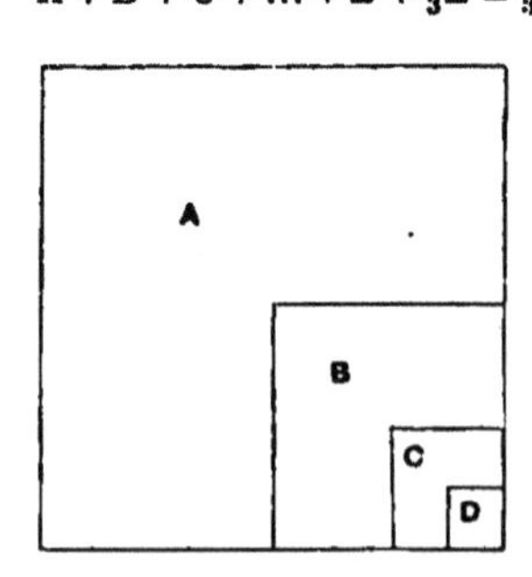

Let $b = \frac{1}{3}B,$

$c = \frac{1}{3}C,$

$d = \frac{1}{3}D,$ and so on.

Then, since $b = \frac{1}{3}B,$

and $B = \frac{1}{4}A,$

$B + b = \frac{1}{3}A.$

Similarly $C + c = \frac{1}{3}B,$

....

Therefore $B + C + D + \ldots + Z + b + c + d + \ldots + z$

$= \frac{1}{3}(A + B + C + D + \ldots + Y).$

But $b + c + d + \ldots + y = \frac{1}{3}(B + C + D + \ldots + Y);$

$\therefore B + C + D + \ldots + Z + z = \frac{1}{3}A,$

or $A + B + C + D + \ldots + Z + \frac{1}{3}Z = \frac{4}{3}A.$

7. *Every segment bounded by a parabola and a chord is four-thirds of the triangle which has the same base and equal altitude.*

Let $K = \frac{4}{3} . \triangle PQq,$

and we have then to prove that the segment is equal to K.

Now, if the segment is not equal to K, it must be either greater or less.

First, suppose it greater. Then, continuing the construction indicated in No. 4, we shall finally have segments remaining whose sum is less than the area by which the segment PQq exceeds K [No. 3, Cor.].

Therefore the polygon must exceed K: which is impossible, for, by the last proposition,

$$A + B + C + \ldots + Z < \frac{4}{3}A,$$

where $A = \triangle PQq.$

Secondly, suppose the segment less than K.

If $\triangle PQq = A, \quad B = \frac{1}{4}A, \quad C = \frac{1}{4}B,$

and so on, until we arrive at an area X such that X is less than the difference between K and the segment,

$$A + B + C + \ldots + X + \frac{1}{3}X = \frac{4}{3}A$$
$$= K.$$

Now, since K exceeds $A + B + C + \ldots + X$ by an area less than X, and the segment by an area greater than X, it follows that

$$A + B + C + \ldots + X$$

is greater than the segment: which is impossible, by No. 4 above.

Thus, since the segment is neither greater nor less than K, it follows that

$$\text{the segment} = K = \tfrac{4}{3} \cdot \triangle PQq.$$

8. The second proposition of the second Book of the treatise *On the equilibrium of planes* (ἐπιπέδων ἰσορροπιῶν) gives a special term for the construction of a polygon in a parabolic segment after the manner indicated in Nos. 2, 4 and 5 above, and enunciates certain theorems connected with it, in the following passage:

"If in a segment bounded by a straight line and a section of a right-angled cone a triangle be inscribed having the same base as the segment and equal altitude, if again triangles be inscribed in the remaining segments having the same bases as those segments and equal altitude, and if in the remaining segments triangles be continually inscribed in the same manner, let the figure so produced be said *to be inscribed in the recognised manner* (γνωρίμως ἐγγράφεσθαι) in the segment.

And it is plain

(1) *that the lines joining the two angles of the figure so inscribed which are nearest to the vertex of the segment, and the next pairs of angles in order, will be parallel to the base of the segment,*

(2) *that the said lines will be bisected by the diameter of the segment, and*

(3) *that they will cut the diameter in the proportions of the successive odd numbers, the number one having reference to [the length adjacent to] the vertex of the segment.*

And these properties will have to be proved in their proper places (ἐν ταῖς τάξεσιν)."

These propositions were no doubt established by Archimedes by means of the above-mentioned properties of parabolic segments; and the last words indicate an intention to collect the propositions in systematic order with proofs. But the intention does not appear to have been carried out, or at least we know of no lost work of Archimedes in which they could have been included. Eutocius proves them by means of Apollonius' *Conics*, as he does not appear to have seen the work on the area of a parabolic segment; but the first two are easily derived from No. 2 above (p. lxi).

The third may be proved as follows.

If $Q_4Q_3Q_2Q_1Pq_1q_2q_3q_4$ be a figure γνωρίμως ἐγγεγραμμένον, we have, since Q_1q_1, Q_2q_2 ... are all parallel and bisected by PV_1,

$$PV_1 : PV_2 : PV_3 : PV_4 \ldots$$
$$= Q_1V_1^2 : Q_2V_2^2 : Q_3V_3^2 : Q_4V_4^2 \ldots$$
$$= 1 : 4 : 9 : 16 \ldots\ldots;$$

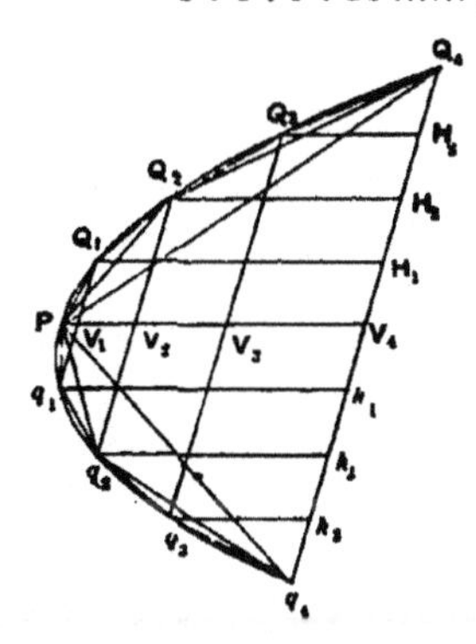

whence it follows that

$$PV_1 : V_1V_2 : V_2V_3 : V_3V_4 \ldots$$
$$= 1 : 3 : 5 : 7 \ldots.$$

9. *If QQ′ be a chord of a parabola bisected in V by the diameter PV, and if PV is of constant length, then the areas of the triangle PQQ′ and of the segment PQQ′ are both constant whatever be the direction of QQ′.*

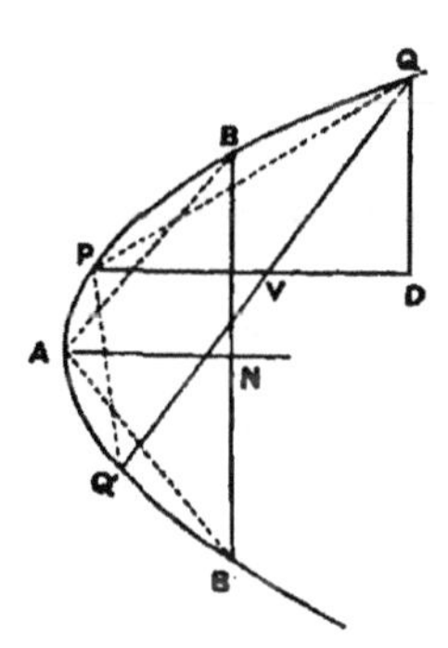

If BAB' be the particular segment whose vertex is A, so that BB' is bisected perpendicularly by the axis at the point N where $AN = PV$, and if QD be drawn perpendicular to PV, we have (by No. 3 on p. liii)

$$QV^2 : QD^2 = p : p_a.$$

Also, since $$AN = PV,$$

$$QV^2 : BN^2 = p : p_a;$$

$$\therefore BN = QD.$$

Hence $$BN . AN = QD . PV,$$

and $$\triangle ABB' = \triangle PQQ'.$$

Therefore the triangle PQQ' is of constant area provided that PV is of given length.

Also the area of the segment PQQ' is equal to $\frac{4}{3} . \triangle PQQ'$;

[No. 7, p. lxiii].

therefore the area of the segment is also constant under the same conditions.

10. *The area of any ellipse is to that of a circle whose diameter is equal to the major axis of the ellipse as the minor axis is to the major (or the diameter of the circle).*

[This is proved in Prop. 4 of the book *On Conoids and Spheroids.*]

11. *The area of an ellipse whose axes are a, b is to that of a circle whose diameter is d, as ab to d^2.*

[*On Conoids and Spheroids*, Prop. 5.]

12. *The areas of ellipses are to one another as the rectangles under their axes; and hence similar ellipses are to one another as the squares of corresponding axes.*

[*On Conoids and Spheroids*, Prop. 6 and Cor.]

It is not within the scope of the present work to give an account of the *applications* of conic sections, by Archimedes and others, e.g. for the purpose of solving equations of a degree higher than the second or in the problems known as νεύσεις*. The former application

* The word νεῦσις, commonly *inclinatio* in Latin, is difficult to translate satisfactorily. Its meaning is best gathered from Pappus' explanation. He says (p. 670): "A line is said to verge (νεύειν) towards a point if, being produced, it reaches the point." As particular cases of the general form of the problem he gives the following:

"Two lines being given in position, to place between them a straight line given in length and verging towards a given point."

"A semicircle and a straight line at right angles to the base being given in

is involved in Prop. 4 of Book II. *On the Sphere and Cylinder*, where the problem is to cut a given sphere (by a plane) so that the segments may bear to one another a given ratio. The book *On Spirals* contains propositions which assume the solution of certain νεύσεις, e.g. Props. 8 and 9, in which Archimedes assumes the following problem to be effected: If AB be any chord of a circle and O any point on the circumference, to draw through O a straight line ODP meeting AB in D and the circle again in P and such that DP is equal to a given length. Though Archimedes does not give the solution, we may infer that he obtained it by means of conic sections*.

A full account of these applications of conic sections by the Greeks will be found in the 11th and 12th chapters of Zeuthen's work, *Die Lehre von den Kegelschnitten im Altertum.*

position, or two semicircles with their bases in a straight line, to place between the two lines a straight line given in length and verging towards a corner of the semicircle."

Thus a line has to be laid across two given lines or curves so that it passes through a given point and the portion intercepted between the lines or curves is equal to a given length.

Zeuthen translates the word νεῦσις by "Einschiebung," or as we might say, "interpolation"; but this fails to express the condition that the required line must pass through a given point, just as the Latin *inclinatio* (and for that matter the Greek term itself) does not explicitly express the other requirement that the intercepted portion of the line shall be of given length.

* Cf. Pappus, pp. 298—302.

PART II.

INTRODUCTION TO THE *CONICS* OF APOLLONIUS.

CHAPTER I.

THE AUTHOR AND HIS OWN ACCOUNT OF THE *CONICS*.

We possess only the most meagre information about Apollonius, viz. that he was born at Perga, in Pamphylia, in the reign of Ptolemy Euergetes (247–222 B.C.), that he flourished under Ptolemy Philopator, and that he went when quite young to Alexandria, where he studied under the successors of Euclid. We also hear of a visit to Pergamum, where he made the acquaintance of Eudemus, to whom he dedicated the first three of the eight Books of the *Conics*. According to the testimony of Geminus, quoted by Eutocius, he was greatly held in honour by his contemporaries, who, in admiration of his marvellous treatise on conics, called him the "great geometer*."

Seven Books only out of the eight have survived, four in the original Greek, and three in an Arabic translation. They were edited by Halley in 1710, the first four Books being given in Greek with a Latin translation, and the remaining three in a Latin translation from the Arabic, to which Halley added a conjectural restoration of the eighth Book.

The first four Books have recently appeared in a new edition by J. L. Heiberg (Teubner, Leipzig, 1891 and 1893), which contains, in addition to the Greek text and a Latin translation, the fragments of the other works of Apollonius which are still extant in Greek, the commentaries and lemmas of Pappus, and the commentaries of Eutocius.

* The quotation is from the sixth Book of Geminus' τῶν μαθημάτων θεωρία. See Apollonius (ed. Heiberg) Vol. II. p. 170.

No additional light has been thrown on the Arabic text of Books V. to VII. since the monumental edition of Halley, except as regards the preface and the first few propositions of Book V., of which L. M. Ludwig Nix published a German translation in 1889*.

For fuller details relating to the MSS. and editions of the *Conics* reference should be made to the Prolegomena to the second volume of Heiberg's edition.

The following is a literal translation of the dedicatory letters in which Apollonius introduces the various Books of his *Conics* to Eudemus and Attalus respectively.

1. Book I. General preface.

"Apollonius to Eudemus, greeting.

"If you are in good health and circumstances are in other respects as you wish, it is well; I too am tolerably well. When I was with you in Pergamum, I observed that you were eager to become acquainted with my work in conics; therefore I send you the first book which I have corrected, and the remaining books I will forward when I have finished them to my satisfaction. I daresay you have not forgotten my telling you that I undertook the investigation of this subject at the request of Naucrates the geometer at the time when he came to Alexandria and stayed with me, and that, after working it out in eight books, I communicated them to him at once, somewhat too hurriedly, without a thorough revision (as he was on the point of sailing), but putting down all that occurred to me, with the intention of returning to them later. Wherefore I now take the opportunity of publishing each portion from time to time, as it is gradually corrected. But, since it has chanced that some other persons also who have been with me have got the first and second books before they were corrected, do not be surprised if you find them in a different shape.

* This appeared in a dissertation entitled *Das fünfte Buch der Conica des Apollonius von Perga in der arabischen Uebersetzung des Thabit ibn Corrah* (Leipzig, 1889), which however goes no further than the middle of the 7th proposition of Book v. and ends on p. 32 in the middle of a sentence with the words "gleich dem Quadrat von"! The fragment is nevertheless valuable in that it gives a new translation of the important preface to Book v., part of which Halley appears to have misunderstood.

"Now of the eight books the first four form an elementary introduction; the first contains the modes of producing the three sections and the opposite branches [of the hyperbola] (τῶν ἀντικειμένων) and their fundamental properties worked out more fully and generally than in the writings of other authors; the second treats of the properties of the diameters and axes of the sections as well as the asymptotes and other things of general importance and necessary for determining limits of possibility (πρὸς τοὺς διορισμούς)*, and what I mean by diameters and axes you will learn from this book. The third book contains many remarkable theorems useful for the synthesis of solid loci and determinations of limits; the most and

* It is not possible to express in one word the meaning of διορισμός here. In explanation of it it will perhaps be best to quote Eutocius who speaks of "that [διορισμός] which does not admit that the proposition is general, but says when and how and in how many ways it is possible to make the required construction, like that which occurs in the twenty-second proposition of Euclid's Elements, *From three straight lines, which are equal to three given straight lines, to construct a triangle*; for in this case it is of course a necessary condition that any two of the straight lines taken together must be greater than the remaining one," [*Comm. on Apoll.* p. 178]. In like manner Pappus [p. 30], in explaining the distinction between a 'theorem' and a 'problem,' says: "But he who propounds a problem, even though he requires what is for some reason impossible of realisation, may be pardoned and held free from blame; for it is the business of the man who seeks a solution to determine at the same time [καὶ τοῦτο διορίσαι] the question of the possible and the impossible, and, if the solution be possible, when and how and in how many ways it is possible." Instances of the διορισμός are common enough. Cf. Euclid VI. 27, which gives the criterion for the possibility of a real solution of the proposition immediately following; the διορισμός there expresses the fact that, for a real solution of the equation $x(a-x)=b^2$, it is a necessary condition that $b^2 \ngtr \left(\frac{a}{2}\right)^2$.

Again, we find in Archimedes, *On the Sphere and Cylinder* [p. 214], the remark that a certain problem "stated thus absolutely requires a διορισμός, but, if certain conditions here existing are added, it does not require a διορισμός."

Many instances will be found in Apollonius' work; but it is to be observed that, as he uses the term, it frequently involves, not only a necessary condition, as in the cases just quoted, but, closely connected therewith, the determination of the *number* of solutions. This can be readily understood when the use of the word in the preface to Book IV. is considered. That Book deals with the number of possible points of intersection of two conics; it follows that, when e.g. in the fifth Book hyperbolas are used for determining by their intersections with given conics the feet of normals to the latter, the number of solutions comes to light at the same time as the conditions necessary to admit of a solution.

prettiest of these theorems are new, and, when I had discovered them, I observed that Euclid had not worked out the synthesis of the locus with respect to three and four lines, but only a chance portion of it and that not successfully: for it was not possible that the synthesis could have been completed without my additional discoveries. The fourth book shows in how many ways the sections of cones meet one another and the circumference of a circle; it contains other matters in addition, none of which has been discussed by earlier writers, concerning the number of points in which a section of a cone or the circumference of a circle meets [the opposite branches of a hyperbola]*.

"The rest [of the books] are more by way of surplusage† (*περιουσιαστικώτερα*): one of them deals somewhat fully (*ἐπὶ πλέον*) with *minima* and *maxima*, one with equal and similar sections of cones, one with theorems involving determination of limits (*διοριστικῶν θεωρημάτων*), and the last with determinate conic problems.

* The reading here translated is Heiberg's *κώνου τομὴ ἢ κύκλου περιφέρεια <ταῖς ἀντικειμέναις> κατὰ πόσα σημεῖα συμβάλλουσι.* Halley had read *κώνου τομὴ ἢ κύκλου περιφέρεια καὶ ἔτι ἀντικείμεναι ἀντικειμέναις κατὰ πόσα σημεῖα συμβάλλουσι.* Heiberg thinks Halley's longer interpolation unnecessary, but I cannot help thinking that Halley gives the truer reading, for the following reasons. (1) The contents of Book IV. show that the sense is not really complete without the mention of the number of intersections of a double-branch hyperbola with another double-branch hyperbola as well as with any of the single-branch conics; and it is scarcely conceivable that Apollonius, in describing what was new in his work, should have mentioned only the less complicated question. (2) If Heiberg's reading is right we should hardly have the plural *συμβάλλουσι* after the disjunctive expression "a section of a cone *or* the circumference of a circle." (3) There is positive evidence for *καὶ ἀντικείμεναι* in Pappus' quotation from this preface [ed. Hultsch, p. 676], where the words are *κώνου τομὴ κύκλου περιφερείᾳ καὶ ἀντικείμεναι ἀντικειμέναις*, "a section of a cone with the circumference of a circle and opposite branches with opposite branches." Thus to combine the reading of our text and that of Pappus would give a satisfactory sense as follows: "in how many points a section of a cone or a circumference of a circle, as well as opposite branches, may [respectively] intersect opposite branches." See, in addition, the note on the corresponding passage in the preface to Book IV. given below.

† *περιουσιαστικώτερα* has been translated "more advanced," but literally it implies extensions of the subject beyond the mere essentials. Hultsch translates "ad abundantiorem scientiam pertinent," and Heiberg less precisely "ulterius progrediuntur."

"When all the books are published it will of course be open to those who read them to judge them as they individually please. Farewell."

2. Preface to Book II.

"Apollonius to Eudemus, greeting.

"If you are in good health, it is well; I too am moderately well. I have sent my son Apollonius to you with the second book of my collected conics. Peruse it carefully and communicate it to those who are worthy to take part in such studies. And if Philonides the geometer, whom I introduced to you in Ephesus, should at any time visit the neighbourhood of Pergamum, communicate the book to him. Take care of your health. Farewell."

3. Preface to Book IV.

"Apollonius to Attalus, greeting.

"Some time ago, I expounded and sent to Eudemus of Pergamum the first three books of my conics collected in eight books; but, as he has passed away, I have resolved to send the remaining books to you because of your earnest desire to possess my works. Accordingly I now send you the fourth book. It contains a discussion of the question, in how many points at most it is possible for the sections of cones to meet one another and the circumference of a circle, on the supposition that they do not coincide throughout, and further in how many points at most a section of a cone and the circumference of a circle meet the opposite branches [of a hyperbola]*

* Here again Halley adds to the text as above translated the words καὶ ἔτι ἀντικείμεναι ἀντικειμέναις. Heiberg thinks the addition unnecessary as in the similar passage in the first preface above. I cannot but think that Halley is right both for the reasons given in the note on the earlier passage, and because, without the added words, it seems to me impossible to explain satisfactorily the distinction between the three separate questions referred to in the next sentence. Heiberg thinks that these refer to the intersections

(1) of conic sections with one another or with a circle,

(2) of sections of a cone with the double-branch hyperbola,

(3) of circles with the double-branch hyperbola.

But to specify separately, as essentially distinct questions, Heiberg's (2) and

and, besides these questions, not a few others of a similar character. Now the first-named question Conon expounded to Thrasydaeus, without however showing proper mastery of the proofs, for which cause Nicoteles of Cyrene with some reason fell foul of him. The second matter has merely been mentioned by Nicoteles, in connexion with his attack upon Conon, as one capable of demonstration; but I have not found it so demonstrated either by himself or by any one else. The third question and the others akin to it I have not found so much as noticed by any one. And all the matters alluded to, which I have not found proved hitherto, needed many and various novel theorems, most of which I have already expounded in the first three books, while the rest are contained in the present one. The investigation of these theorems is of great service both for the synthesis of problems and the determinations of limits of possibility (*πρός τε τὰς τῶν προβλημάτων συνθέσεις καὶ τοὺς διορισμούς*). On the other hand Nicoteles, on account of his controversy with Conon, will not have it that any use can be made of the discoveries of Conon for determinations of limits: in which opinion he is mistaken, for, even if it is possible, without using them at all, to arrive at results relating to such determinations, yet they at all events afford a more ready means of observing some things, e.g. that several

(3) is altogether inconsistent with the scientific method of Apollonius. When he mentions a circle, it is always as a mere appendage to the other curves (*ὑπερβολὴ ἢ ἔλλειψις ἢ κύκλου περιφέρεια* is his usual phrase), and it is impossible, I think, to imagine him drawing a serious distinction between (2) and (3) or treating the omission of Nicoteles to mention (3) as a matter worth noting. *τὸ τρίτον* should surely be something essentially distinct from, not a particular case of, *τὸ δεύτερον*. I think it certain, therefore, that *τὸ τρίτον* is the case of the intersection of two double-branch hyperbolas with one another; and the adoption of Halley's reading would make the passage intelligible. We should then have the following three distinct cases,

(1) the intersections of single-branch conics with one another or with a circle,

(2) the intersections of a single-branch conic or a circle with the double-branch hyperbola,

(3) the intersections of two double-branch hyperbolas;

and *ἄλλα οὐκ ὀλίγα ὅμοια τούτοις* may naturally be taken as referring to those cases *e.g.* where the curves *touch* at one or two points.

solutions are possible or that they are so many in number, and again that no solution is possible; and such previous knowledge secures a satisfactory basis for investigations, while the theorems in question are further useful for the analyses of determinations of limits (πρὸς τὰς ἀναλύσεις δὲ τῶν διορισμῶν). Moreover, apart from such usefulness, they are worthy of acceptance for the sake of the demonstrations themselves, in the same way as we accept many other things in mathematics for this and for no other reason."

4. Preface to Book V*.

"Apollonius to Attalus, greeting.

"In this fifth book I have laid down propositions relating to *maximum* and *minimum* straight lines. You must know that our predecessors and contemporaries have only superficially touched upon the investigation of the shortest lines, and have only proved what straight lines touch the sections and, conversely, what properties they have in virtue of which they are tangents. For my part, I have proved these properties in the first book (without however making any use, in the proofs, of the doctrine of the shortest lines) inasmuch as I wished to place them in close connexion with that part of the subject in which I treated of the production of the three conic sections, in order to show at the same time that in each of the three sections numberless properties and necessary results appear, as they do with reference to the original (transverse) diameter. The propositions in which I discuss the shortest lines I have separated into classes, and dealt with each individual case by careful demonstration; I have also connected the investigation of them with the investigation of the greatest lines above mentioned, because I considered that those who cultivate this science needed them for obtaining a knowledge of the analysis and determination of problems as well as for their synthesis, irrespective of the fact that the subject is one of those which seem worthy of study for their own sake. Farewell."

* In the translation of this preface I have followed pretty closely the German translation of L. M. L. Nix above referred to [p. lxix, note]. The prefaces to Books VI. and VII. are translated from Halley.

5. Preface to Book VI.

"Apollonius to Attalus, greeting.

"I send you the sixth book of the conics, which embraces propositions about conic sections and segments of conics equal and unequal, similar and dissimilar, besides some other matters left out by those who have preceded me. In particular, you will find in this book how, in a given right cone, a section is to be cut equal to a given section, and how a right cone is to be described similar to a given cone and so as to contain a given conic section. And these matters in truth I have treated somewhat more fully and clearly than those who wrote before our time on these subjects. Farewell."

6. Preface to Book VII.

"Apollonius to Attalus, greeting.

"I send to you with this letter the seventh book on conic sections. In it are contained very many new propositions concerning diameters of sections and the figures described upon them; and all these have their use in many kinds of problems, and especially in the determination of the conditions of their possibility. Several examples of these occur in the determinate conic problems solved and demonstrated by me in the eighth book, which is by way of an appendix, and which I will take care to send you as speedily as possible. Farewell."

The first point to be noted in the above account by Apollonius of his own work is the explicit distinction which he draws between the two main divisions of it. The first four Books contain matters which fall within the range of an elementary introduction (*πέπτωκεν εἰς ἀγωγὴν στοιχειώδη*), while the second four are extensions beyond the mere essentials (*περιουσιαστικώτερα*), or (as we may say) more "advanced," provided that we are careful not to understand the relative terms "elementary" and "advanced" in the sense which we should attach to them in speaking of a modern mathematical work. Thus it would be wrong to regard the investigations of the fifth Book as more advanced than the earlier Books on the ground that the results, leading to the determination of the evolute of any conic, are such as are now generally obtained by the aid of the differential calculus; for the investigation of the limiting conditions for the possibility of drawing a certain number of normals to a given conic from a given point is essentially similar in character to many other διορισμοί found in other writers. The only difference is that, while in the case of the parabola the investigation is not very difficult, the corresponding propositions for the hyperbola and ellipse make exceptionally large demands on a geometer's acuteness and grasp. The real distinction between the first four Books and the fifth consists rather in the fact that the former contain a connected and scientific exposition of the general theory of conic sections as the indispensable basis for further extensions of the subject in certain special directions, while the fifth Book is an instance of such specialisation; and the same is true of the sixth and seventh Books. Thus the first four Books were limited to what were considered the essential principles; and their scope was that prescribed by tradition for treatises intended to form an accepted groundwork for such special applications as were found e.g. in the kindred theory of *solid loci* developed by Aristaeus. It would follow that the subject-matter would be for the most part the same as that of earlier treatises, though it would naturally be the object of Apollonius to introduce such improvements of method as the state of knowledge at the time suggested, with a view to securing greater generality and establishing a more thoroughly scientific, and therefore more definitive, system. One effect of the repeated working-up, by successive authors, of for the most part existing material would be to produce crystallisation, so to speak; and therefore we should expect to find in the first four Books of Apollonius greater conciseness than would be possible in a treatise where new ground was being broken. In the latter case the advance would be more gradual, precautions would have to be taken with a view to securing the absolute impregnability of each successive position, and one result would naturally be a certain diffuseness and an apparently excessive attention to minute detail. We find this contrast in the two divisions of Apollonius' *Conics*; in fact, if we except the somewhat lengthy treatment of a small proportion of new matter (such as the properties of the hyperbola with two branches regarded as one conic), the first four Books are concisely put together in comparison with Books V.—VII.

The distinction, therefore, between the two divisions of the work is the distinction between what may be called a text-book or compendium of conic sections and a series of monographs on special portions of the subject.

For the first four Books it will be seen that Apollonius does not claim originality except as regards a number of theorems in the third Book and the investigations in the fourth Book about intersecting conics; for the rest he only claims that the treatment is more full and general than that contained in the earlier works on conics. This statement is quite consistent with that of Pappus that in his first four Books Apollonius incorporated and completed (ἀναπληρώσας) the four Books of Euclid on the same subject.

Eutocius, however, at the beginning of his commentary claims more for Apollonius than he claims for himself. After quoting Geminus' account of the old method of producing the three conics from right cones with different vertical angles by means of plane sections in every case perpendicular to a generator, he says (still purporting to quote Geminus), "But afterwards Apollonius of Perga investigated the general proposition that in every cone, whether right or scalene, all the sections are found, according as the plane [of section] meets the cone in different ways." Again he says, "Apollonius supposed the cone to be either right or scalene, and made the sections different by giving different inclinations to the plane." It can only be inferred that, according to Eutocius, Apollonius was the first discoverer of the fact that other sections than those perpendicular to a generator, and sections of cones other than right cones, had the same properties as the curves produced in the old way. But, as has already been pointed out, we find (1) that Euclid had already declared in the *Phaenomena* that, if a cone (presumably right) or a cylinder be cut by a plane not parallel to the base, the resulting section is a "section of an acute-angled cone," and Archimedes states expressly that all sections of a cone which meet all the generators (and here the cone may be oblique) are either circles or "sections of an acute-angled cone." And it cannot be supposed that Archimedes, or whoever discovered this proposition, could have discovered it otherwise than by a method which would equally show that hyperbolic and parabolic sections could be produced in the same general manner as elliptic sections, which Archimedes singles out for mention because he makes special use of them. Nor (2) can any different conclusion be drawn from the continued use of the old names of the curves even after the more general method of producing them was known; there is nothing unnatural in this because, *first*, hesitation might well be felt in giving up a traditional definition associated with certain standard propositions, determinations of constants, &c., and *secondly*, it is not thought strange, e.g. in a modern text-book of analytical geometry, to define conic sections by means of simple properties and equations, and to adhere to the definitions after it is proved that the curves represented by the general equation of the second degree are none other than the identical curves of the definitions. Hence we must conclude that the statement of Eutocius (which is in any case too general, in that it might lead to the supposition that *every* hyperbola could be produced as a section of any cone) rests on a misapprehension, though perhaps a natural one considering that to him, living so much later, conics probably meant the treatise of Apollonius only, so that he might easily lose sight of the extent of the knowledge possessed by earlier writers*.

At the same time it seems clear that, in the generality of his treatment of the subject from the very beginning, Apollonius was making an entirely new departure. Though Archimedes was aware of the possibility of producing the three conics by means of sections of an oblique or scalene cone, we find no sign of his having used sections other than those which are perpendicular to the plane of symmetry; in other words, he only derives directly from the cone the fundamental property referred to an *axis*, i.e. the relation

$$PN^2 : AN . A'N = P'N'^2 : AN' . A'N',$$

and we must assume that it was by means of the equation referred to the *axes* that the more general property

$$QV^2 : PV . P'V = \text{(const.)}$$

was proved. Apollonius on the other hand starts at once with

* There seems also to have been some confusion in Eutocius' mind about the exact basis of the names *parabola*, *ellipse* and *hyperbola*, though, as we shall see, Apollonius makes this clear enough by connecting them immediately with the method of *application of areas*. Thus Eutocius speaks of the hyperbola as being so called because a certain pair of angles (the vertical angle of an obtuse-angled right cone and the right angle at which the section, made in the old way, is inclined to a generator) together exceed (ὑπερβάλλειν) two right angles, or because the plane of the section passes beyond (ὑπερβάλλειν) the apex of the cone and meets the half of the double cone beyond the apex; and he gives similar explanations of the other two names. But on this interpretation the nomenclature would have no significance; for in each case we could choose different angles in the figure with equal reason, and so vary the names.

the most general section of an oblique cone, and proves directly from the cone that the conic has the latter general property with reference to a particular diameter arising out of his construction, which however is not in general one of the principal diameters. Then, in truly scientific fashion, he proceeds to show directly that the same property which was proved true with reference to the original diameter is equally true with reference to *any* other diameter, and the axes do not appear at all until they appear as particular cases of the new (and arbitrary) diameter. Another indication of the originality of this fuller and more general working-out of the principal properties (*τὰ ἀρχικὰ συμπτώματα ἐπὶ πλέον καὶ καθόλου μᾶλλον ἐξειργασμένα*) is, I think, to be found in the preface to Book V. as newly translated from the Arabic. Apollonius seems there to imply that *minimum* straight lines (i.e. normals) had only been discussed by previous writers in connexion with the properties of tangents, whereas his own order of exposition necessitated an early introduction of the tangent properties, independently of any questions about normals, for the purpose of effecting the transition from the original diameter of reference to any other diameter. This is easily understood when it is remembered that the ordinary properties of normals are expressed with reference to the *axes*, and Apollonius was not in a position to use the axes until they could be brought in as particular cases of the new and arbitrary diameter of reference. Hence he had to adopt a different order from that of earlier works and to postpone the investigation of normals for separate and later treatment.

All authorities agree in attributing to Apollonius the designation of the three conics by the names *parabola*, *ellipse* and *hyperbola*; but it remains a question whether the exact *form* in which their fundamental properties were stated by him, and which suggested the new names, represented a new discovery or may have been known to earlier writers of whom we may take Archimedes as the representative.

It will be seen from Apollonius I. 11 [Prop. 1] that the fundamental property proved from the cone for the parabola is that expressed by the Cartesian equation $y^2 = px$, where the axes of coordinates are any diameter (as the axis of x) and the tangent at its extremity (as the axis of y). Let it be assumed in like manner for the ellipse and hyperbola that y is the ordinate drawn from any point to the original diameter of the conic, x the abscissa measured from one extremity of the diameter, while x_1 is the abscissa measured from the other extremity. Apollonius' procedure is then to take a certain length (p, say) determined in a certain manner with reference to the cone, and to prove, *first*, that

$$y^2 : x . x_1 = p : d \quad \text{..................(1)},$$

where d is the length of the original diameter, and, *secondly*, that, if a perpendicular be erected to the diameter at that extremity of it from which x is measured and of length p, then y^2 is equal to a rectangle of breadth x and "applied" to the perpendicular of length p, but falling short (or exceeding) by a rectangle similar and similarly situated to that contained by p and d; in other words,

$$y^2 = px \mp \frac{x^2}{d^2} . pd$$

or

$$y^2 = px \mp \frac{p}{d} . x^2 \quad \text{..................(2)}.$$

Thus for the ellipse or hyperbola an equation is obtained which differs from that of the parabola in that it contains another term, and y^2 is less or greater than px instead of being equal to it. The line p is called, for all three curves alike, the *parameter* or *latus rectum* corresponding to the original diameter, and the characteristics expressed by the respective equations suggested the three names. Thus the *parabola* is the curve in which the rectangle which is equal to y^2 is applied to p and neither falls short of it nor overlaps it, the *ellipse* and *hyperbola* are those in which the rectangle is applied to p but falls short of it, or overlaps it, respectively.

In Archimedes, on the other hand, while the parameter duly appears with reference to the parabola, no such line is anywhere mentioned in connexion with the ellipse or hyperbola, but the fundamental property of the two latter curves is given in the form

$$\frac{y^2}{x . x_1} = \frac{y'^2}{x' . x_1'},$$

it being further noted that, in the ellipse, either of the equal ratios is equal to $\frac{b^2}{a^2}$ in the case where the equation is referred to the axes and a, b are the major and minor semi-axes respectively.

Thus Apollonius' equation expressed the equality of two *areas*, while Archimedes' equation expressed the equality of two *propor-*

tions; and the question is whether Archimedes and his predecessors were acquainted with the equation of the central conic in the form in which Apollonius gives it, in other words, whether the special use of the *parameter* or *latus rectum* for the purpose of graphically constructing a rectangle having x for one side and equal in area to y^2 was new in Apollonius or not.

On this question Zeuthen makes the following observations.

(1) The equation of the conic in the form

$$\frac{y^2}{x \cdot x_1} = (\text{const.})$$

had the advantage that the constant could be expressed in any shape which might be useful in a particular case, e.g. it might be expressed either as the ratio of one area to another or as the ratio of one straight line to another, in which latter case, if the consequent in the ratio were assumed to be the diameter d, the antecedent would be the parameter p.

(2) Although Archimedes does not, as a rule, connect his description of conics with the technical expressions used in the well-known method of *application of areas*, yet the practical use of that method stood in the same close relation to the formula of Archimedes as it did to that of Apollonius. Thus, where the axes of reference are the axes of the conic and a represents the major or transverse axis, the equation

$$\frac{y^2}{x \cdot x_1} = (\text{const.}) = \lambda \text{ (say)}$$

is equivalent to the equation

$$\frac{y^2}{ax \mp x^2} = \lambda \text{ ....(3)},$$

and, in one place (*On Conoids and Spheroids*, 25, p. 420) where Archimedes uses the property that $\frac{y^2}{x \cdot x_1}$ has the same value for all points on a hyperbola, he actually expresses the denominator of the ratio in the form in which it appears in (3), speaking of it as an area applied to a line equal to a but *exceeding by a square figure* (ὑπερβάλλον εἴδει τετραγώνῳ), in other words, as the area denoted by $ax + x^2$.

(3) The equation $\frac{y^2}{x \cdot x_1} = (\text{const.})$ represents y as a mean proportional between x and a certain constant multiple of x_1, which

last can easily be expressed as the ordinate Y, corresponding to the abscissa x, of a point on a certain straight line passing through the other extremity of the diameter (i.e. the extremity from which x_1 is measured). Whether this particular line appeared as an auxiliary line in the figures used by the predecessors of Apollonius (of which there is no sign), or the well-known constructions were somewhat differently made, is immaterial.

(4) The differences between the two modes of presenting the fundamental properties are so slight that we may regard Apollonius as in reality the typical representative of the Greek theory of conics and as giving indications in his proofs of the train of thought which had led his predecessors no less than himself to the formulation of the various propositions.

Thus, where Archimedes chooses to use *proportions* in investigations for which Apollonius prefers the method of *application of areas* which is more akin to our algebra, Zeuthen is most inclined to think that it is Archimedes who is showing individual peculiarities rather than Apollonius, who kept closer to his Alexandrine predecessors: a view which (he thinks) is supported by the circumstance that the system of applying areas as found in Euclid Book II. is decidedly older than the Euclidean doctrine of proportions.

I cannot but think that the argument just stated leaves out of account the important fact that, as will be seen, the Archimedean form of the equation actually *appears as an intermediate step* in the proof which Apollonius gives of his own fundamental equation. Therefore, as a matter of fact, the Archimedean form can hardly be regarded as a personal variant from the normal statement of the property according to the Alexandrine method. Further, to represent Archimedes' equation in the form

$$\frac{y^2}{x \cdot x_1} = (\text{const.}),$$

and to speak of this as having the advantage that the constant may be expressed differently for different purposes, implies rather more than we actually find in Archimedes, who never uses the constant at all when the hyperbola is in question, and uses it for the ellipse only in the case where the axes of reference are the axes of the ellipse, and then only in the single form $\frac{b^2}{a^2}$.

Now the equation

$$\frac{y^2}{ax - x^2} = \frac{b^2}{a^2},$$

or

$$y^2 = \frac{b^2}{a} \cdot x - \frac{b^2}{a^2} \cdot x^2,$$

does not give an easy means of exhibiting the area y^2 as a simple rectangle applied to a straight line but falling short by another rectangle of equal breadth, *unless* we take some line equal to $\frac{b^2}{a}$ and erect it perpendicularly to the abscissa x at that extremity of it which is on the curve. Therefore, for the purpose of arriving at an expression for y^2 corresponding to those obtained by means of the principle of application of areas, the essential thing was the determination of the parameter p and the expression of the constant in the particular form $\frac{p}{d}$, which however does not appear in Archimedes.

Again, it is to be noted that, though Apollonius actually supplies the proof of the Archimedean form of the fundamental property in the course of the propositions I. 12, 13 [Props. 2, 3] establishing the basis of his definitions of the hyperbola and ellipse, he *retraces his steps* in I. 21 [Prop. 8], and proves it again as a deduction from those definitions: a procedure which suggests a somewhat forced adherence to the latter at the cost of some repetition. This slight awkwardness is easily accounted for if it is assumed that Apollonius was deliberately supplanting an old form of the fundamental property by a new one; but the facts are more difficult to explain on any other assumption. The idea that the form of the equation as given by Apollonius was new is not inconsistent with the fact that the principle of *application of areas* was older than the Euclidean theory of proportions; indeed there would be no cause for surprise if so orthodox a geometer as Apollonius intentionally harked back and sought to connect his new system of conics with the most ancient traditional methods.

It is curious that Pappus, in explaining the new definitions of Apollonius, says (p. 674): "For a certain rectangle applied to a certain line in the section of an acute-angled cone becomes *deficient by a square* (ἐλλεῖπον τετραγώνῳ), in the section of an obtuse-angled cone *exceeding by a square*, and in that of a right-angled cone neither deficient nor exceeding." There is evidently some confusion here, because in the definitions of Apollonius there is no question of exceeding or falling-short *by a square*, but the rectangle which is equal to y^2 exceeds or falls short by a *rectangle* similar and similarly situated to that contained by the diameter and the latus rectum. The description "deficient, or exceeding, by a square" recalls Archimedes' description of the rectangle $x \cdot x_1$ appearing in the equation of the hyperbola as *ὑπερβάλλον εἴδει τετραγώνῳ*; so that it would appear that Pappus somehow confused the two forms in which the two writers give the fundamental property.

It will be observed that the "opposites," by which are meant the opposite branches of a hyperbola, are specially mentioned as distinct from the three sections (the words used by Apollonius being *τῶν τριῶν τομῶν καὶ τῶν ἀντικειμένων*). They are first introduced in the proposition I. 14 [Prop. 4], but it is in I. 16 [Prop. 6] that they are for the first time regarded as together forming one curve. It is true that the preface to Book IV. shows that other writers had already noticed the two opposite branches of a hyperbola, but there can be no doubt that the complete investigation of their properties was reserved for Apollonius. This view is supported by the following evidence. (1) The words of the first preface promise something new and more perfect with reference to the double-branch hyperbola as well as the three single-branch curves; and a comparison between the works of Apollonius and Archimedes (who does not mention the two branches of a hyperbola) would lead us to expect that the greater generality claimed by Apollonius for his treatment of the subject would show itself, if anywhere, in the discussion of the complete hyperbola. The words, too, about the "new and remarkable theorems" in the third Book point unmistakeably to the extension to the case of the complete hyperbola of such properties as that of the rectangles under the segments of intersecting chords. (2) That the treatment of the two branches as one curve was somewhat new in Apollonius is attested by the fact that, notwithstanding the completeness with which he establishes the correspondence between their properties and those of the single branch, he yet continues throughout to speak of them as two independent curves and to prove each proposition with regard to them separately and subsequently to the demonstration of it for the single curves, the result being a certain diffuseness which might have been avoided if the first propositions had been so combined as to prove each property at one and the same time for both double-branch and single-branch conics, and if the further developments had then taken as their basis the generalised property. As it is, the diffuseness marking the separate treatment of the double hyperbola contrasts strongly with the remarkable ingenuity shown by Apollonius in compressing into one proposition the proof of a property common to all three conics. This facility in treating the three curves together is to be explained by the fact that, as successive discoveries in conics were handed down by tradition, the general notion of a conic had been gradually evolved; whereas, if Apollonius had to add new matter with reference to the double hyperbola, it would naturally take the form of propositions supplementary to those affecting the three single-branch curves.

It may be noted in this connexion that the proposition I. 38 [Prop. 15] makes use for the first time of the *secondary diameter* (d') of a hyperbola regarded as a line of definite length determined by the relation

$$\frac{d'^2}{d^2} = \frac{p}{d},$$

where d is the transverse diameter and p the parameter of the ordinates to it. The actual definition of the *secondary diameter* in this sense occurs earlier in the Book, namely between I. 16 and I. 17. The idea may be assumed to have been new, as also the determination of the conjugate hyperbola with two branches as the complete hyperbola which has a pair of conjugate diameters common with the original hyperbola, with the difference that the secondary diameter of the original hyperbola is the transverse diameter of the conjugate hyperbola and *vice versâ*.

The reference to Book II. in the preface does not call for any special remark except as regards the meaning given by Apollonius to the terms *diameter* and *axis*. The words of the preface suggest that the terms were used in a new sense, and this supposition agrees with the observation made above (p. xlix) that with Archimedes only the *axes* are diameters.

The preface speaks of the "many remarkable theorems" contained in Book III. as being useful for "the synthesis of solid loci," and goes on to refer more particularly to the "locus with respect to three and four lines." It is strange that in the Book itself we do not find any theorem stating in terms that a particular geometrical locus is a conic section, though of course we find theorems stating conversely that all points on a conic have a certain property. The explanation of this is probably to be found in the fact that the determination of a locus, even when it was a conic section, was not regarded as belonging to a synthetic treatise on *conics*, and the ground for this may have been that the subject of such loci was extensive enough to require a separate book. This conjecture is supported by the analogy of the treatises of Euclid and Aristaeus on *conics* and *solid loci* respectively, where, so far as we can judge, a very definite line of demarcation appears to have been drawn between the determination of the loci themselves and the theorems in conics which were useful for that end.

There can be no doubt that the brilliant investigations in Book V. with reference to normals regarded as *maximum* and *minimum* straight lines from certain points to the curve were mostly, if not altogether, new. It will be seen that they lead directly to the determination of the Cartesian equation to the evolute of any conic.

Book VI. is about similar conics for the most part, and Book VII. contains an elaborate series of propositions about the magnitude of various functions of the lengths of conjugate diameters, including the determination of their *maximum* and *minimum* values. A comparison of the contents of Book VII. with the remarks about Book VII. and VIII. in the preface to the former suggests that the lost Book VIII. contained a number of problems having for their object the finding of conjugate diameters in a given conic such that certain functions of their lengths have given values. These problems would be solved by means of the results of Book VII., and it is probable that Halley's restoration of Book VIII. represents the nearest conjecture as to their contents which is possible in the present state of our knowledge.

CHAPTER II.

GENERAL CHARACTERISTICS.

§ 1. Adherence to Euclidean form, conceptions and language.

The accepted form of geometrical proposition with which Euclid's *Elements* more than any other book has made mathematicians familiar, and the regular division of each proposition into its component parts or stages, cannot be better described than in the words of Proclus. He says*: "Every problem and every theorem which is complete with all its parts perfect purports to contain in itself all of the following elements: *enunciation* (πρότασις), *setting-out* (ἔκθεσις), *definition*† (διορισμός), *construction* (κατασκευή), *proof* (ἀπόδειξις), *conclusion* (συμπέρασμα). Now of these the *enunciation* states what is given and what is that which is sought, the perfect *enunciation* consisting of both these parts. The *setting-out* marks off what is given, by itself, and adapts it beforehand for use in the investigation. The *definition* states separately and makes clear what the particular thing is which is sought. The *construction* adds what is wanting to the datum for the purpose of finding what is sought. The *proof* draws the required inference by reasoning scientifically from acknowledged facts. The *conclusion* reverts again to the *enunciation*, confirming what has been demonstrated. These are all the parts of problems and theorems, but the most essential and those which are found in all are *enunciation*, *proof*, *conclusion*. For it is equally necessary to know beforehand what is sought, and that this should be demonstrated by means of the intermediate steps and the demonstrated fact should be inferred; it is impossible to dispense

* Proclus (ed. Friedlein), p. 203.

† The word *definition* is used for want of a better. As will appear from what follows, διορισμὸς really means a *closer description*, by means of a concrete figure, of what the enunciation states in general terms as the property to be proved or the problem to be solved.

with any of these three things. The remaining parts are often brought in, but are often left out as serving no purpose. Thus there is neither *setting-out* nor *definition* in the problem of constructing an isosceles triangle having each of the angles at the base double of the remaining angle, and in most theorems there is no *construction* because the *setting-out* suffices without any addition for demonstrating the required property from the data. When then do we say that the *setting-out* is wanting? The answer is, when there is nothing *given* in the *enunciation*; for, though the *enunciation* is in general divided into what is given and what is sought, this is not always the case, but sometimes it states only what is sought, i.e. what must be known or found, as in the case of the problem just mentioned. That problem does not, in fact, state beforehand with what datum we are to construct the isosceles triangle having each of the equal angles double of the remaining one, but (simply) that we are to find such a triangle.... When, then, the enunciation contains both (what is given and what is sought), in that case we find both *definition* and *setting-out*, but, whenever the datum is wanting, they too are wanting. For not only is the *setting-out* concerned with the datum but so is the *definition* also, as, in the absence of the datum, the *definition* will be identical with the *enunciation*. In fact, what could you say in defining the object of the aforesaid problem except that it is required to find an isosceles triangle of the kind referred to? But that is what the *enunciation* stated. If then the *enunciation* does not include, on the one hand, what is given and, on the other, what is sought, there is no *setting-out* in virtue of there being no datum, and the *definition* is left out in order to avoid a mere repetition of the *enunciation*."

The constituent parts of an Euclidean proposition will be readily identified by means of the above description without further details. It will be observed that the word διορισμός has here a different signification from that described in the note to p. lxx above. Here it means a closer definition or description of the object aimed at, by means of the concrete lines or figures set out in the ἔκθεσις instead of the general terms used in the enunciation; and its purpose is to rivet the attention better, as indicated by Proclus in a later passage, τρόπον τινὰ προσεχείας ἐστὶν αἴτιος ὁ διορισμός.

The other technical use of the word to signify the limitations to which the possible solutions of a problem are subject is also described by Proclus, who speaks of διορισμοί determining "whether what is sought is impossible or possible, and how far it is practicable and in how many ways*"; and the διορισμός in this sense appears in the same form in Euclid as in Archimedes and Apollonius. In Apollonius it is sometimes inserted in the body of a problem as in the instance II. 50 [Prop. 50] given below; in another case it forms the subject of a separate preliminary theorem, II. 52 [Prop. 51], the result being quoted in the succeeding proposition II. 53 [Prop. 52] in the same way as the διορισμός in Eucl. VI. 27 is quoted in the enunciation of VI. 28 (*see* p. cviii).

Lastly, the orthodox division of a problem into *analysis* and *synthesis* appears regularly in Apollonius as in Archimedes. Proclus speaks of the preliminary *analysis* as a way of investigating the more recondite problems (τὰ ἀσαφέστερα τῶν προβλημάτων); thus it happens that in this respect Apollonius is often even more formal than Euclid, who, in the *Elements*, is generally able to leave out all the preliminary analysis in consequence of the comparative simplicity of the problems solved, though the *Data* exhibit the method as clearly as possible.

In order to illustrate the foregoing remarks, it is only necessary to reproduce a theorem and a problem in the exact form in which they appear in Apollonius, and accordingly the following propositions are given in full as typical specimens, the translation on the right-hand side following the Greek exactly, except that the *letters* are changed in order to facilitate comparison with the same propositions as reproduced in this work and with the corresponding figures.

III. 54 [Prop. 75 with the first figure].

Ἐὰν κώνου τομῆς ἢ κύκλου περιφερείας δύο εὐθεῖαι ἐφαπτόμεναι συμπίπτωσι, διὰ δὲ τῶν ἁφῶν παράλληλοι ἀχθῶσι ταῖς ἐφαπτομέναις, καὶ ἀπὸ τῶν ἁφῶν πρὸς τὸ αὐτὸ σημεῖον τῆς γραμμῆς διαχθῶσιν εὐθεῖαι τέμνουσαι τὰς παραλλήλους, τὸ περιεχόμενον ὀρθογώνιον ὑπὸ τῶν ἀποτεμνομένων πρὸς τὸ ἀπὸ τῆς ἐπιζευγνυούσης τὰς ἁφὰς τετράγωνον λόγον ἔχει τὸν συγκείμενον ἔκ τε τοῦ, ὃν ἔχει τῆς ἐπιζευγνυούσης τὴν σύμπτωσιν τῶν ἐφαπτομένων καὶ τὴν διχοτομίαν τῆς τὰς ἁφὰς ἐπιζευγνυούσης

If two straight lines touching a section of a cone or the circumference of a circle meet, and through the points of contact parallels be drawn to the tangents, and from the points of contact straight lines be drawn through the same point of the curve cutting the parallels, the rectangle contained by the intercepts bears to the square on the line joining the points of contact the ratio compounded [1] of that which the square of the inner seg-

* Proclus, p. 202.

τὸ ἐντὸς τμῆμα πρὸς τὸ λοιπὸν δυνάμει, καὶ τοῦ, ὃν ἔχει τὸ ὑπὸ τῶν ἐφαπτομένων περιεχόμενον ὀρθογώνιον πρὸς τὸ τέταρτον μέρος τοῦ ἀπὸ τῆς τὰς ἁφὰς ἐπιζευγνυούσης τετραγώνου.

ment of the line joining the point of concourse of the tangents and the point of bisection of the line joining the points of contact bears to the square of the remaining segment, and [2] of that which the rectangle contained by the tangents bears to the fourth part of the square on the line joining the points of contact.

ἔστω κώνου τομὴ ἢ κύκλου περιφέρεια ἡ ΑΒΓ καὶ ἐφαπτόμεναι αἱ ΑΔ, ΓΔ, καὶ ἐπεζεύχθω ἡ ΑΓ καὶ δίχα τετμήσθω κατὰ τὸ Ε, καὶ ἐπεζεύχθω ἡ ΔΒΕ, καὶ ἤχθω ἀπὸ μὲν τοῦ Α παρὰ τὴν ΓΔ ἡ ΑΖ, ἀπὸ δὲ τοῦ Γ παρὰ τὴν ΑΔ ἡ ΓΗ, καὶ εἰλήφθω τι σημεῖον ἐπὶ τῆς γραμμῆς τὸ Θ, καὶ ἐπιζευχθεῖσαι αἱ ΑΘ, ΓΘ ἐκβεβλήσθωσαν ἐπὶ τὰ Η, Ζ. λέγω, ὅτι τὸ ὑπὸ ΑΖ, ΓΗ πρὸς τὸ ἀπὸ ΑΓ τὸν συγκείμενον ἔχει λόγον ἐκ τοῦ, ὃν ἔχει τὸ ἀπὸ ΕΒ πρὸς τὸ ἀπὸ ΒΔ καὶ τὸ ὑπὸ ΑΔΓ πρὸς τὸ τέταρτον τοῦ ἀπὸ ΑΓ, τουτέστι τὸ ὑπὸ ΑΕΓ.

Let QPQ' be a section of a cone or the circumference of a circle and QT, $Q'T$ tangents, and let QQ' be joined and bisected at V, and let TPV be joined, and let there be drawn, from Q, Qr parallel to $Q'T$ and, from Q', $Q'r'$ parallel to QT, and let any point R be taken on the curve, and let QR, $Q'R$ be joined and produced to r', r. I say that the rectangle contained by Qr, $Q'r'$ has to the square on QQ' the ratio compounded of that which the square on VP has to the square on PT and that which the rectangle under QTQ'* has to the fourth part of the square on QQ', i.e. the rectangle under QVQ'.

ἤχθω γὰρ ἀπὸ μὲν τοῦ Θ παρὰ τὴν ΑΓ ἡ ΚΘΟΞΛ, ἀπὸ δὲ τοῦ Β ἡ ΜΒΝ· φανερὸν δή, ὅτι ἐφάπτεται ἡ ΜΝ. ἐπεὶ οὖν ἴση ἐστὶν ἡ ΑΕ τῇ ΕΓ, ἴση ἐστὶ καὶ ἡ ΜΒ τῇ ΒΝ καὶ ἡ ΚΟ τῇ ΟΛ καὶ ἡ ΘΟ τῇ ΟΞ καὶ ἡ ΚΘ τῇ ΞΛ. ἐπεὶ οὖν ἐφάπτονται αἱ ΜΒ, ΜΑ, καὶ παρὰ τὴν ΜΒ ἦκται ἡ ΚΘΛ, ἔστιν, ὡς τὸ ἀπὸ ΑΜ πρὸς τὸ ἀπὸ ΜΒ, τουτέστι τὸ ὑπὸ ΜΒΝ, τὸ ἀπὸ ΑΚ πρὸς τὸ ὑπὸ ΞΚΘ, τουτέστι τὸ ὑπὸ ΛΘΚ. ὡς δὲ τὸ ὑπὸ ΝΓ, ΜΑ πρὸς τὸ ἀπὸ ΜΑ, τὸ ὑπὸ ΛΓ, ΚΑ πρὸς τὸ ἀπὸ ΚΑ· δι' ἴσου ἄρα, ὡς τὸ ὑπὸ ΝΓ, ΜΑ πρὸς τὸ ὑπὸ ΝΒΜ, τὸ ὑπὸ ΛΓ, ΚΑ πρὸς τὸ ὑπὸ

For let there be drawn, from R, $KRWR'K'$, and, from P, LPL' parallel to QQ'; it is then clear that LL' is a tangent. Now, since QV is equal to VQ', LP is also equal to PL' and KW to WK' and RW to WR' and KR to $R'K'$. Since therefore LP, LQ are tangents, and KRK' is drawn parallel to LP, as the square on QL is to the square on LP, that is, the rectangle under LPL', so is the square on QK to the rectangle under $R'KR$, that is, the rectangle under $K'RK$. And, as the rectangle under $L'Q'$,

* τὸ ὑπὸ ΑΔΓ, "the rect. under QTQ'," means the rectangle $QT.TQ'$, and similarly in other cases.

ΛΘΚ. τὸ δὲ ὑπὸ ΑΓ, ΚΑ πρὸς τὸ ὑπὸ ΛΘΚ τὸν συγκείμενον ἔχει λόγον ἐκ τοῦ τῆς ΓΛ πρὸς ΛΘ, τουτέστι τῆς ΖΑ πρὸς ΑΓ, καὶ τοῦ τῆς ΑΚ πρὸς ΚΘ, τουτέστι τῆς ΗΓ πρὸς ΓΑ, ὅς ἐστιν ὁ αὐτὸς τῷ, ὃν ἔχει τὸ ὑπὸ ΗΓ, ΖΑ πρὸς τὸ ἀπὸ ΓΑ· ὡς ἄρα τὸ ὑπὸ ΝΓ, ΜΑ πρὸς τὸ ὑπὸ ΝΒΜ, τὸ ὑπὸ ΗΓ, ΖΑ πρὸς τὸ ἀπὸ ΓΑ. τὸ δὲ ὑπὸ ΓΝ, ΜΑ πρὸς τὸ ὑπὸ ΝΒΜ τοῦ ὑπὸ ΝΔΜ μέσου λαμβανομένου τὸν συγκείμενον ἔχει λόγον ἐκ τοῦ, ὃν ἔχει τὸ ὑπὸ ΓΝ, ΑΜ προς τὸ ὑπὸ ΝΔΜ καὶ τὸ ὑπὸ ΝΔΜ πρὸς τὸ ὑπὸ ΝΒΜ· τὸ ἄρα ὑπὸ ΗΓ, ΖΑ πρὸς τὸ ἀπὸ ΓΑ τὸν συγκείμενον ἔχει λόγον ἐκ τοῦ τοῦ ὑπὸ ΓΝ, ΑΜ πρὸς τὸ ὑπὸ ΝΔΜ καὶ τοῦ ὑπὸ ΝΔΜ πρὸς τὸ ὑπὸ ΝΒΜ. ἀλλ' ὡς μὲν τὸ ὑπὸ ΝΓ, ΑΜ πρὸς τὸ ὑπὸ ΝΔΜ, τὸ ἀπὸ ΕΒ πρὸς τὸ ἀπὸ ΒΔ, ὡς δὲ τὸ ὑπὸ ΝΔΜ πρὸς τὸ ὑπὸ ΝΒΜ, τὸ ὑπὸ ΓΔΑ πρὸς τὸ ὑπὸ ΓΕΑ· τὸ ἄρα ὑπὸ ΗΓ, ΑΖ πρὸς τὸ ἀπὸ ΑΓ τὸν συγκείμενον ἔχει λόγον ἐκ τοῦ τοῦ ἀπὸ ΒΕ πρὸς τὸ ἀπὸ ΒΔ καὶ τοῦ ὑπὸ ΓΔΑ πρὸς τὸ ὑπὸ ΓΕΑ.

LQ is to the square on LQ, so is the rectangle under $K'Q'$, KQ to the square on KQ; therefore *ex aequo* as the rectangle under $L'Q'$, LQ is to the rectangle under $L'PL$, so is the rectangle under $K'Q'$, KQ to the rectangle under $K'RK$. But the rectangle under $K'Q'$, KQ has to the rectangle under $K'RK$ the ratio compounded of that of $Q'K'$ to $K'R$, that is, of rQ to QQ', and of that of QK to KR, that is, of $r'Q'$ to $Q'Q$, which is the same as the ratio which the rectangle under $r'Q'$, rQ has to the square on $Q'Q$; hence, as the rectangle under $L'Q'$, LQ is to the rectangle under $L'PL$, so is the rectangle under $r'Q'$, rQ to the square on $Q'Q$. But the rectangle under $Q'L'$, LQ has to the rectangle under $L'PL$ (if the rectangle under $L'TL$ be taken as a mean) the ratio compounded of that which the rectangle under $Q'L'$, QL has to the rectangle under $L'TL$ and the rectangle under $L'TL$ to the rectangle under $L'PL$; hence the rectangle under $r'Q'$, rQ has to the square on $Q'Q$ the ratio compounded of that of the rectangle under $Q'L'$, QL to the rectangle under $L'TL$ and of the rectangle under $L'TL$ to the rectangle under $L'PL$. But, as the rectangle under $L'Q'$, QL is to the rectangle under $L'TL$, so is the square on VP to the square on PT, and, as the rectangle under $L'TL$ is to the rectangle under $L'PL$, so is the rectangle under $Q'TQ$ to the rectangle under $Q'VQ$; therefore the rectangle under $r'Q'$, rQ has to the square on QQ' the ratio compounded of that of the square on PV to the square on PT and of the rectangle under $Q'TQ$ to the rectangle under $Q'VQ$.

II. 50 [Prop. 50 (Problem)].

(So far as relating to the *hyperbola*.)

Τῆς δοθείσης κώνου τομῆς ἐφαπτομένην ἀγαγεῖν, ἥτις πρὸς τῷ ἄξονι γωνίαν ποιήσει ἐπὶ ταὐτὰ τῇ τομῇ ἴσην τῇ δοθείσῃ ὀξείᾳ γωνίᾳ.

To draw a tangent to a given section of a cone which shall make with the axis towards the same parts with the section an angle equal to a given acute angle.

* * * *

Ἔστω ἡ τομὴ ὑπερβολή, καὶ γεγονέτω, καὶ ἔστω ἐφαπτομένη ἡ ΓΔ, καὶ εἰλήφθω τὸ κέντρον τῆς τομῆς τὸ Χ, καὶ ἐπεζεύχθω ἡ ΓΧ καὶ κάθετος ἡ ΓΕ· λόγος ἄρα τοῦ ὑπὸ τῶν ΧΕΔ πρὸς τὸ ἀπὸ τῆς ΕΓ δοθείς· ὁ αὐτὸς γάρ ἐστι τῷ τῆς πλαγίας πρὸς τὴν ὀρθίαν. τοῦ δὲ ἀπὸ τῆς ΓΕ πρὸς τὸ ἀπὸ τῆς ΕΔ λόγος ἐστὶ δοθείς· δοθεῖσα γὰρ ἑκατέρα τῶν ὑπὸ ΓΔΕ, ΔΕΓ. λόγος ἄρα καὶ τοῦ ὑπὸ ΧΕΔ πρὸς τὸ ἀπὸ τῆς ΕΔ δοθείς· ὥστε καὶ τῆς ΧΕ πρὸς ΕΔ λόγος ἐστὶ δοθείς. καὶ δοθεῖσα ἡ πρὸς τῷ Ε· δοθεῖσα ἄρα καὶ ἡ πρὸς τῷ Χ. πρὸς δὴ θέσει εὐθείᾳ τῇ ΧΕ καὶ δοθέντι τῷ Χ διῆκταί τις ἡ ΓΧ ἐν δεδομένῃ γωνίᾳ· θέσει ἄρα ἡ ΓΧ. θέσει δὲ καὶ ἡ τομή· δοθὲν ἄρα τὸ Γ. καὶ διῆκται ἐφαπτομένη ἡ ΓΔ· θέσει ἄρα ἡ ΓΔ.

Let the section be a hyperbola, and suppose it done, and let PT be the tangent, and let the centre C of the section be taken and let PC be joined and PN be perpendicular; therefore the ratio of the rectangle contained by CNT to the square on NP is given, for it is the same as that of the transverse to the erect. And the ratio of the square PN to the square on NT is given, for each of the angles PTN, TNP is given. Therefore also the ratio of the rectangle under CNT to the square on NT is given; so that the ratio of CN to NT is also given. And the angle at N is given; therefore also the angle at C is given. Thus with the straight line CN [given] in position and at the given point C a certain straight line PC has been drawn at a given angle; therefore PC is [given] in position. Also the section is [given] in position; therefore P is given. And the tangent PT has been drawn; therefore PT is [given] in position.

ἤχθω ἀσύμπτωτος τῆς τομῆς ἡ ΖΧ· ἡ ΓΔ ἄρα ἐκβληθεῖσα συμπεσεῖται τῇ ἀσυμπτώτῳ. συμπιπτέτω κατὰ τὸ Ζ. μείζων ἄρα ἔσται ἡ ὑπὸ ΖΔΕ γωνία τῆς ὑπὸ ΖΧΔ. δεήσει ἄρα εἰς τὴν σύνθεσιν τὴν δεδομένην ὀξεῖαν γωνίαν μείζονα εἶναι τῆς ἡμισείας τῆς περιεχομένης ὑπὸ τῶν ἀσυμπτώτων.

Let the asymptote LC of the section be drawn; then PT produced will meet the asymptote. Let it meet it in L; then the angle LTN will be greater than the angle LCT. Therefore it will be necessary for the synthesis that the given acute angle should be greater than

συντεθήσεται δὴ τὸ πρόβλημα οὕτως· ἔστω ἡ μὲν δοθεῖσα ὑπερβολή, ἧς ἄξων ὁ ΑΒ, ἀσύμπτωτος δὲ ἡ ΧΖ, ἡ δὲ δοθεῖσα γωνία ὀξεῖα μείζων οὖσα τῆς ὑπὸ τῶν ΑΧΖ ἡ ὑπὸ ΚΘΗ, καὶ ἔστω τῇ ὑπὸ τῶν ΑΧΖ ἴση ἡ ὑπὸ ΚΘΛ, καὶ ἤχθω ἀπὸ τοῦ Α τῇ ΑΒ πρὸς ὀρθὰς ἡ ΑΖ, εἰλήφθω δέ τι σημεῖον ἐπὶ τῆς ΗΘ τὸ Η, καὶ ἤχθω ἀπ' αὐτοῦ ἐπὶ τὴν ΘΚ κάθετος ἡ ΗΚ. ἐπεὶ οὖν ἴση ἐστὶν ἡ ὑπὸ ΖΧΑ τῇ ὑπὸ ΛΘΚ, εἰσὶ δὲ καὶ αἱ πρὸς τοῖς Α, Κ γωνίαι ὀρθαί, ἔστιν ἄρα, ὡς ἡ ΧΑ πρὸς ΑΖ, ἡ ΘΚ πρὸς ΚΛ. ἡ δὲ ΘΚ πρὸς ΚΛ μείζονα λόγον ἔχει ἤπερ πρὸς τὴν ΗΚ· καὶ ἡ ΧΑ πρὸς ΑΖ ἄρα μείζονα λόγον ἔχει ἤπερ ἡ ΘΚ πρὸς ΚΗ. ὥστε καὶ τὸ ἀπὸ ΧΑ πρὸς τὸ ἀπὸ ΑΖ μείζονα λόγον ἔχει ἤπερ τὸ ἀπὸ ΘΚ πρὸς τὸ ἀπὸ ΚΗ. ὡς δὲ τὸ ἀπὸ ΧΑ πρὸς τὸ ἀπὸ ΑΖ, ἡ πλαγία πρὸς τὴν ὀρθίαν· καὶ ἡ πλαγία ἄρα πρὸς τὴν ὀρθίαν μείζονα λόγον ἔχει ἤπερ τὸ ἀπὸ ΘΚ πρὸς τὸ ἀπὸ ΚΗ. ἐὰν δὴ ποιήσωμεν, ὡς τὸ ἀπὸ ΧΑ πρὸς τὸ ἀπὸ ΑΖ, οὕτως ἄλλο τι πρὸς τὸ ἀπὸ ΚΗ, μεῖζον ἔσται τοῦ ἀπὸ ΘΚ. ἔστω τὸ ὑπὸ ΜΚΘ· καὶ ἐπεζεύχθω ἡ ΗΜ. ἐπεὶ οὖν μεῖζόν ἐστι τὸ ἀπὸ ΜΚ τοῦ ὑπὸ ΜΚΘ, τὸ ἄρα ἀπὸ ΜΚ πρὸς τὸ ἀπὸ ΚΗ μείζονα λόγον ἔχει ἤπερ τὸ ὑπὸ ΜΚΘ πρὸς τὸ ἀπὸ ΚΗ, τουτέστι τὸ ἀπὸ ΧΑ πρὸς τὸ ἀπὸ ΑΖ. καὶ ἐὰν ποιήσωμεν, ὡς τὸ ἀπὸ ΜΚ πρὸς τὸ ἀπὸ ΚΗ, οὕτως τὸ ἀπὸ ΧΑ πρὸς ἄλλο τι, ἔσται πρὸς ἔλαττον τοῦ ἀπὸ ΑΖ· καὶ ἡ ἀπὸ τοῦ Χ ἐπὶ τὸ ληφθὲν σημεῖον ἐπιζευγνυμένη εὐθεῖα ὅμοια ποιήσει τὰ τρίγωνα, καὶ διὰ τοῦτο μείζων ἐστὶν ἡ ὑπὸ ΖΧΑ τῆς ὑπὸ ΗΜΚ. κείσθω δὴ τῇ ὑπὸ ΗΜΚ ἴση ἡ ὑπὸ ΑΧΓ· ἡ ἄρα ΧΓ τεμεῖ τὴν τομήν. τεμνέτω κατὰ τὸ Γ, καὶ ἀπὸ τοῦ Γ ἐφαπτομένη τῆς τομῆς ἤχθω ἡ ΓΔ, καὶ κάθετος ἡ ΓΕ· ὅμοιον

the half of that contained by the asymptotes.

Thus the synthesis of the problem will proceed as follows: let the given hyperbola be that of which AA' is the axis and CZ an asymptote, and the given acute angle (being greater than the angle ACZ) the angle FED, and let the angle FEH be equal to the angle ACZ, and let AZ be drawn from A at right angles to AA', and let any point D be taken on DE, and let a perpendicular DF be drawn from it upon EF. Then, since the angle ZCA is equal to the angle HEF, and also the angles at A, F are right, as CA is to AZ, so is EF to FH. But EF has to FH a greater ratio than it has to FD; therefore also CA has to AZ a greater ratio than EF has to FD. Hence also the square on CA has to the square on AZ a greater ratio than the square on EF has to the square on FD. And, as the square on CA is to the square on AZ, so is the transverse to the erect; therefore also the transverse has to the erect a greater ratio than the square on EF has to the square on FD. If then we make, as the square on CA to the square on AZ, so some other area to the square on FD, that area will be greater than the square on EF. Let it be the rectangle under KFE; and let DK be joined. Then, since the square on KF is greater than the rectangle under KFE, the square on KF has to the square on FD a greater ratio than the rectangle under KFE has to the square on FD, that is, the square on CA to the square on AZ. And if we make, as the square on KF to the square on FD, so the square on CA to

ἄρα ἐστὶ τὸ ΓΧΕ τρίγωνον τῷ ΗΜΚ. ἔστιν ἄρα, ὡς τὸ ἀπὸ ΧΕ πρὸς τὸ ἀπὸ ΕΓ, τὸ ἀπὸ ΜΚ πρὸς τὸ ἀπὸ ΚΗ. ἔστι δὲ καί, ὡς ἡ πλαγία πρὸς τὴν ὀρθίαν, τό τε ὑπὸ ΧΕΔ πρὸς τὸ ἀπὸ ΕΓ καὶ τὸ ὑπὸ ΜΚΘ πρὸς τὸ ἀπὸ ΚΗ. καὶ ἀνάπαλιν, ὡς τὸ ἀπὸ ΓΕ πρὸς τὸ ὑπὸ ΧΕΔ, τὸ ἀπὸ ΗΚ πρὸς τὸ ὑπὸ ΜΚΘ· δι' ἴσου ἄρα, ὡς τὸ ἀπὸ ΧΕ πρὸς τὸ ὑπὸ ΧΕΔ, τὸ ἀπὸ ΜΚ πρὸς τὸ ὑπὸ ΜΚΘ. καὶ ὡς ἄρα ἡ ΧΕ πρὸς ΕΔ, ἡ ΜΚ πρὸς ΚΘ. ἦν δὲ καί, ὡς ἡ ΓΕ πρὸς ΕΧ, ἡ ΗΚ πρὸς ΚΜ· δι' ἴσου ἄρα, ὡς ἡ ΓΕ πρὸς ΕΔ, ἡ ΗΚ πρὸς ΚΘ. καὶ εἰσὶν ὀρθαὶ αἱ πρὸς τοῖς Ε, Κ γωνίαι· ἴση ἄρα ἡ πρὸς τῷ Δ γωνία τῇ ὑπὸ ΗΘΚ.

another area, [the ratio] will be to a smaller area than the square on *AZ*; and the straight line joining *C* to the point taken will make the triangles similar, and for this reason the angle *ZCA* is greater than the angle *DKF*. Let the angle *ACP* be made equal to the angle *DKF*; therefore *CP* will cut the section. Let it cut it at *P*, and from *P* let *PT* be drawn touching the section, and *PN* perpendicular; therefore the triangle *PCN* is similar to *DKF*. Therefore, as is the square on *CN* to the square on *NP*, so is the square on *KF* to the square on *FD*. Also, as the transverse is to the erect, so is both the rectangle under *CNT* to the square on *NP* and the rectangle under *KFE* to the square on *FD*. And conversely, as the square on *PN* is to the rectangle under *CNT*, so is the square on *DF* to the rectangle under *KFE*; therefore *ex aequo*, as the square on *CN* is to the rectangle under *CNT*, so is the square on *KF* to the rectangle under *KFE*. Therefore, as *CN* is to *NT*, so is *KF* to *FE*. But also, as *PN* is to *NC*, so was *DF* to *FK*; therefore *ex aequo*, as *PN* is to *NT*, so is *DF* to *FE*. And the angles at *N*, *F* are right; therefore the angle at *T* is equal to the angle *DEF*.

In connexion with the propositions just quoted, it may not be out of place to remark upon some peculiar advantages of the Greek language as a vehicle for geometrical investigations. Its richness in grammatical forms is, from this point of view, of extreme importance. For instance, nothing could be more elegant than the regular use of the perfect imperative passive in constructions; thus, where we should have to say "let a perpendicular be drawn" or, more peremptorily, "draw a perpendicular," the Greek expression is ἤχθω κάθετος, the former word expressing in itself the meaning "let it *have been* drawn" or "suppose it drawn," and similarly in all other cases, e.g. γεγράφθω, ἐπεζεύχθω, ἐκβεβλήσθω, τετμήσθω, εἰλήφθω, ἀφῃρήσθω and the like. Neatest of all is the word γεγονέτω with which the *analysis* of a problem begins, "suppose it done." The same form is used very effectively along with the usual expression for a proportion, e.g. πεποιήσθω, ὡς ἡ HK πρὸς KE, ἡ NΞ πρὸς ΞM, which can hardly be translated in English by anything shorter than "Let NΞ be so taken that NΞ is to ΞM as HK to KE."

Again, the existence of the separate masculine, feminine and neuter forms of the definite article makes it possible to abbreviate the expressions for straight lines, angles, rectangles and squares by leaving the particular substantive to be understood. Thus ἡ HK is ἡ HK (γραμμή), *the line* HK; in ἡ ὑπὸ ΑΒΓ or ἡ ὑπὸ τῶν ΑΒΓ the word understood is γωνία and the meaning is *the angle* ΑΒΓ (i.e. the angle *contained by* AB and ΒΓ); τὸ ὑπὸ ΑΒΓ or τὸ ὑπὸ τῶν ΑΒΓ is τὸ ὑπὸ ΑΒΓ (χωρίον or ὀρθογώνιον), *the rectangle contained by* AB, ΒΓ; τὸ ἀπὸ AB is τὸ ἀπὸ AB (τετράγωνον), *the square on* AB. The result is that much of the language of Greek geometry is scarcely less concise than the most modern notation.

The closeness with which Apollonius followed the Euclidean tradition is further illustrated by the exact similarity of language between the enunciations of Apollonius' propositions about the conic and the corresponding propositions in Euclid's third Book about circles. The following are some obvious examples.

Eucl. III. 1.

Τοῦ δοθέντος κύκλου τὸ κέντρον εὑρεῖν.

Ap. II. 45.

Τῆς δοθείσης ἐλλείψεως ἢ ὑπερβολῆς τὸ κέντρον εὑρεῖν.

Eucl. III. 2.

Ἐὰν κύκλου ἐπὶ τῆς περιφερείας ληφθῇ δύο τυχόντα σημεῖα, ἡ ἐπὶ τὰ σημεῖα ἐπιζευγνυμένη εὐθεῖα ἐντὸς πεσεῖται τοῦ κύκλου.

Ap. I. 10.

Ἐὰν ἐπὶ κώνου τομῆς ληφθῇ δύο σημεῖα, ἡ μὲν ἐπὶ τα σημεῖα ἐπιζευγνυμένη εὐθεῖα ἐντὸς πεσεῖται τῆς τομῆς, ἡ δὲ ἐπ' εὐθείας αὐτῇ ἐκτός.

Eucl. III. 4.

Ἐὰν ἐν κύκλῳ δύο εὐθεῖαι τέμνωσιν ἀλλήλας μὴ διὰ τοῦ κέντρου οὖσαι, οὐ τέμνουσιν ἀλλήλας δίχα.

Ap. II. 26.

Ἐὰν ἐν ἐλλείψει ἢ κύκλου περιφερείᾳ δύο εὐθεῖαι τέμνωσιν ἀλλήλας μὴ διὰ τοῦ κέντρου οὖσαι, οὐ τέμνουσιν ἀλλήλας δίχα.

Eucl. III. 7.

Ἐὰν κύκλου ἐπὶ τῆς διαμέτρου ληφθῇ τι σημεῖον, ὃ μή ἐστι κέντρον τοῦ κύκλου, ἀπὸ δὲ τοῦ σημείου πρὸς τὸν κύκλον προσπίπτωσιν εὐθεῖαί τινες, μεγίστη μὲν ἔσται, ἐφ' ἧς τὸ κέντρον, ἐλαχίστη δὲ ἡ λοιπή, τῶν δὲ ἄλλων ἀεὶ ἡ ἔγγιον τῆς διὰ τοῦ κέντρου τῆς ἀπώτερον μείζων ἐστίν, δύο δὲ μόνον ἴσαι ἀπὸ τοῦ σημείου προσπεσοῦνται πρὸς τὸν κύκλον ἐφ' ἑκάτερα τῆς ἐλαχίστης.

Ap. v. 4 and 6.
(Translated from Halley.)

If a point be taken on the axis of an *ellipse* whose distance from the vertex of the section is equal to half the latus rectum, and if from the point any straight lines whatever be drawn to the section, the least of all the straight lines drawn from the given point will be that which is equal to half the latus rectum, the greatest the remaining part of the axis, and of the rest those which are nearer to the least will be less than those more remote......

As an instance of Apollonius' adherence to the *conceptions* of Euclid's *Elements*, those propositions of the first Book of the *Conics* may be mentioned which first introduce the notion of a tangent. Thus in I. 17 we have the proposition that, if in a conic a straight line be drawn through the extremity of the diameter parallel to the ordinates to that diameter, the said straight line will fall without the conic; and the conclusion is drawn that it is a *tangent*. This argument recalls the Euclidean definition of a tangent to a circle as "any straight line which meets the circle and being produced does not cut the circle." We have also in Apollonius as well as in Euclid the proof that no straight line can fall between the tangent and the curve. Compare the following enunciations:

Eucl. III. 16.

Ἡ τῇ διαμέτρῳ τοῦ κύκλου πρὸς ὀρθὰς ἀπ' ἄκρας ἀγομένη ἐκτὸς πεσεῖται τοῦ κύκλου, καὶ εἰς τὸν μεταξὺ τόπον τῆς τε εὐθείας καὶ τῆς περιφερείας ἑτέρα εὐθεῖα οὐ παρεμπεσεῖται.

Ap. I. 32.

Ἐὰν κώνου τομῆς διὰ τῆς κορυφῆς εὐθεῖα παρὰ τεταγμένως κατηγμένην ἀχθῇ, ἐφάπτεται τῆς τομῆς, καὶ εἰς τὸν μεταξὺ τόπον τῆς τε κώνου τομῆς καὶ τῆς εὐθείας ἑτέρα εὐθεῖα οὐ παρεμπεσεῖται.

Another instance of the orthodoxy of Apollonius is found in the fact that, when enunciating propositions as holding good of a circle as well as a conic, he speaks of "a hyperbola or an ellipse or the *circumference of a* circle," not of a *circle* simply. In this he follows the practice of Euclid based upon his definition of a circle as "a plane figure bounded by one line." It is only very exceptionally that the word *circle* alone is used to denote the *circumference* of the circle, e.g. in Euclid IV. 16 and Apollonius I. 37.

§ 2. Planimetric character of the treatise.

Apollonius, like all the Greek geometers whose works have come down to us, uses the stereometric origin of the three conics as sections of the cone only so far as is necessary in order to deduce a single fundamental plane property for each curve. This plane property is then made the basis of the further development of the theory, which proceeds without further reference to the cone, except indeed when, by way of rounding-off the subject, it is considered necessary to prove that a cone can be found which will contain any given conic. As pointed out above (p. xxi), it is probable that the discovery of the conic sections was the outcome of the attempt of Menaechmus to solve the problem of the two mean proportionals by constructing the plane loci represented by the equations

$$x^2 = ay, \quad y^2 = bx, \quad xy = ab,$$

and, in like manner, the Greek geometers in general seem to have connected the conic sections with the cone only because it was in their view necessary to give the curves a geometrical definition expressive of their relation to other known geometrical figures, as distinct from an abstract definition as the loci of points satisfying certain conditions. Hence *finding* a particular conic was understood as being synonymous with *localising* it in a cone, and we actually meet with this idea in Apollonius I. 52—58 [Props. 24, 25, 27], where the problem of "finding" a parabola, an ellipse, and a hyperbola satisfying certain conditions takes the form of finding a cone of which the required curves are sections. Menaechmus and his contemporaries would perhaps hardly have ventured, without such a geometrical definition, to regard the loci represented by the three equations as being really curves. When however they were found to be producible by cutting a cone in a particular manner, this fact was a sort of guarantee that they were genuine curves; and there was no longer any hesitation in proceeding with the further investigation of their properties in a plane, without reference to their origin in the cone.

There is no reason to suppose that the method adopted in the *Solid Loci* of Aristaeus was different. We know from Pappus that Aristaeus called the conics by their original names; whereas, if (as the title might be thought to imply) he had used in his book the methods of solid geometry, he would hardly have failed to discover a more general method of producing the curves than that implied by their old names. We may also assume that the other predecessors of Apollonius used, equally with him, the planimetric method; for (1) among the properties of conics which were well-known before his time there are many, e.g. the asymptote-properties of the hyperbola, which could not have been evolved in any natural way from the consideration of the cone, (2) there are practically no traces of the deduction of the plane properties of a conic from other stereometric investigations, even in the few instances where it would have been easy. Thus it would have been easy to regard an ellipse as a section of a right cylinder and then to prove the property of conjugate diameters, or to find the area of the ellipse, by projection from the circular sections; but this method does not appear to have been used.

§ 3. Definite order and aim.

Some writers have regarded the *Conics* as wanting in system and containing merely a bundle of propositions thrown together in a hap-hazard way without any definite plan having taken shape in the author's mind. This idea may have been partly due to the words used at the beginning of the preface, where Apollonius speaks of having put down everything as it occurred to him; but it is clear that the reference is to the imperfect copies of the Books which had been communicated to various persons before they took their final form. Again, to a superficial observer the order adopted in the first Book might seem strange, and so tend to produce the same impression; for the investigation begins with the properties of the conics derived from the cone itself, then it passes to the properties of conjugate diameters, tangents, etc., and returns at the end of the Book to the connexion of particular conics with the cone, which is immediately dropped again. But, if the Book is examined more closely, it is apparent that from the beginning to the end a definite object is aimed at, and only such propositions are given as are necessary for the attainment of that object. It is true that they contain plane properties which are constantly made use of afterwards; but for the time being they are simply links in a chain of proof leading to the conclusion that the parabolas, ellipses and hyperbolas which Apollonius obtains by any possible section of any kind of circular cone are identical with those which are produced from sections of cones of revolution.

The order of procedure (leaving out unnecessary details) is as follows. First, we have the property of the conic which is the equivalent of the Cartesian equation referred to the particular diameter which emerges from the process of cutting the cone, and the tangent at its extremity, as axes of coordinates. Next, we are introduced to the conjugate diameter and the reciprocal relation between it and the original diameter. Then follow properties of tangents (1) at the extremity of the original diameter and (2) at any other point of the curve which is not on the diameter. After these come a series of propositions leading up to the conclusion that *any* new diameter, the tangent at its extremity, and the chords parallel to the tangent (in other words, the ordinates to the new diameter) have to one another the same relation as that subsisting between the original diameter, the tangent at its extremity, and the ordinates to it, and hence that the equation of the conic when referred to the new diameter and the tangent at its extremity is of the same *form* as the equation referred to the original diameter and tangent*. Apollonius is now in a position to pass to the proof of the proposition that the curves represented by his original definitions can be represented by equations of the same form with reference to *rectangular* axes, and can be produced by means of sections of *right* cones. He proceeds to propose the problem "to find" a parabola, ellipse, or hyperbola, when a diameter, the angle of inclination of its ordinates, and the corresponding parameter are given, or, in other words, when the curve is given by its equation referred to given axes. "Finding" the curve is, as stated above, regarded as synonymous with determining it as a section of a right circular cone. This Apollonius does in two steps: he first assumes that the ordinates are at right angles to the diameter and solves the problem for this particular case, going back to the method followed in his original derivation of the curves from the cone, and not using any of the results obtained in the intervening plane investigations; then, secondly, he reduces the case where the ordinates are not perpen-

* The definiteness of the design up to this point is attested by a formal recapitulation introduced by Apollonius himself at the end of I. 51 and concluding with the statement that "all the properties which have been shown to be true with regard to the sections by reference to the original diameters will equally result when the other diameters are taken."

dicular to the diameter to the former case, proving by his procedure that it is always possible to draw a diameter which is at right angles to the chords bisected by it. Thus what is proved here is not the mere converse of the first propositions of the Book. If that had been all that was intended, the problems would more naturally have followed directly after those propositions. It is clear, however, that the solution of the problems as given is not possible without the help of the intermediate propositions, and that Apollonius does in fact succeed in proving, concurrently with the solution of the problems, that there cannot be obtained from oblique cones any other curves than can be derived from right cones, and that all conics have axes.

The contents of the first Book, therefore, so far from being a fortuitous collection of propositions, constitute a complete section of the treatise arranged and elaborated with a definite intention throughout.

In like manner it will be seen that the other Books follow, generally, an intelligible plan; as, however, it is not the object of this introduction to give an abstract of the work, the remaining Books shall speak for themselves.

CHAPTER III.

THE METHODS OF APOLLONIUS.

As a preliminary to the consideration in detail of the methods employed in the *Conics*, it may be stated generally that they follow steadily the accepted principles of geometrical investigation which found their definitive expression in the *Elements* of Euclid. Any one who has mastered the *Elements* can, if he remembers what he gradually learns as he proceeds in his reading of the *Conics*, understand every argument of which Apollonius makes use. In order, however, to thoroughly appreciate the whole course of his thought, it is necessary to bear in mind that some of the methods employed by the Greek geometers were much more extensively used than they are in modern geometry, and were consequently handled by Apollonius and his contemporary readers with much greater deftness and facility than would be possible, without special study, to a modern mathematician. Hence it frequently happens that Apollonius omits an intermediate step such as a practised mathematician would now omit in a piece of algebraical work which was not intended for the mere beginner. In several such instances Pappus and Eutocius think it necessary to supply the omission by a lemma.

§ 1. The principal machinery used by Apollonius as well as by the earlier geometers comes under the head of what has been not inappropriately called a **geometrical Algebra**; and it will be convenient to exhibit the part which this plays in the *Conics* under the following important subdivisions.

(1) The theory of proportions.

This theory in its most complete form, as expounded in the fifth and sixth Books of Euclid, lies at the very root of the system of

Apollonius; and a very short consideration suffices to show how far it is capable of being used as a substitute for algebraical operations. Thus it is obvious that it supplies a ready method of effecting the operations of *multiplication* and *division*. Again, suppose, for example, that we have a series in geometrical progression consisting of the terms $a_0, a_1, a_2 \ldots a_n$, so that

$$\frac{a_0}{a_1} = \frac{a_1}{a_2} = \frac{a_2}{a_3} = \ldots = \frac{a_{n-1}}{a_n}.$$

We have then $\frac{a_n}{a_0} = \left(\frac{a_1}{a_0}\right)^n$, or $\frac{a_1}{a_0} = \sqrt[n]{\frac{a_n}{a_0}}$.

Thus the continued use of the method of proportions enables an expression to be given for the sum of the geometrical series (cf. the summation in Eucl. IX. 35).

(2) **The application of areas.**

Whether the theory of proportions in the form in which Euclid presents it is due to Eudoxus of Cnidus (408—355 B.C.) or not, there is no doubt that the method of *application of areas*, to which allusion has already been made, was used much earlier still. We have the authority of the pupils of Eudemus (quoted by Proclus on Euclid I. 44) for the statement that "these propositions are the discoveries of the Pythagorean muse, the application of areas, their exceeding, and their falling short" (ἥ τε παραβολὴ τῶν χωρίων καὶ ἡ ὑπερβολὴ καὶ ἡ ἔλλειψις), where we find the very terms afterwards applied by Apollonius to the three conic sections on the ground of the corresponding distinction between their respective fundamental properties as presented by him. The problem in Euclid I. 44 is "to apply to a given straight line a parallelogram which shall be equal to a given triangle and have one of its angles equal to a given rectilineal angle." The solution of this clearly gives the means of *adding together* or *subtracting* any triangles, parallelograms, or other figures which can be decomposed into triangles.

Next, the second Book of Euclid (with an extension which is found in VI. 27—29) supplies means for solving the problems of modern algebra so long as they do not involve expressions above the second degree, and provided, so far as the solution of quadratic equations is concerned, that negative and imaginary solutions are excluded; the only further qualification to be borne in mind is that, since negative magnitudes are not used in Greek geometry, it is often necessary to solve a problem in two parts, with different figures, where one solution by algebra would cover both cases.

It is readily seen that Book II. of the *Elements* makes it possible to *multiply* two factors with any number of linear terms in each; and the compression of the result into a single product follows by the aid of the *application*-theorem. That theorem itself supplies a method of *dividing* the product of any two linear factors by a third. The remaining operations for which the second Book affords the means are, however, the most important of all, namely,

(*a*) the finding of a square whose area is equal to that of a given rectangle [II. 14], which problem is the equivalent of extracting the square root, or of the solution of a pure quadratic equation,

(*b*) the geometrical solution of a mixed quadratic equation, which can be derived from II. 5, 6.

In the first case (*a*) we produce the side AB of the rectangle to E, making BE equal to BC; then we bisect AE in F, and, with F as centre and radius FE, draw a circle meeting CB produced in G.

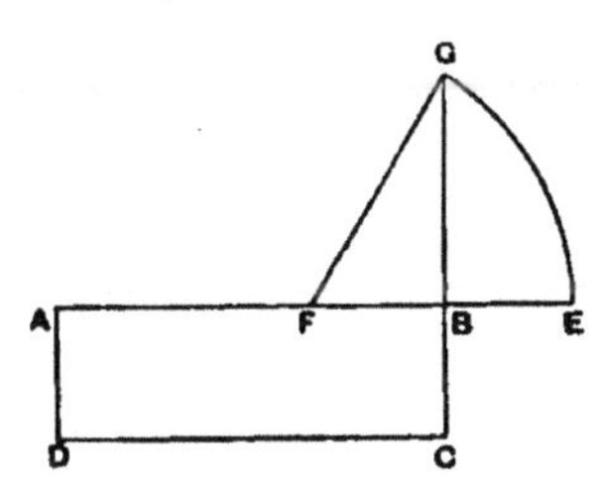

Then $$FG^2 = FB^2 + BG^2.$$

Also $$FG^2 = FE^2 = AB \cdot BE + FB^2,$$

whence, taking away the common FB^2,

$$BG^2 = AB \cdot BE.$$

This corresponds to the equation

$$x^2 = ab \quad \text{..............................(1)},$$

and BG or x is found.

In the second case (*b*) we have, if AB is divided equally at C and unequally at D,

$$AD \cdot DB + CD^2 = CB^2. \qquad \text{[Eucl. II. 5.]}$$

Now suppose $AB = a$, $DB = x$.

Then
$$ax - x^2 = \text{rect. } AH$$
$$= \text{the gnomon } CMF.$$
Thus, if the area of the gnomon is given ($=b^2$, say), and if a is given ($=AB$), the problem of solving the equation
$$ax - x^2 = b^2$$
is, in the language of geometry, "To a given straight line (a) to apply a rectangle which shall be equal to a given square (b^2) and *deficient by a square*," i.e. to construct the rectangle AH.

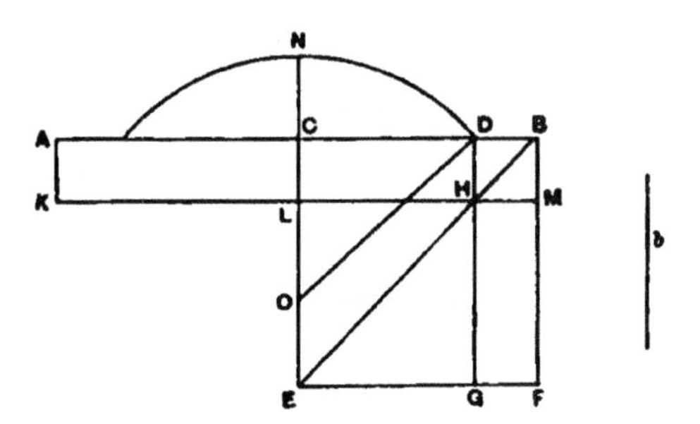

This simply requires the construction of a gnomon, equal in area to b^2, of which each of the outer sides is given $\left(CB, \text{ or } \frac{a}{2}\right)$. Now we know the area $\frac{a^2}{4}$ (i.e. the square CF), and we know the area of part of it, the required gnomon CMF ($=b^2$); hence we have only to find the difference between the two, namely the area of the square LG, in order to find CD which is equal to its side. This can be done by applying the Pythagorean proposition, I. 47.

Simson gives the following easy solution in his note on VI. 28–29. Measure CO perpendicular to AB and equal to b, produce OC to N so that $ON = CB$ $\left(\text{or } \frac{a}{2}\right)$, and with O as centre and radius ON describe a circle cutting CB in D.

Then DB (or x) is found, and therefore the rectangle AH.

For
$$AD \,.\, DB + CD^2 = CB^2$$
$$= OD^2$$
$$= OC^2 + CD^2,$$
whence
$$AD \,.\, DB = OC^2,$$
or
$$ax - x^2 = b^2 \quad \ldots\ldots\ldots\ldots\ldots\ldots\ldots\ldots(2).$$

It is clear that it is a necessary condition of the possibility of a real solution that b^2 must not be greater than $\left(\frac{a}{2}\right)^2$, and that the geometrical solution derived from Euclid does not differ from our practice of solving a quadratic by completing the square on the side containing the terms in x^2 and x*.

To show how closely Apollonius keeps to this method and to the old terminology connected therewith, we have only to compare his way of describing the *foci* of a hyperbola or an ellipse. He says, "Let a rectangle equal to one fourth part of the 'figure' [i.e. equal to CB'^2] be applied to the axis at either end, for the hyperbola or the opposite branches exceeding, but for the ellipse deficient, by a square"; and the case of the *ellipse* corresponds exactly to the solution of the equation just given.

* It will be observed that, while in this case there are two geometrically real solutions, Euclid gives only one. It must not however be understood from this that he was unaware that there are two solutions. The contrary may be inferred from the proposition VI. 27, in which he gives the *διορισμός* stating the necessary condition corresponding to $b^2 \ngtr \left(\frac{a}{2}\right)^2$; for, although the separate treatment, in the text translated by Simson, of the two cases where the base of the applied parallelogram is greater and less than half the given line appears to be the result of interpolations (see Heiberg's edition, Vol. II. p. 161), the distinction is perfectly obvious, and we must therefore assume that, in the case given above in the text, Euclid was aware that $x=AD$ satisfies the equation as well as $x=BD$. The reason why he omitted to specify the former solution is no doubt that the rectangle so found would simply be an equal rectangle but on BD as base instead of AD, and therefore there is no real object in distinguishing two solutions. This is easily understood when we regard the equation as a statement of the problem of finding two quantities whose sum (a) and product (b^2) are given, i.e. as equivalent to the simultaneous equations

$$x+y=a,$$
$$xy=b^2.$$

These symmetrical equations have really only one solution, as the two apparent solutions are simply the result of interchanging the values of x and y. This form of the problem was known to Euclid, as appears from Prop. 86 of the *Data* (as translated by Simson): "If two straight lines contain a parallelogram given in magnitude, in a given angle; if both of them together be given, they shall each of them be given."

From Euclid's point of view the equations next referred to in the text

$$x^2 \pm ax = b^2$$

have of course only *one* solution.

Again, from the proposition in Euclid II. 6, we have, if AB is bisected at C and produced to D,

$$AD \,.\, DB + CB^2 = CD^2.$$

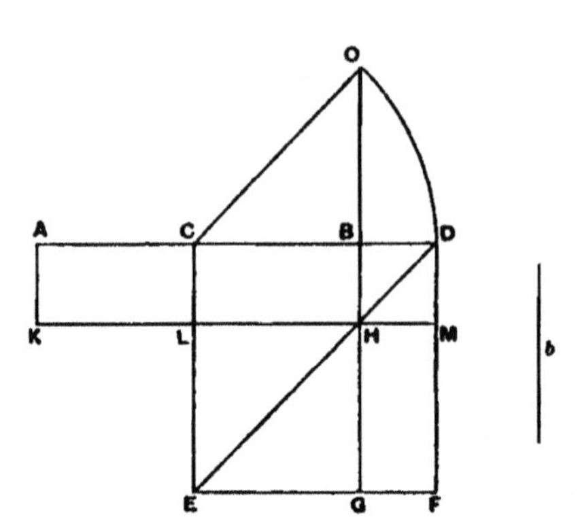

Let us suppose that, in Euclid's figure, $AB = a$, $BD = x$.

Then $$AD \,.\, DB = ax + x^2,$$

and, if this is equal to b^2 (a given area), the solution of the equation

$$ax + x^2 = b^2$$

is equivalent to finding a gnomon equal in area to b^2 and having as one of the sides containing the inner right angle a straight line equal to the given length CB or $\frac{a}{2}$. Thus we know $\left(\frac{a}{2}\right)^2$ and b^2, and we have to find, by the Pythagorean proposition, a square equal to the sum of two given squares.

To do this Simson draws BO at right angles to AB and equal to b, joins CO, and describes with centre C and radius CO a circle meeting AB produced in D. Thus BD, or x, is found.

Now $$AD \,.\, DB + CB^2 = CD^2$$
$$= CO^2$$
$$= CB^2 + BO^2,$$

whence $$AD \,.\, DB = BO^2,$$

or $$ax + x^2 = b^2.$$

This solution corresponds exactly to Apollonius' determination of the foci of the *hyperbola*.

The equation $$x^2 - ax = b^2$$
can be dealt with in a similar manner.

If $AB = a$, and if we suppose the problem solved, so that $AD = x$, then

$$x^2 - ax = AM = \text{the gnomon } CMF,$$

and, to find the gnomon, we have its area (b^2), and the area CB^2 or $\left(\frac{a}{2}\right)^2$ by which the gnomon differs from CD^2. Thus we can find D (and therefore AD, or x) by the same construction as in the case immediately preceding.

Hence Euclid has no need to treat this case separately, because it is the same as the preceding except that here x is equal to AD instead of BD, and one solution can be derived from the other.

So far Euclid has not put his propositions in the form of an actual solution of the quadratic equations referred to, though he has in II. 5, 6 supplied the means of solving them. In VI. 28, 29 however he has not only made the problem more general by substituting for the *square* by which the required rectangle is to exceed or fall short a *parallelogram* similar and similarly situated to a given parallelogram, but he has put the propositions in the form of an actual solution of the general quadratic, and has prefixed to the first case (the deficiency by a parallelogram) the necessary condition of possibility [VI. 27] corresponding to the obvious διορισμός referred to above in connection with the equation

$$ax - x^2 = b^2.$$

Of the problems in VI. 28, 29 Simson rightly says "These two problems, to the first of which the 27th prop. is necessary, are the most general and useful of all in the elements, and are most frequently made use of by the ancient geometers in the solution of other problems; and therefore are very ignorantly left out by Tacquet and Dechales in their editions of the *Elements*, who pretend that they are scarce of any use.*"

* It is strange that, notwithstanding this observation of Simson's, the three propositions VI. 27, 28, 29 are omitted from Todhunter's Euclid, which contains a note to this effect: "We have omitted in the sixth Book Propositions 27, 28, 29 and the first solution which Euclid gives of Proposition 30, as they appear now to be never required and have been condemned as useless by various modern commentators; see Austin, Walker, and Lardner."

I would suggest that all three propositions should be at once restored to the text-books of Euclid with a note explaining their mathematical significance.

The enunciations of these propositions are as follows* :

VI. 27. "*Of all the parallelograms applied to the same straight line and deficient by parallelograms similar and similarly situated to that which is described upon the half of the line, that which is applied to the half, and is similar to its defect, is greatest.*

VI. 28. "*To a given straight line to apply a parallelogram equal to a given rectilineal figure and deficient by a parallelogram similar to a given parallelogram: But the given rectilineal figure must not be greater than the parallelogram applied to half of the given line and similar to the defect.*

VI. 29. "*To a given straight line to apply a parallelogram equal to a given rectilineal figure and exceeding by a parallelogram similar to a given one.*"

Corresponding propositions are found among the *Data* of Euclid. Thus Prop. 83 states that, "*If a parallelogram equal to a given space be applied to a given straight line, deficient by a parallelogram given in species, the sides of the defect are given,*" and Prop. 84 states the same fact in the case of an *excess*.

It is worth while to give shortly Euclid's proof of one of these propositions, and VI. 28 is accordingly selected.

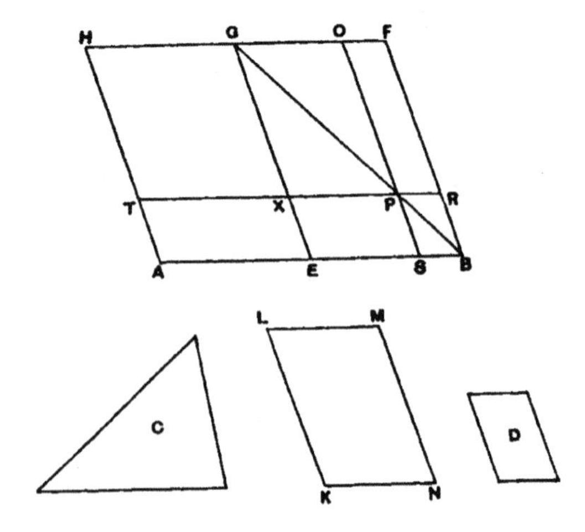

* The translation follows the text of Heiberg's edition of Euclid (Teubner, 1883–8).

Let AB be the given straight line, C the given area, D the parallelogram to which the *defect* of the required parallelogram is to be similar.

Bisect AB at E, and on EB describe a parallelogram $GEBF$ similar and similarly situated to D [by VI. 18]. Then, by the διορισμός [VI. 27], AG must be either equal to C or greater than it. If the former, the problem is solved; if the latter, it follows that the parallelogram EF is greater than C.

Now construct a parallelogram $LKNM$ equal to the excess of EF over C and similar and similarly situated to D [VI. 25].

Therefore $LKNM$ is similar and similarly situated to EF, while, if GE, LK, and GF, LM, are homologous sides respectively,

$$GE > LK, \text{ and } GF > LM.$$

Make GX (along GE) and GO (along GF) equal respectively to LK, LM, and complete the parallelogram $XGOP$.

Then GPB must be the diagonal of the parallelogram GB [VI. 26]. Complete the figure, and we have

$$EF = C + KM, \text{ by construction,}$$

and $$XO = KM.$$

Therefore the difference, the gnomon ERO, is equal to C.

Hence the parallelogram TS, which is equal to the gnomon, is equal to C.

Suppose now that $AB = a$, $SP = x$, and that $b : c$ is the ratio of the sides KN, LK of the parallelogram $LKNM$ to one another; we then have, if m is a certain constant,

$$TB = m \, . \, ax,$$

$$SR = m \, . \, \frac{x^2}{c^2} \, . \, bc$$

$$= m \, . \, \frac{b}{c} x^2,$$

so that $$ax - \frac{b}{c} x^2 = \frac{C}{m}.$$

Proposition 28 in like manner solves the equation

$$ax + \frac{b}{c} x^2 = \frac{C}{m}.$$

If we compare these equations with those by which Apollonius expresses the fundamental property of a central conic, viz.

$$px \mp \frac{p}{d}x^2 = y^2,$$

it is seen that the only difference is that p takes the place of a and, instead of *any* parallelogram whose sides are in a certain ratio, that particular similar parallelogram is taken whose sides are p, d. Further, Apollonius draws p at right angles to d. Subject to these differences, the phraseology of the *Conics* is similar to that of Euclid: the square of the ordinate is said to be equal to a rectangle "applied to" a certain straight line (i.e. p), "having as its width" (πλάτος ἔχον) the abscissa, and "falling short (or exceeding) by a figure similar and similarly situated to that contained by the diameter and the parameter."

It will be seen from what has been said, and from the book itself, that Apollonius is nothing if not orthodox in his adherence to the traditional method of *application of areas*, and in his manipulation of equations between areas such as are exemplified in the second Book of Euclid. From the extensive use which is made of these principles we may conclude that, where equations between areas are stated by Apollonius without proof, though they are not immediately obvious, the explanation is to be found in the fact that his readers as well as himself were so imbued with the methods of geometrical algebra that they were naturally expected to be able to work out any necessary intermediate step for themselves. And, with regard to the manner of establishing the results assumed by Apollonius, we may safely infer, with Zeuthen, that it was the practice to prove them directly by using the *procedure* of the second Book of the *Elements* rather than by such combinations and transformations of the *results* obtained in that Book as we find in the lemmas of Pappus to the propositions of Apollonius. The kind of result most frequently assumed by Apollonius is some relation between the products of pairs of segments of a straight line divided by points on it into a number of parts, and Pappus' method of proving such a relation amounts practically to the procedure of modern algebra, whereas it is more likely that Apollonius and his contemporaries would, after the manner of *geometrical* algebra, draw a figure showing the various rectangles and squares, and thence, in many cases by simple inspection, conclude e.g. that one rectangle is equal to the sum of two others, and so on.

An instance will make this clear. In Apollonius III. 26 [Prop. 60] it is assumed that, if E, A, B, C, D be points on a line in the order named, and if $AB = CD$, then

$$EC \,.\, EB = AB \,.\, BD + ED \,.\, EA.$$

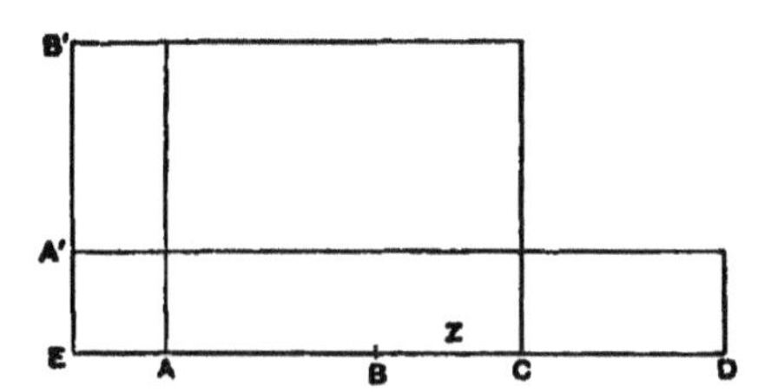

This appears at once if we set off EB' perpendicular and equal to EB, and EA' along EB' equal to EA, and if we complete the parallelograms as in the figure*.

Similarly Eutocius' lemma to III. 29 [Prop. 61] is more likely to represent Apollonius' method of proof than is Pappus' 6th lemma to Book III. (ed. Hultsch, p. 949).

(3) Graphic representation of areas by means of auxiliary lines.

The Greek geometers were fruitful in devices for the compression of the sum or difference of the areas of any rectilineal figures into a single area; and in fact the *Elements* of Euclid furnish the means of effecting such compression generally. The *Conics* of Apollonius contain some instances of similar procedure which deserve mention for their elegance. There is, *first*, the representation of the area of the square on the ordinate y in the form of a rectangle whose base is the abscissa x. While the procedure for this purpose is, in

* On the other hand Pappus' method is simply to draw a line with points on it, and to proceed semi-algebraically. Thus in this case [Lemma 4 to Book III., p. 947] he proceeds as follows, first bisecting BC in Z.

$$CE \,.\, EB + BZ^2 = EZ^2,$$

and

$$DE \,.\, EA + AZ^2 = EZ^2,$$

while

$$AZ^2 = CA \,.\, AB + BZ^2.$$

It follows that

$$CE \,.\, EB + BZ^2 = DE \,.\, EA + CA \,.\, AB + BZ^2,$$

whence

$$CE \,.\, EB = DE \,.\, EA + CA \,.\, AB,$$

(and $CA = BD$).

form, closely connected with the traditional *application of areas*, its special neatness is due to the use of a certain *auxiliary line*. The Cartesian equation of a central conic referred to any diameter of length d and the tangent at its extremity is (if d' be the length of the conjugate diameter)

$$y^2 = \frac{d'^2}{d} x \mp \frac{d'^2}{d^2} \cdot x^2,$$

and the problem is to express the right hand side of the equation in the form of a single rectangle xY, in other words, to find a simple construction for Y where

$$Y = \frac{d'^2}{d} \mp \frac{d'^2}{d^2} \cdot x.$$

Apollonius' device is to take a length p such that

$$\frac{p}{d} = \frac{d'^2}{d^2},$$

(so that p is the parameter of the ordinates to the diameter of length d). If PP' be the diameter taken as the axis of x, and P the origin of coordinates, he draws PL perpendicular to PP' and of length p, and joins $P'L$. Then, if $PV = x$, and if VR drawn parallel to PL meets $P'L$ in R we have (using the figures of Props. 2, 3), by similar triangles,

$$\frac{p}{d} = \frac{VR}{P'V} = \frac{VR}{d \mp x},$$

so that

$$VR = p \mp \frac{p}{d} x$$

$$= \frac{d'^2}{d} \mp \frac{d'^2}{d^2} \cdot x$$

$$= Y,$$

and the construction for Y is therefore effected.

Again, in v. 1–3 [Prop. 81], another auxiliary line is used for expressing y^2 in the form of an area standing on x as base in the particular case where y is an ordinate to the axis. AM is drawn perpendicular to AA' and of length equal to $\frac{p_a}{2}$ (where p_a is the parameter corresponding to the axis AA'), and CM is joined. If the ordinate PN meets CM in H, it is then proved that

$$y^2 = 2 \text{ (quadrilateral } MANH).$$

Apollonius then proceeds in v. 9, 10 [Prop. 86] to give, by means of a second auxiliary line, an extremely elegant construction for an area equal to the difference between the square on a normal PG and the square on $P'G$, where P' is any other point on the curve than P'. The method is as follows. If PN is the ordinate of P, measure NG along the axis away from the nearer vertex so that

$$NG : CN = p_a : AA' [= CB^2 : CA^2].$$

In the figures of Prop. 86 let PN produced meet CM in H, as before. GH is now joined and produced if necessary, forming the second auxiliary line. It is then proved at once that $NG = NH$, and therefore that

$$NG^2 = 2 \triangle NGH,$$

and similarly that $\qquad N'G^2 = 2 \triangle N'GH'$.

Hence, by the aid of the expression for y^2 above, the areas PG^2 and $P'G^2$ are exhibited in the figures, and it is proved that

$$P'G^2 - PG^2 = 2 \triangle HKH',$$

so that we have in the figures a graphic representation of the difference between the areas of the two squares effected by means of the two fixed auxiliary lines CM, GH.

(4) Special use of auxiliary points in Book VII.

The seventh Book investigates the values of certain quadratic functions of the lengths of any two conjugate diameters PP', DD' in central conics of different excentricities, with particular reference to the *maximum* and *minimum* values of those functions. The whole procedure of Apollonius depends upon the reduction of the ratio $CP^2 : CD^2$ to a ratio between *straight lines* MH' and MH, where H, H' are fixed points on the transverse axis of the hyperbola or on either axis of the ellipse, and M is a variable point on the same axis determined in a certain manner with reference to the position of the point P. The proposition that

$$PP'^2 : DD'^2 = MH' : MH$$

appears in VII. 6, 7 [Prop. 127], and the remainder of the Book is a sufficient proof of the effectiveness of this formula as the geometrical substitute for algebraical operations.

The bearing of the proposition may be exhibited as follows, with the help of the notation of analytical geometry. If the axes of coordinates are the principal axes of the conic, and if a, b are the lengths of the axes, we have, e.g., in the case of the *hyperbola*,

$$\frac{CP^2+CD^2}{CP^2-CD^2}=\frac{2(x^2+y^2)-\left\{\left(\frac{a}{2}\right)^2-\left(\frac{b}{2}\right)^2\right\}}{\left(\frac{a}{2}\right)^2-\left(\frac{b}{2}\right)^2},$$

where x, y are the coordinates of P.

Eliminating y by means of the equation of the curve, we obtain

$$\frac{CP^2+CD^2}{CP^2-CD^2}=\frac{2x^2\cdot\frac{a^2+b^2}{a^2}-\left(\frac{a}{2}\right)^2-\left(\frac{b}{2}\right)^2}{\left(\frac{a}{2}\right)^2-\left(\frac{b}{2}\right)^2}$$

$$=\frac{4\left(2x^2-\frac{a^2}{4}\right)}{a^2\cdot\frac{a^2-b^2}{a^2+b^2}}.$$

Apollonius' procedure is to take a certain fixed point H on the axis whose coordinates are $(h, 0)$, and a variable point M whose coordinates are $(x', 0)$, such that the numerator and denominator of the last expression are respectively equal to $2ax'$, $2ah$; whence the fraction is itself equal to $\frac{x'}{h}$, and we have

$$\frac{h}{\left(\frac{a}{2}\right)}=\frac{a^2-b^2}{a^2+b^2}\quad\text{.........................}(1),$$

and

$$ax'=2\left(2x^2-\frac{a^2}{4}\right),$$

or

$$a\left(x'+\frac{a}{2}\right)=4x^2\quad\text{.........................}(2).$$

From (1) we derive at once

$$\frac{\frac{a}{2}-h}{\frac{a}{2}+h}=\frac{b^2}{a^2},$$

whence

$$AH:A'H=b^2:a^2$$
$$=p_a:AA'.$$

Thus, to find H, we have only to divide AA' in the ratio $p_a : AA'$. This is what is done in VII. 2, 3 [Prop. 124].

H' is similarly found by dividing $A'A$ in the same ratio $p_a : AA'$, and clearly $AH = A'H'$, $A'H = AH'$.

Again, from (2), we have

$$a : \left(x' + \frac{a}{2}\right) = \frac{a^2}{4x} : x.$$

In other words, $AA' : A'M = CT : CN$

or $A'M : AM = CN : TN$ .... (3).

If now, as in the figures of Prop. 127, we draw AQ parallel to the tangent at P meeting the curve again in Q, AQ is bisected by CP; and, since AA' is bisected at C, it follows that $A'Q$ is parallel to CP.

Hence, if QM' be the ordinate of Q, the triangles $A'QM'$, CPN are similar, as also are the triangles AQM', TPN;

$$\therefore A'M' : AM' = CN : TN.$$

Thus, on comparison with (3), it appears that M coincides with M'; or, in other words, the determination of Q by the construction described gives the position of M.

Since now H, H', M are found, and x', h were so determined that

$$\frac{CP^2 + CD^2}{CP^2 - CD^2} = \frac{x'}{h},$$

it follows that $CP^2 : CD^2 = x' + h : x' - h$,

or $PP'^2 : DD'^2 = MH' : MH$.

The construction is similar for the ellipse except that in that case AA' is divided *externally* at H, H' in the ratio described.

§ 2. The use of coordinates.

We have here one of the most characteristic features of the Greek treatment of conic sections. The use of coordinates is not peculiar to Apollonius, but it will have been observed that the same point of view appears also in the earlier works on the subject. Thus Menaechmus used the characteristic property of the parabola which we now express by the equation $y^2 = px$ referred to rectangular axes. He used also the property of the rectangular hyperbola which is expressed in our notation by the equation $xy = c^2$, where the axes of coordinates are the asymptotes.

Archimedes too used the same form of equation for the parabola, while his mode of representing the fundamental property of a central conic

$$\frac{y^2}{x \cdot x_1} = (\text{const.})$$

can easily be put into the form of the Cartesian equation.

So Apollonius, in deriving the three conics from any cone cut in the most general manner, seeks to find the relation between the coordinates of any point on the curve referred to the original diameter and the tangent at its extremity as axes (in general oblique), and proceeds to deduce from this relation, when found, the other properties of the curves. His method does not essentially differ from that of modern analytical geometry except that in Apollonius geometrical operations take the place of algebraical calculations.

We have seen that the graphic representation of the area of y^2 in the form of a rectangle on x as base, where (x, y) is any point on a central conic, was effected by means of an auxiliary fixed line $P'L$ whose equation referred to PP', PL as *rectangular* axes is

$$Y = p \mp \frac{p}{d} x.$$

That an equation of this form between the coordinates x, Y represents a straight line we must assume Apollonius to have been aware, because we find in Pappus' account of the contents of the first Book of his separate work on *plane loci* the following proposition:

"If straight lines be drawn from a point meeting at given angles two straight lines given in position, and if the former lines are in a given ratio, or if the sum of one of them and of such a line as bears a given ratio to the second is given, then the point will lie on a given straight line"; in other words, the equation

$$x + ay = b$$

represents a straight line, where a, b are positive.

The altitude of the rectangle whose base is x and whose area is equal to y^2 is thus determined by a procedure like that of analytical geometry except that Y is found by a geometrical construction instead of being calculated algebraically from the equation of the auxiliary line

$$Y = p \mp \frac{p}{d} x.$$

If it should seem curious that the auxiliary line is determined with reference to an independent (rectangular) pair of coordinate axes different from the oblique axes to which the conic is itself referred, it has only to be borne in mind that, in order to show the area y^2 as a *rectangle*, it was necessary that the angle between x and Y should be right. But, as soon as the line $P'L$ was once drawn, the object was gained, and the subsidiary axes of coordinates were forthwith dropped, so that there was no danger of confusion in the further development of the theory.

Another neat example of the use of an auxiliary line regarded from the point of view of coordinate geometry occurs in I. 32 [Prop. 11], where it is proved that, if a straight line be drawn from the end of a diameter parallel to its ordinates (in other words, a tangent), no straight line can fall between the parallel and the curve. Apollonius first supposes that such a line can be drawn from P passing through K, a point outside the curve, and the ordinate KQV is drawn. Then, if y', y be the ordinates of K, Q respectively, and x their common abscissa, referred to the diameter and tangent as axes, we have for the central conic (figures on pp. 23, 24)

$$y'^2 > y^2 \text{ or } xY,$$

where Y represents the ordinate of the point on the auxiliary line $P'L$ before referred to corresponding to the abscissa x (with PP', PL as independent rectangular axes).

Let y'^2 be equal to xY', so that $Y' > Y$, and let Y' be measured along Y (so that, in the figures referred to, $VR = Y$, and $VS = Y'$).

Then the locus of the extremity of Y for different values of x is the straight line $P'L$, and the locus of the extremity of Y' for *different* points K on PK is the straight line PS. It follows, since the lines $P'L$, PS intersect, that there is one point (their intersection R') where $Y = Y'$, and therefore that, for the corresponding points Q', M on the conic and the supposed line PK respectively, $y = y'$, so that Q', M are coincident, and accordingly PK must meet the curve between P and K. Hence PK cannot lie between the tangent and the curve in the manner supposed.

Here then we have two auxiliary lines used, viz.

$$Y = p \mp \frac{p}{d}x,$$

and $$Y = mx,$$

where m is some constant; and the point of intersection of PK and the conic is determined by the point of intersection of the two auxiliary lines; only here again the latter point is found by a geometrical construction and not by an algebraical calculation.

In seeking in the various propositions of Apollonius for the equivalent of the Cartesian equation of a conic referred to other axes different from those originally taken, it is necessary to bear in mind what has already been illustrated by the original equation which forms the basis of the respective definitions, viz. that, where the equivalents of Cartesian equations occur, they appear in the guise of simple equations between areas. The book contains several such equations between areas which can either be directly expressed as, or split up into parts which are seen to be, constant multiples of x^2, xy, y^2, x, and y, where x, y are the coordinates of any point on the curve referred to different coordinate axes; and we have therefore the equivalent of so many different Cartesian equations.

Further, the essential difference between the Greek and the modern method is that the Greeks did not direct their efforts to making the fixed lines of the figure as few as possible, but rather to expressing their equations between areas in as short and simple a form as possible. Accordingly they did not hesitate to use a number of auxiliary fixed lines, provided only that by that means the areas corresponding to the various terms in x^2, xy, ... forming the Cartesian equation could be brought together and combined into a smaller number of terms. Instances have already been given in which such compression is effected by means of one or two auxiliary lines. In the case, then, where two auxiliary fixed lines are used in addition to the original axes of coordinates, and it appears that the properties of the conic (in the form of equations between areas) can be equally well expressed relatively to the two auxiliary lines and to the two original axes of reference, we have clearly what amounts to a *transformation of coordinates*.

§ 3. Transformation of coordinates.

A simple case is found as early as I. 15 [Prop. 5], where, for the ellipse, the axes of reference are changed from the original diameter and the tangent at its extremity to the diameter *conjugate* to the first and the corresponding tangent. This transformation may with sufficient accuracy be said to be effected, first, by a simple transference of the *origin* of coordinates from the extremity of the original diameter to the centre of the ellipse, and, secondly, by moving the origin a second time from the centre to D, the end of the conjugate diameter. We find in fact, as an intermediate step in the proof, the statement of the property that (d being the original diameter and d' its conjugate in the figure of Prop. 5)

$$\left(\frac{d'}{2}\right)^2 - y^2 = \text{the rectangle } RT \,.\, TE$$

$$= \frac{d'^2}{d^2} \cdot x^2,$$

where x, y are the coordinates of the point Q with reference to the diameter and its conjugate as axes and the centre as origin; and ultimately the equation is expressed in the old form, only with d' for diameter and p' for the corresponding parameter, where

$$\frac{p'}{d'} = \frac{d}{p}.$$

The equation of the hyperbola as well as of the ellipse referred to the centre as origin and the original diameter and its conjugate as axes is at once seen to be included as a particular case in I. 41 [Prop. 16], which proposition proves generally that, if two similar parallelograms be described on CP, CV respectively, and an equiangular parallelogram be described on QV such that QV is to the other side of the parallelogram on it in the ratio compounded of the ratio of CP to the other side of the parallelogram on CP and of the ratio $p : d$, then the parallelogram on QV is equal to the difference between the parallelograms on CP, CV. Suppose now that the parallelograms on CP, CV are squares, and therefore that the parallelogram on QV is a rectangle; it follows that

$$x^2 \sim \left(\frac{d}{2}\right)^2 = \frac{d}{p} \cdot y^2$$

$$= \frac{d^2}{d'^2} \cdot y^2 \qquad \text{.......................(1)}.$$

Apollonius is now in a position to undertake the transformation to a different pair of axes consisting of any diameter whatever and the tangent at its extremity. The method which he adopts is to use the new diameter as what has been termed an auxiliary fixed line.

It will be best to keep to the case of the *ellipse* throughout, in order to avoid ambiguities of sign. Suppose that the new diameter CQ meets the tangent at P in E, as in the figure of I. 47 [Prop. 21];

then, if from any point R on the curve the ordinate RW is drawn to PP', it is parallel to the tangent PE, and, if it meets CQ in F, the triangles CPE, CWF are similar, and one angle in each is that between the old and the new diameters.

Also, as the triangles CPE, CWF are the halves of two similar parallelograms on CP, CW, we can use the relation proved in I. 41 [Prop. 16] for parallelograms, provided that we take a triangle on RW as base such that RWP is one angle, and the side WU lying along WP is determined by the relation

$$\frac{RW}{WU} = \frac{CP}{PE} \cdot \frac{p}{d}.$$

Apollonius satisfies this condition by drawing RU parallel to QT, the tangent at Q. The proof is as follows.

From the property of the tangent, I. 37 [Prop. 14],

$$\frac{QV^2}{CV \cdot VT} = \frac{p}{d}.$$

Also, by similar triangles,

$$\frac{QV}{VT} = \frac{RW}{WU}, \text{ and } \frac{QV}{CV} = \frac{PE}{CP}.$$

Therefore $$\frac{RW}{WU} \cdot \frac{PE}{CP} = \frac{p}{d},$$

or $$\frac{RW}{WU} = \frac{CP}{PE} \cdot \frac{p}{d} \text{ (the required relation).}$$

Thus it is clear that the proposition I. 41 [Prop. 16] is true of the three triangles CPE, CFW, RUW; that is,

$$\triangle CPE - \triangle CFW = \triangle RUW \quad\text{.................(2).}$$

It is now necessary to prove, as is done in I. 47 [Prop. 21], that the chord RR' parallel to the tangent at Q is bisected by CQ*, in order to show that RM is the ordinate to CQ in the same way as

* This is proved in I. 47 [Prop. 21] as follows:

$$\triangle CPE - \triangle CFW = \triangle RUW.$$

Similarly, $$\triangle CPE - \triangle CF'W' = \triangle R'UW'.$$

By subtraction, $$F'W'WF = R'W'WR,$$

whence, taking away the figure $R'W'WFM$ from each side,

$$\triangle R'F'M = \triangle RFM,$$

and it follows that $$RM = R'M.$$

RW is to CP. It then follows that the two triangles RUW, CFW have the same relation to the original axes, and to the diameter QQ', as the triangles RFM, CUM have to the new axes, consisting of QQ' and the tangent at Q, and to the diameter PP', respectively.

Also the triangle CPE has the same relation to the old axes that the triangle CQT has to the new.

Therefore, in order to prove that a like relation to that in (2) above holds between three triangles similarly determined with reference to CQ, the tangent at Q and the diameter PP', it has to be shown that

$$\triangle CQT - \triangle CUM = \triangle RMF.$$

The first step is to prove the equality of the triangles CPE, CQT, as to which see note on I. 50 [Prop. 23] and III. 1 [Prop. 53].

We have then, from (2) above,

$$\triangle CQT - \triangle CFW = \triangle RUW,$$

or the quadrilateral $QTWF = \triangle RUW$,

therefore, subtracting the quadrilateral $MUWF$ from each side,

$$\triangle CQT - \triangle CUM = \triangle RMF,$$

the property which it was required to prove.

Thus a relation between areas has been found in exactly the same form as that in (2), but with QQ' as the diameter of reference in place of PP'. Hence, by reversing the process, we can determine the parameter q corresponding to the diameter QQ', and so obtain the equation of the conic with reference to the new axes in the same form as the equation (1) above (p. cxix) referred to PP' and its conjugate; and, when this is done, we have only to move the origin from C to Q in order to effect the complete transformation to the new axes of coordinates consisting of QQ' and the tangent at Q, and to obtain the equation

$$y^2 = qx - \frac{q}{QQ'} \cdot x^2.$$

Now the original parameter p is determined with reference to the length (d) of PP' by the relation

$$\frac{p}{d} = \frac{QV^2}{CV \cdot VT} = \frac{PE}{CP} \cdot \frac{OP}{PT} = \frac{OP}{PT} \cdot \frac{2PE}{d},$$

so that
$$p = \frac{OP}{PT} \cdot 2PE;$$

and the corresponding value for q should accordingly be given by the equation

$$q = \frac{OQ}{QE} \cdot 2QT,$$

which Apollonius proves to be the case in I. 50 [Prop. 23].

No mention of the parabola has been made in the above, because the proof of the corresponding transformation is essentially the same; but it may be noted here that Archimedes was familiar with a method of effecting the same transformation for the parabola. This has been already alluded to (p. liii) as easily deducible from the proposition of Apollonius.

There is another result, and that perhaps the most interesting of all, which can be derived from the foregoing equations between areas. We have seen that

$$\triangle RUW = \triangle CPE - \triangle CFW,$$

so that $$\triangle RUW + \triangle CFW = \triangle CPE,$$

i.e. the quadrilateral $CFRU = \triangle CPE$.

Now, if PP', QQ' are *fixed* diameters, and R a variable point on the curve, we observe that RU, RF are drawn always in fixed directions (parallel to the tangents at Q, P respectively), while the area of the triangle CPE is constant.

It follows therefore that, *if PP', QQ' are two fixed diameters and if from any point R on the curve ordinates be drawn to PP', QQ' meeting QQ', PP' in F, U respectively, then*

the area of the quadrilateral $CFRU$ is constant.

Conversely, *if in a quadrilateral $CFRU$ the two sides CU, CF lie along fixed straight lines, while the two other sides are drawn from a moveable point R in given directions and meeting the fixed lines, and if the quadrilateral has a constant area, then the locus of the point R is an ellipse or a hyperbola.*

Apollonius does not specifically give this converse proposition, nor in fact any proposition stating that this or that locus is a conic. But, as he says in his preface that his work contains "remarkable theorems which are useful for the synthesis of solid loci," we must conclude that among them was the proposition which in effect states that the area of the quadrilateral $CFRU$ is constant, and that the converse way of stating it was perfectly well known to him.

It will be seen from the note to Prop. 18 that the proposition that the area of $CFRU$ is constant is the equivalent of saying that *the equation of a central conic referred to any two diameters as axes is*

$$\alpha x^2 + \beta xy + \gamma y^2 = A,$$

where α, β, γ, A are constants.

It is also interesting to observe that this equation is the equivalent of the intermediate step in the transformation from one diameter and tangent to another diameter and tangent as axes; in other words, *Apollonius passes from the equation referred to one pair of conjugate diameters to the equation referred to a second pair of conjugate diameters by means of the more general equation of the curve referred to axes consisting of one of each pair of conjugates.*

Other forms of the equation of the conic can be obtained, e.g. by regarding RF, RU as fixed coordinate axes and expressing the constancy of the area of the quadrilateral $CF'R'U'$ for any point R' with reference to RF, RU as axes. The axes of reference may then be *any* axes meeting in a point on the curve.

For obtaining the equation we may use the formula

$$CFRU = CF'R'U',$$

or the other relations derived immediately from it, viz.

$$F'IRF = IUU'R',$$

or

$$FJR'F' = JU'UR,$$

which are proved in III. 3 [Prop. 55].

The coordinates of R' would in this case be $R'I$, $R'J$.

Similarly an equation can be found corresponding to the property in III. 2 [Prop. 54] that

$$\triangle HFQ = \text{quadrilateral } HTUR.$$

Again, III. 54, 56 [Prop. 75] lead at once to the "locus with respect to three lines," and from this we obtain the well-known equation to a conic with reference to two tangents as axes, where the lengths of the tangents are h, k, viz.

$$\left(\frac{x}{h} + \frac{y}{k} - 1\right) = 2\lambda \left(\frac{xy}{hk}\right)^{\frac{1}{2}}$$

and, in the particular case of the parabola,

$$\left(\frac{x}{h}\right)^{\frac{1}{2}} + \left(\frac{y}{k}\right)^{\frac{1}{2}} = 1.$$

The latter equation can also be derived directly from III. 41 [Prop. 65], which proves that three tangents to a parabola forming a triangle are divided in the same proportion.

Thus, if x, y be the coordinates of Q with reference to qR, qP as axes, and if $qp = x_1$, $rq = y_1$ (cf. the figure of Prop. 65), we have, by the proposition,

$$\frac{x}{x_1 - x} = \frac{rQ}{Qp} = \frac{y_1 - y}{y} = \frac{k - y_1}{y_1} = \frac{x_1}{h - x_1}.$$

From these equations we find

$$\left.\begin{aligned} \frac{x_1}{x} - 1 &= \frac{h}{x_1} - 1, \quad \text{or} \quad x_1^2 = hx \\ \frac{y_1}{y} - 1 &= \frac{k}{y_1} - 1, \quad \text{or} \quad y_1^2 = ky \end{aligned}\right\} \text{..............(1).}$$

Also, since $\qquad \dfrac{x_1}{x} = \dfrac{y_1}{y_1 - y},$

$$\frac{x}{x_1} + \frac{y}{y_1} = 1 \text{ .... (2),}$$

therefore by combining (1) and (2) we obtain

$$\left(\frac{x}{h}\right)^{\frac{1}{2}} + \left(\frac{y}{k}\right)^{\frac{1}{2}} = 1.$$

The same equation can equally be derived from the property proved by Archimedes (pp. lix, lx).

Lastly, we find of course the equation of the hyperbola referred to its asymptotes

$$xy = c^2,$$

and, if Apollonius had had a relation between the coordinates of a point (x, y) represented to him in a geometrical form equivalent to the equation

$$xy + ax + by + C = 0,$$

he would certainly not have failed to see that the locus was a hyperbola; for the nature of the equation would immediately have suggested the compression of it into a form which would show that the product of the distances of the point (reckoned in fixed directions) from two fixed straight lines is constant.

§ 4. Method of finding two mean proportionals.

It will be remembered that Menaechmus' solution of the problem of the two mean proportionals was effected by finding the points of intersection between any two of the curves

$$x^2 = ay,\ y^2 = bx,\ xy = ab.$$

It is clear that the points of intersection of the first two curves lie on the circle

$$x^2 + y^2 - bx - ay = 0,$$

and therefore that the two mean proportionals can be determined by means of the intersection of this circle with any one of the three curves.

Now, in the construction for two mean proportionals which is attributed to Apollonius, we find this very circle used, and we must therefore assume that he had discovered that the points of intersection of the two parabolas lay on the circle.

We have it on the authority of Ioannes Philoponus* (who quotes one Parmenio) that Apollonius solved the problem thus.

Let the two given unequal straight lines be placed at right angles, as OA, OB.

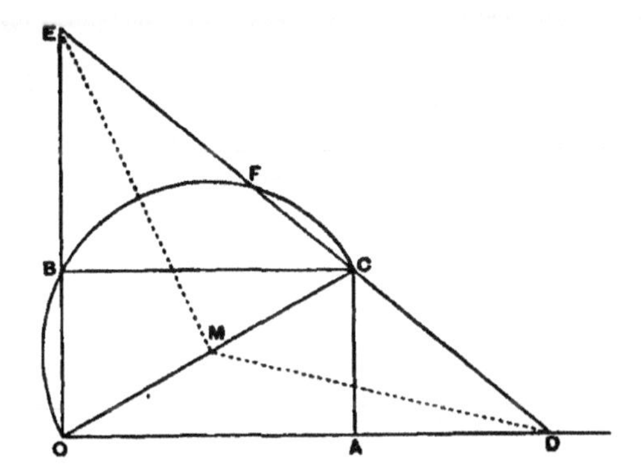

Complete the parallelogram and draw the diagonal OC. On OC as diameter describe the semicircle OBC, produce OA, OB, and through C draw $DCFE$ (meeting OA in D, the circle again in F, and OB in E) so that $DC = FE$. "*And this is assumed as a postulate unproved.*"

Now $DC = FE$, and therefore $DF = CE$.

* On the *Anal. post.* I. The passage is quoted in Heiberg's *Apollonius*, Vol. II. p. 105.

And, since the circle on OC as diameter passes through A,

$$\begin{aligned} OD \,.\, DA &= FD \,.\, DC \\ &= CE \,.\, EF \\ &= OE \,.\, EB; \end{aligned}$$

$$\therefore\ OD : OE = BE : AD \quad \ldots\ldots\ldots\ldots(1).$$

But, by similar triangles,

$$\begin{aligned} OD : OE &= CB : BE \\ &= OA : BE \quad \ldots\ldots\ldots\ldots(2). \end{aligned}$$

Also, by similar triangles,

$$\begin{aligned} OD : OE &= DA : AC \\ &= DA : OB \quad \ldots\ldots\ldots\ldots(3). \end{aligned}$$

It follows from (1), (2) and (3) that

$$OA : BE = BE : AD = AD : OB;$$

hence BE, AD are the two required mean proportionals.

The important step in the above is the *assumed* step of drawing DE through C so that $DC = FE$.

If we compare with this the passage in Pappus which says that Apollonius "has also contrived the resolution of it by means of the sections of the cone*," we may conclude that the point F in the above figure was determined by drawing a rectangular hyperbola with OA, OB as asymptotes and passing through C. And this is the actual procedure of the Arabian scholiast in expounding this solution. Hence it is sufficiently clear that Apollonius' solution was obtained by means of the intersection of the circle on OC as diameter with the rectangular hyperbola referred to, i.e. by the intersection of the curves

$$\left.\begin{aligned} x^2 + y^2 - bx - ay &= 0 \\ xy &= ab \end{aligned}\right\}.$$

The *mechanical* solution attributed to Apollonius is given by Eutocius†. In this solution M, the middle point of OC, is taken, and with M as centre a circle has to be described cutting OA, OB produced in points D, E such that the line DE passes through C; and this, the writer says, can be done *by moving a ruler about C as a fixed point until the distances of D, E* (the points in which it crosses OA, OB) *from M are equal.*

* Pappus III. p. 56. Οὗτοι γὰρ ὁμολογοῦντες στερεὸν εἶναι τὸ πρόβλημα τὴν κατασκευὴν αὐτοῦ μόνον ὀργανικῶς πεποίηνται συμφώνως Ἀπολλωνίῳ τῷ Περγαίῳ, ὃς καὶ τὴν ἀνάλυσιν αὐτοῦ πεποίηται διὰ τῶν τοῦ κώνου τομῶν.

† Archimedes, Vol. III. pp. 76—78.

It is clear that this solution is essentially the same as the other, because, if DC be made equal to FE as in the former case, the line from M perpendicular to DE must bisect it, and therefore $MD = ME$. This coincidence is noticed in Eutocius' description of the solution of the problem by Philo Byzantinus. This latter solution is the same as that attributed by Ioannes Philoponus to Apollonius except that Philo obtains the required position for DE by moving the ruler about C until DC, FE become equal. Eutocius adds that this solution is almost the same as Heron's (given just before and identical with the *mechanical* solution of Apollonius), but that Philo's method is more convenient in practice (πρὸς χρῆσιν εὐθετώτερον), because it is, by dividing the ruler into equal and continuous parts, possible to watch the equality of the lines DC, FE with much greater ease than to make trial with a pair of compasses (καρκίνῳ διαπειράζειν) whether MD, ME are equal*.

It may be mentioned here that, when Apollonius uses the problem of the two mean proportionals in the *Conics*, it is for the purpose of connecting the coordinates of a point on a central conic with the coordinates of the corresponding centre of curvature, i.e. of the corresponding point on the evolute. The propositions on the subject are v. 51, 52 [Prop. 99].

§ 5. Method of constructing normals passing through a given point.

Without entering into details, for which reference should be made to v. 58–63 [Props. 102, 103], it may be stated generally that Apollonius' method of finding the feet of the various normals passing through a given point is by the construction of a certain rectangular hyperbola which determines, by its intersections with the conic, the required points.

The analytical equivalent of Apollonius' procedure is as follows. Suppose O to be the fixed point through which the normals are to pass, and PGO to be one of those normals, meeting the major or transverse axis of a central conic, or the axis of a parabola, in G. Let PN be the ordinate of P, and OM the perpendicular from O on the axis.

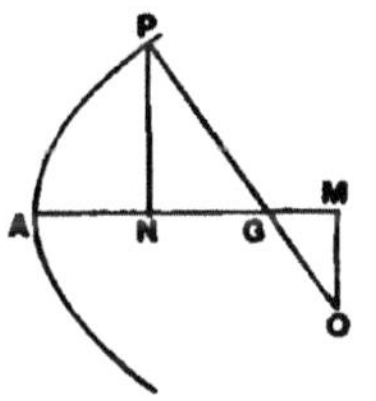

Then, if we take as axes of coordinates the axes of the central conic, and, for the parabola,

* Archimedes, Vol. III. p. 76.

the axis and the tangent at the vertex, and if (x, y), (x_1, y_1) be the coordinates of P, O respectively, we have

$$\frac{y}{-y_1} = \frac{NG}{x_1 - x - NG}.$$

Therefore, (1) for the *parabola*,

$$\frac{y}{-y_1} = \frac{\frac{p_a}{2}}{x_1 - x - \frac{p_a}{2}},$$

or $$xy - \left(x_1 - \frac{p_a}{2}\right)y - y_1 \cdot \frac{p_a}{2} = 0 \text{..................}(1);$$

(2) for the *ellipse* or *hyperbola*,

$$xy\left(1 \mp \frac{b^2}{a^2}\right) - x_1 y \pm \frac{b^2}{a^2} \cdot y_1 x = 0.$$

The intersections of these rectangular hyperbolas with the respective conics give the feet of the various normals passing through O.

Now Pappus criticises this procedure, so far as applied to the *parabola*, as being unorthodox. He is speaking (p. 270) of the distinction between the three classes of "plane" (ἐπίπεδα), "solid" (στερεά), and the still more complicated "linear" problems (γραμμικὰ προβλήματα), and says, "Such procedure seems a serious error on the part of geometers when the solution of a plane problem is discovered by means of conics or higher curves, and generally when it is solved by means of a foreign kind (ἐξ ἀνοικείου γένους), as, for example, the problem in the fifth Book of the *Conics* of Apollonius in the case of the parabola, and the solid νεῦσις with reference to a circle assumed in the book about the spiral by Archimedes; for it is possible without the use of anything solid to discover the theorem propounded by the latter...." The first allusion must clearly be to the use of the intersections of a rectangular *hyperbola* with the parabola when the same points could be obtained by means of the intersections of the latter with a certain *circle*. Presumably Pappus regarded the parabola itself as being completely drawn and given, so that its character as a "solid locus" was not considered to affect the order of the problem. On this assumption the criticism has no doubt some force, because it is a clear advantage to be able to effect the construction by means of the line and circle only.

The circle in this case can of course be obtained by combining the equation of the rectangular hyperbola (1) above with that of the parabola $y^2 = p_a x$.

Multiply (1) by $\frac{y}{p_a}$, and we have

$$\frac{x}{p_a} y^2 - \left(x_1 - \frac{p_a}{2}\right) \frac{y^2}{p_a} - \frac{yy_1}{2} = 0,$$

and, substituting $p_a x$ for y^2,

$$x^2 - \left(x_1 - \frac{p_a}{2}\right) x - \frac{yy_1}{2} = 0,$$

whence, by adding the equation of the parabola, we have

$$x^2 + y^2 - \left(x_1 + \frac{p_a}{2}\right) x - \frac{y_1}{2} . y = 0.$$

But there is nothing in the operations leading to this result which could not have been expressed in the geometrical language which the Greeks used. Moreover we have seen that in Apollonius' solution of the problem of the two mean proportionals the same reduction of the intersections between two conics to the intersections of a conic and a circle is found. We must therefore assume that Apollonius could have reduced the problem of the normals to a parabola in the same way, but that he purposely refrained from doing so. Two explanations of this are possible; either (1) he may have been unwilling to sacrifice to a pedantic orthodoxy the convenience of using one uniform method for all three conics alike, or (2) he may have regarded the presence of one "solid locus" (the given parabola) in his figure as determinative of the class of problem, and may have considered that to solve it with the help of a circle only would not, in the circumstances, have the effect of making it a "plane" problem.

CHAPTER IV.

THE CONSTRUCTION OF A CONIC BY MEANS OF TANGENTS.

IN Book III. 41–43 [Props. 65, 66, 67] Apollonius gives three theorems which may be enunciated as follows:

41. *If three straight lines, each of which touches a parabola, meet one another, they will be cut in the same proportion.*

42. *If in a central conic parallel tangents be drawn at the extremities of a fixed diameter, and if both tangents be met by any variable tangent, the rectangle under the intercepts on the parallel tangents is constant, being equal to the square on half the parallel diameter, i.e. the diameter conjugate to that joining the points of contact.*

43. *Any tangent to a hyperbola cuts off lengths from the asymptotes whose product is constant.*

There is an obvious family likeness between these three consecutive propositions, and their arrangement in this manner can hardly have been the result of mere accident. It is true that III. 42 [Prop. 66] is used almost directly afterwards for determining the foci of a central conic, and it might be supposed that it had its place in the book for this reason only; but, if this were the case, we should have expected that the propositions about the foci would follow directly after it instead of being separated from it by III. 43, 44 [Props. 67, 68]. We have also a strong positive reason for supposing that the arrangement was due to set purpose rather than to chance, namely the fact that all three propositions can be used for describing a conic by means of tangents. Thus, if two tangents to a parabola are given, the first of the three propositions gives a general method of drawing any number of other tangents; while the second and third give the simplest cases of the construction of an ellipse and a hyperbola by the same means, those cases, namely, in which the fixed tangents employed are chosen in a special manner.

As therefore the three propositions taken together contain the essentials for the construction of all three conics by this method, it becomes important to inquire whether Apollonius possessed the means of drawing any number of tangents satisfying the given conditions in each case. That Apollonius was in a position to solve this problem is proved by the contents of two of his smaller treatises. One of these, λόγου ἀποτομῆς β′ (two Books *On cutting off a proportion*), we possess in a translation by Halley from the Arabic under the title *De sectione rationis*; the other, now lost, was χωρίου ἀποτομῆς β′ (two Books *On cutting off a space*, which means cutting off from two fixed lines lengths, measured from fixed points on the lines respectively, such that they contain a rectangle of constant area). Now the very problem just mentioned of drawing any number of tangents to a parabola reduces precisely to that which is discussed with great fulness in the former of the two treatises, while the construction of any number of tangents to the ellipse and hyperbola in accordance with the conditions of III. 42, 43 [Props. 66, 67] reduces to two important cases of the general problem discussed in the second treatise.

I. In the case of the *parabola*, if two tangents qP, qR and the points of contact P, R are given, we have to draw through any point a straight line which will intersect the given tangents (in r, p respectively) in such a way that

$$Pr : rq = qp : pR,$$

or

$$Pr : Pq = qp : qR;$$

that is, we must have

$$Pr : qp = Pq : qR \text{ (a constant ratio).}$$

In fact, we have to draw a line such that the intercept on one tangent measured from the point of contact is to the intercept on the other tangent measured from the intersection of the tangents in a given ratio. How to do this is shown in the greatest detail in the first Book λόγου ἀποτομῆς.

If, again, instead of the points of contact, two other tangents are given meeting the fixed tangent qP in r_1, r_2 and the fixed tangent qR in p_1, p_2, we have to draw a straight line rp cutting off along the tangents qP, qR parts measured from r_1, p_1 respectively which are in a given proportion, i.e. such that

$$r_1r : p_1p = r_1r_2 : p_1p_2 \text{ (a fixed ratio)};$$

and this problem is solved in the second Book λόγου ἀποτομῆς.

The general problem discussed in that treatise is, to draw from a point O a straight line which shall cut off from two given straight lines portions, measured from two fixed points A, B, which are in a given proportion, e.g., in the accompanying figure, ONM is to be drawn so that $AM : BN$ is a given ratio. In the second Book of

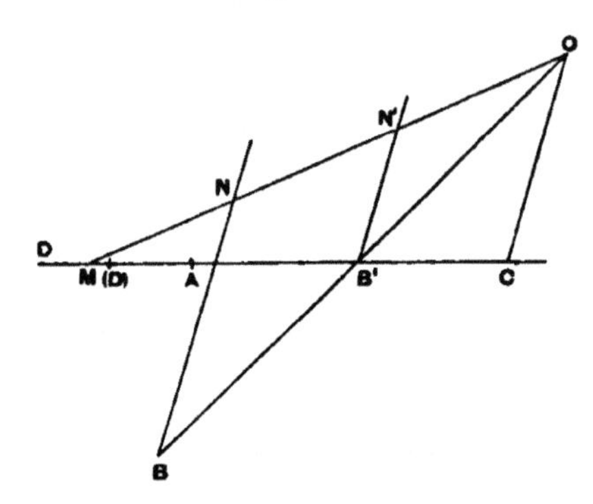

the treatise this general case is reduced to a more special one in which the fixed point B occupies a position B' on the first line AM, so that one of the intercepts is measured from the intersection of the two lines. The reduction is made by joining OB and drawing $B'N'$ parallel to BN from the point B' in which OB, MA intersect.

Then clearly $B'N' : BN$ is a given ratio, and therefore the ratio $B'N' : AM$ is given.

We have now to draw a straight line $ON'M$ cutting MAB', $B'N'$ in points M, N' such that

$$\frac{B'N'}{AM} = \text{a given ratio, } \lambda \text{ suppose.}$$

This problem is solved in the first Book, and the solution is substantially as follows.

Draw OC parallel to $N'B'$ meeting MA produced in C. Now suppose a point D found on AM such that

$$\lambda = \frac{OC}{AD}.$$

Then, supposing that the ratio $\frac{B'N'}{AM}$ is made equal to λ, we have

$$\frac{AM}{AD} = \frac{B'N'}{OC} = \frac{B'M}{CM},$$

whence

$$\frac{MD}{AD} = \frac{CB'}{CM},$$

and therefore $CM \,.\, MD = AD \,.\, CB'$ (a given rectangle).

Thus a given line CD has to be divided at M so that $CM \,.\, MD$ has a given value; and this is the Euclidean problem of *applying to a given straight line a rectangle equal to a given area but falling short, or exceeding, by a square.*

In the absence of algebraical signs, it was of course necessary for Apollonius to investigate a large number of separate cases, and also to find the limiting conditions of possibility and the number of the possible solutions between each set of limits. In the case represented in the above figure the solution is always possible for any value of the given ratio, because the given value $AD \,.\, CB'$, to which $CM \,.\, MD$ is to be equal, is always less than $CA \,.\, AD$, and therefore always less than $\left(\frac{CD}{2}\right)^2$, the maximum value of the rectangle whose sides are together equal to CD. As the *application* of the rectangle would give two positions of M, it remains to be proved that only one of them falls on AD and so gives a solution such as the figure requires; and this is so because $CM \,.\, MD$ must be less than $CA \,.\, AD$.

The application to the parabola has more significance in the cases where the given ratio must be subject to certain limits in order that the solution of the problem may be possible. This will be so, e.g. in the annexed figure, where the letters have the same meaning as before, and the particular case is taken in which one

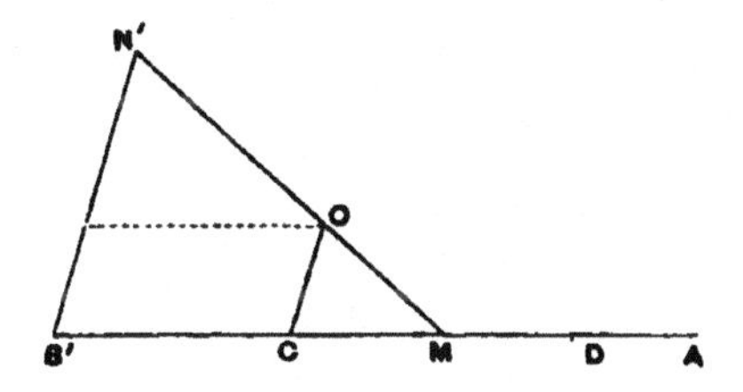

intercept $B'N'$ is measured from B', the intersection of the two fixed lines. Apollonius begins by stating the limiting case, saying that we obtain a solution in a special manner in the case where M is the middle point of CD, so that the given rectangle $CM . MD$ or $CB' . AD$ has its maximum value.

In order to find the corresponding limiting value of λ, Apollonius seeks the corresponding position of D.

We have $$\frac{B'C}{MD} = \frac{CM}{AD} = \frac{B'M}{MA},$$

whence, since $MD = CM$,

$$\frac{B'C}{B'M} = \frac{CM}{MA} = \frac{B'M}{B'A},$$

and therefore $$B'M^2 = B'C . B'A.$$

Thus M is determined, and therefore D also.

According, therefore, as λ is less or greater than the particular value of $\frac{OC}{AD}$ thus determined, Apollonius finds no solution or two solutions.

At the end we find also the following further determination of the limiting value of λ. We have

$$AD = B'A + B'C - (B'D + B'C)$$
$$= B'A + B'C - 2B'M$$
$$= B'A + B'C - 2\sqrt{B'A . B'C}.$$

Thus, if we refer the various points to a system of coordinates with $B'A$, $B'N'$ as axes, and if we denote the coordinates of O by (x, y) and the length $B'A$ by h, we have

$$\lambda = \frac{OC}{AD} = \frac{y}{h + x - 2\sqrt{hx}}.$$

If we suppose Apollonius to have used these results for the parabola, he cannot have failed to observe that the limiting case described is that in which O is on the parabola, while $N'OM$ is the tangent at O; for, as above,

$$\frac{B'M}{B'A} = \frac{B'C}{B'M}$$
$$= \frac{N'O}{N'M}, \text{ by parallels,}$$

so that $B'A$, $N'M$ are divided at M, O respectively in the same proportion.

Further, if we put for λ the proportion between the lengths of the two fixed tangents, we obtain, if h, k be those lengths,

$$\frac{k}{h} = \frac{y}{h + x - 2\sqrt{hx}},$$

which is the equation of the parabola referred to the fixed tangents as coordinate axes, and which can easily be reduced to the symmetrical form

$$\left(\frac{x}{h}\right)^{\frac{1}{2}} + \left(\frac{y}{k}\right)^{\frac{1}{2}} = 1.$$

II. In the case of the *ellipse* and *hyperbola* the problem is to draw through a given point O a straight line cutting two straight lines in such a way that the intercepts upon them measured from fixed points contain a rectangle of constant area, and for the *ellipse* the straight lines are parallel, while for the *hyperbola* they meet in a point and the intercepts on each are measured from the point of their intersection.

These are particular cases of the general problem which, according to Pappus, was discussed in the treatise entitled χωρίου ἀποτομή; and, as we are told that the propositions in this work corresponded severally to those in the λόγου ἀποτομή, we know that the particular cases now in question were included. We can also form an idea how the general problem was solved. The reduction to the particular case where one of the points from which the intercepts are measured is the intersection of the two fixed lines is effected in the same manner as in the case of proportional section described above. Then, using the same figure (p. cxxxii), we should take the point D (in the position represented by (D) in the figure) such that

$$OC \cdot AD = \text{the given rectangle.}$$

We have then to draw the line $ON'M$ so that

$$B'N' \cdot AM = OC \cdot AD,$$

or

$$\frac{B'N'}{OC} = \frac{AD}{AM}.$$

But, since $B'N'$, OC are parallel,

$$\frac{B'N'}{OC} = \frac{B'M}{CM}.$$

Therefore

$$\frac{AM}{CM} = \frac{AD}{B'M} = \frac{DM}{B'C},$$

and the rectangle $B'M \cdot MD = AD \cdot B'C$, which is given. Hence, as before, the problem is reduced to an *application* of a rectangle in the well-known manner.

The complete treatment of the particular cases of the problem, with their διορισμοί, could present no difficulty to Apollonius.

III. It is not a very great step from what we find in Apollonius to the general theorem that, *if a straight line cuts off from two fixed straight lines intercepts, measured from given points on the lines respectively, which contain a rectangle of given area, the envelope of the first straight line is a conic section touching the two fixed straight lines.*

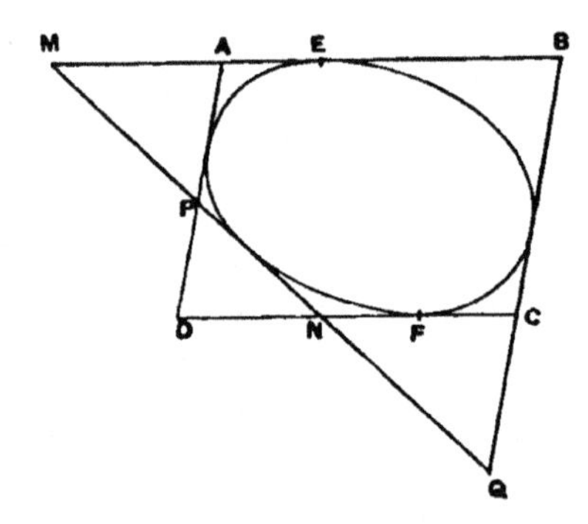

Thus, suppose $ABCD$ to be a parallelogram described about a conic and E, F to be the points of contact of AB, CD. If a fifth tangent MN cuts AB, CD in M, N and AD, CB in P, Q respectively, we have, by the proposition of Apollonius,

$$EA \cdot FD = EM \cdot FN.$$

Therefore
$$\frac{EA}{FN} = \frac{EM}{FD} = \frac{AM}{ND} = \frac{AP}{PD}.$$

Hence, since $EA = CF$,

$$\frac{CF}{AP} = \frac{FN}{PD} = \frac{CN}{AD},$$

and therefore
$$AP \cdot CN = CF \cdot AD,$$

or the rectangle $AP \cdot CN$ has an area independent of the position of the particular fifth tangent MN.

Conversely, if the lines AD, DC are given as well as the points A, C and the area of the rectangle $AP.CN$, we can determine the point F, and therefore also the point E where AB touches the conic. We have then the diameter EF and the direction of the chords bisected by it, as well as the tangent AD; thus we can find the ordinate to EF drawn through the point of contact of AD, and hence we can obtain the equation of the conic referred to the diameter EF and its conjugate as axes of coordinates. Cf. Lemma xxv. of the first Book of Newton's *Principia* and the succeeding investigations.

CHAPTER V.

THE THREE-LINE AND FOUR-LINE LOCUS.

THE so-called τόπος ἐπὶ τρεῖς καὶ τέσσαρας γραμμάς is, as we have seen, specially mentioned in the first preface of Apollonius as a subject which up to his time had not received full treatment. He says that he found that Euclid had not worked out the synthesis of the locus, but only some part of it, and that not successfully, adding that in fact the complete theory of it could not be established without the "new and remarkable theorems" discovered by himself and contained in the third book of his *Conics*. The words used indicate clearly that Apollonius did himself possess a complete solution of the problem of the four-line locus, and the remarks of Pappus on the subject (quoted above, p. xxxi, xxxii), though not friendly to Apollonius, confirm the same inference. We must further assume that the key to Apollonius' solution is to be found in the third Book, and it is therefore necessary to examine the propositions in that Book for indications of the way in which he went to work.

The *three-line locus* need not detain us long, because it is really a particular case of the *four-line locus*. But we have, in fact, in III. 53–56 [Props. 74–76] what amounts to a complete demonstration of the theoretical converse of the three-line locus, viz. the proposition that, *if from any point of a conic there be drawn three straight lines in fixed directions to meet respectively two fixed tangents to the conic and their chord of contact, the ratio of the rectangle contained by the first two lines so drawn to the square on the third line is constant.* The proof of this for the case where the two tangents are parallel is obtained from III. 53 [Prop. 74], and the remaining three propositions, III. 54–56 [Props. 75, 76], give the proof where the tangents are not parallel.

In like manner, we should expect to find the theorem of the *four-line locus* appearing, if at all, in the form of the converse proposition stating that *every conic section has, with reference to any inscribed quadrilateral, the properties of the four-line locus.* It will be seen from the note following Props. 75, 76 that this theorem is easily obtained from that of the three-line locus as presented by Apollonius in those propositions; but there is nowhere in the Book any proposition more directly leading to the former. The explanation may be that the *construction* of the locus, that is, the aspect of the question which would be appropriate to a work on *solid loci* rather than one on *conics*, was considered to be of preponderant importance, and that the theoretical converse was regarded as a mere appendage to it. But, from the nature of the case, that converse must presumably have appeared as an intermediate step in the investigation of the locus, and it could hardly have been unknown even to earlier geometers, such as Euclid and Aristaeus, who had studied the subject thoroughly.

In these circumstances we have to seek for indications of the probable course followed by Greek geometers in their investigation of the four-line locus; and, in doing so, we have to bear in mind that the problem must have been capable of partial solution before the time of Apollonius, and that it could be completely solved by means of the propositions in his third Book.

We observe, in the first place, that III. 54–56 [Props. 75, 76], which lead to the property of the three-line locus, are proved by means of the proposition that the ratio of the rectangles under the segments of any intersecting chords drawn in fixed directions is constant. Also the property of the three-line locus is a particular case of the property of a conic with reference to an inscribed quadrilateral having two of its sides parallel, that case, namely, in which the two parallel sides are coincident; and it will be seen that the proposition relating to the rectangles under the segments of intersecting chords can equally well be used for proving generally that a conic is a four-line locus with reference to any inscribed quadrilateral which has two sides parallel.

For, if AB is a fixed chord of a conic and Rr a chord in a given direction cutting AB in I, we have

$$\frac{RI \, . \, Ir}{AI \, . \, IB} = (\text{const.}).$$

If we measure RK along Rr equal to Ir, the locus of K is a chord

DC meeting the diameter which bisects chords parallel to Rr in the same point in which it is met by AB, and the points D, C lie on lines drawn through A, B respectively parallel to Rr.

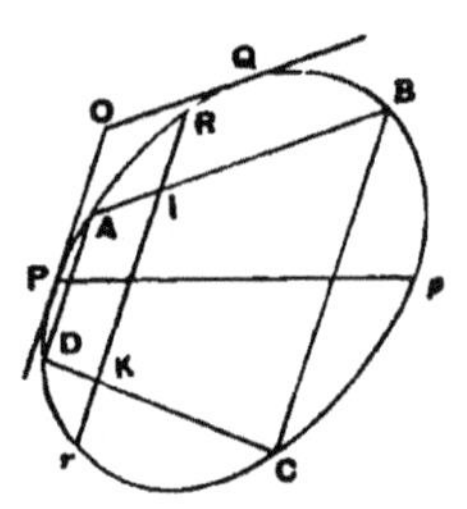

Then, if x, y, z, u be the distances of R from the sides of the quadrilateral $ABCD$, we shall have

$$\frac{xz}{yu} = \text{(const.)}.$$

And, since $ABCD$ may be any inscribed quadrilateral with two sides parallel, or a *trapezium*, the proposition is proved generally for the particular kind of quadrilateral.

If we have, on the other hand, to find the geometrical locus of a point R whose distances x, y, z, u from the sides of such a trapezium are connected by the above relation, we can first manipulate the constants so as to allow the distances to be measured in the directions indicated in the figure, and we shall have

$$\lambda = \frac{RI \,.\, RK}{AI \,.\, IB} = \frac{RI \,.\, Ir}{AI \,.\, IB},$$

where λ is a given constant. We must then try to find a conic whose points R satisfy the given relation, but we must take care to determine it in such a manner as to show synthetically at the same time that the points of the conic so found do really satisfy the given condition; for, of course, we are not yet supposed to know that the locus *is* a conic.

It seems clear, as shown by Zeuthen, that the defective state of knowledge which prevented the predecessors of Apollonius from completing the determination of the four-line locus had reference rather to this first step of finding the locus in the particular case of a trapezium than to the transition from the case of a trapezium to that of a quadrilateral of any form. The transition was in fact, in itself, possible by means which were within the competence of Euclid, as will presently be seen; but the difficulty in the way of the earlier step was apparently due to the fact that the conception of the two branches of a hyperbola as a single curve had not occurred to any one before Apollonius. His predecessors accordingly, in the case where the four-line locus is a complete hyperbola in the modern sense, probably considered only one branch of it; and the question which branch it would be would depend on some further condition determining it as one of the two branches, e.g. the constant might have been determined by means of a given point through which the conic or single-branch hyperbola, which it was required to prove to be the four-line locus, should pass.

To prove that such a single branch of a hyperbola, not passing through all four corners of the quadrilateral, could be the four-line locus, and also to determine the locus corresponding to the value of λ leading to such a hyperbola, it was necessary to know of the connexion of one branch with the other, and the corresponding extensions of all the propositions used in the proof of the property of the inscribed quadrilateral, as well as of the various steps in the converse procedure for determining the locus. These extensions to the case of the complete hyperbola may, as already mentioned (p. lxxxiv *seqq.*), be regarded as due to Apollonius. His predecessors could perfectly well have proved the proposition of the inscribed trapezium for any single-branch conic; and it will be seen that the converse, the construction of the locus, would in the particular case present no difficulty to them. The difficulty would come in where the conic was a hyperbola with two branches.

Assuming, then, that the property of the four-line locus was established with respect to an inscribed trapezium by means of the proposition that the rectangles under the segments of intersecting chords are to one another in the ratio of the squares on the parallel tangents, what was wanted to complete the theory was (1) the extension to the case where the tangents are tangents to opposite branches of a hyperbola, (2) the expression of the constant ratio between the rectangles referred to in those cases where no tangent can be drawn parallel to either of the chords, or where a tangent can be drawn parallel to one of them only. Now we find (1) that Apollonius proves the proposition for the case where the tangents touch opposite branches in III. 19 [Prop. 59, Case I.]. Also (2) the proposition III. 23 [Prop. 59, Case IV.] proves that, where there is no tangent to the hyperbola parallel to either of the chords, the constant ratio of the rectangles is equal to the ratio of the squares of the parallel tangents to the *conjugate* hyperbola; and III. 21 [Prop. 59, Case II.] deals with the case where a tangent can be drawn parallel to one of the chords, while no tangent can be drawn parallel to the other, and proves that, if tQ, the tangent, meets the diameter bisecting the chord to which it is not parallel in t, and if tq is half the chord through t parallel to the same chord, the constant ratio is then $tQ^2 : tq^2$.

Zeuthen suggests (p. 140) that the method adopted for determining the complete conic described about a given trapezium $ABCD$, which is the locus with respect to the four sides of the trapezium corresponding to a given value of the constant ratio λ, may have been to employ an auxiliary figure for the purpose of constructing a conic *similar* to that required to be found, or rather of finding the form of certain rectilineal figures connected with such a similar conic. This procedure is exemplified in Apollonius, II. 50–53 [Props. 50–52], where a certain figure is determined by means of a previous construction of another figure of the same form; and the suggestion that the same procedure was employed in this case has the advantage that it can be successfully applied to each of the separate cases in which Apollonius gives the different expressions for the constant ratio between the rectangles under the segments of intersecting chords in fixed directions.

We have the following data for determining the form of the conic similar to the required conic circumscribing $ABCD$: the value (λ) of the ratio $\dfrac{RI \cdot Ir}{AI \cdot IB}$ between the products of segments of lines in two different directions, and the direction of the diameter Pp bisecting chords in one of the given directions.

I. Suppose that the conic has tangents in both given directions (which is always the case if the conic is a conic in the old sense of the term, i.e. if the double-branch hyperbola is excluded).

Let the points of the auxiliary figure be denoted by accented letters corresponding to those in the figure on p. cxl.

We know the ratio

$$\frac{OP}{OQ} = \sqrt{\lambda}$$

and, if we choose any straight line for $O'P'$, we know (1) the position of a diameter, (2) its extremity P', (3) the direction of the chords bisected by the diameter, (4) a point Q' with the tangent at that point.

Then the intersection of the tangent at Q' with the diameter and the foot of the ordinate to it from Q' determine, with P', three points out of four which are harmonically related, so that the remaining one, the other extremity (p') of the diameter, is found. Hence the conic in the auxiliary figure is determined.

II. Suppose that the conic has no tangent in either direction.

In this case we know the ratio between the tangents to the hyperbola *conjugate* to the required auxiliary hyperbola, and we can therefore determine the conjugate hyperbola in the manner just described; then, by means of the conjugate, the required auxiliary hyperbola is determined.

III. Suppose that the conic has a tangent in the direction of AD, but not in the direction of AB.

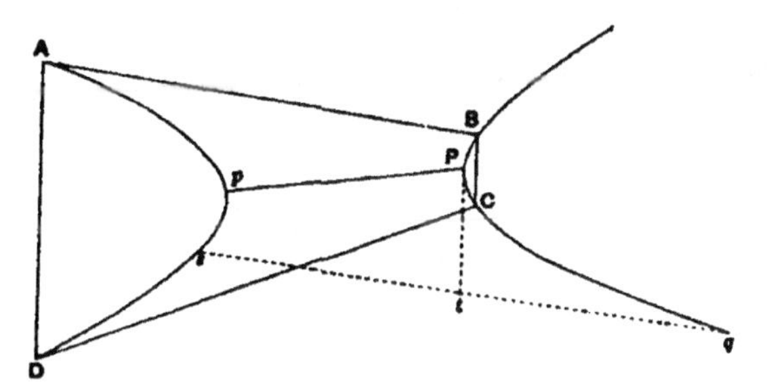

In this case, if the tangent Pt parallel to AD and the diameter bisecting AB meet in t, Apollonius has expressed the constant λ as the ratio between the squares of the tangent tP and of tq, the half of the chord through t parallel to AB. We have then

$$\frac{tP}{tq} = \frac{tP}{ts} = \sqrt{\lambda}.$$

If we now choose $t'P'$ arbitrarily, we have, towards determining the auxiliary similar conic,

(1) a diameter with the direction of chords bisected by it,

(2) one extremity P' of that diameter,

(3) two points q', s' on the curve.

If y_1, y_2 are the ordinates of q', s' with respect to the diameter, x_1, x_2 the distances of the feet of the ordinates from P', and x_1', x_2' their distances from the other (unknown) extremity of the diameter, we have

$$\frac{y_1^2}{x_1 \cdot x_1'} = \frac{y_2^2}{x_2 \cdot x_2'},$$

whence $\frac{x_1'}{x_2'}$ is determined.

The point p' can thus be found by means of the ratio between its distances from two known points on the straight line on which it must lie.

IV. Suppose that the conic has a tangent in the direction of AB, but not in the direction of AD.

Let the tangent at P, parallel to AB, meet the diameter bisecting BC, AD in t, and let tq parallel to AD meet the conic in q; we then have

$$\frac{tq}{tP} = \sqrt{\lambda} = \frac{t'q'}{t'P'}.$$

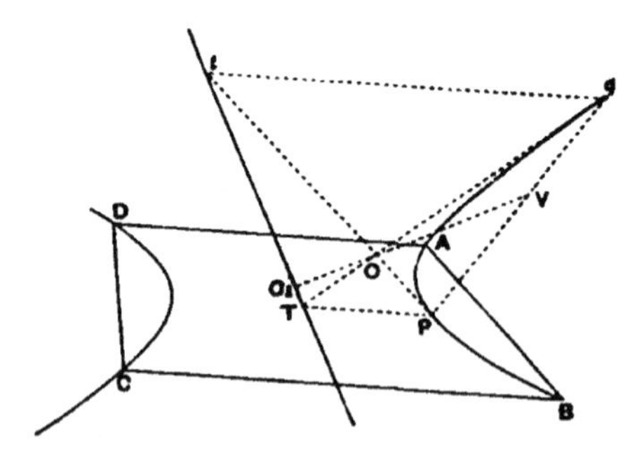

If we choose either $t'q'$ or $t'P'$ arbitrarily, we have

(1) the diameter $t'T'$,

(2) the points P', q' on the curve, the ordinates from which to the diameter meet it in t', T' respectively,

(3) the tangent at P'.

Since $t'P'$ is the tangent at P',

$$C't' \cdot C'T' = \tfrac{1}{4} \cdot a'^2,$$

where C' is the centre, and a' the length of the diameter.

Therefore, by symmetry, $T'q'$ is the tangent at q'. [Prop. 42.]

Hence we can find the centre C' by joining V', the middle point of $P'q'$, to O', the point of intersection of the tangents, since $V'O'$ must be a diameter and therefore meets $t'T'$ in C'.

Thus the auxiliary conic can be readily determined. The relation between the diameter a' and the diameter b' conjugate to it is given by

$$\frac{t'q'^2}{C't' \cdot t'T'} = \frac{b'^2}{a'^2} = \frac{b^2}{a^2}.$$

Thus it is seen that, in all four cases, the propositions of Apollonius supply means for determining an auxiliary figure similar to that which is sought. The transition to the latter can then be made in various ways; e.g. the auxiliary figure gives at once the direction of the diameter bisecting AB, so that the centre is given; and we can effect the transition by means of the ratio between CA and $C'A'$.

There are, however, indications that the auxiliary figures would not in practice be used beyond the point at which the ratio of the diameter (a) bisecting the parallel sides of the trapezium to its conjugate (b) is determined, inasmuch as we find in Apollonius propositions which lead directly to the determination of the absolute values of a and b when the ratio $\frac{a}{b}\left(=\frac{a'}{b'}\right)$ is given. The problem to be solved is, in fact, to describe a conic through two given points A and B such that one diameter of it lies along a given straight line, while the direction of the chords bisected by the diameter is given, as well as the ratio $\left(\frac{a}{b}\right)$ between the length of the diameter and its conjugate.

Suppose that, in the accompanying figure, a straight line is drawn through B parallel to the known direction of the diameter,

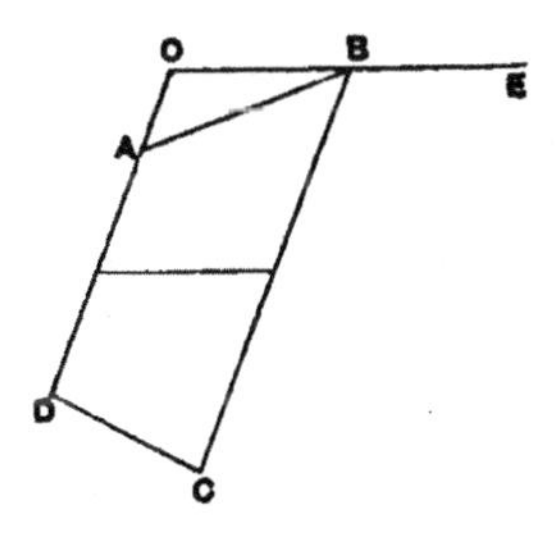

and meeting DA produced in O. Also let OB meet the curve (which we will suppose to be an ellipse) again in E.

Then we must have

$$\frac{OB.OE}{OA.OD}=\frac{a^2}{b^2},$$

whence OE can be found, and therefore the position of E. The line bisecting BE and parallel to AD or BC will determine the centre.

We have now, for the case of the ellipse, a proposition given by Apollonius which determines the value of a^2 directly. By III. 27 [Prop. 61 (1)] we know that

$$OB^2+OE^2+\frac{a^2}{b^2}(OA^2+OD^2)=a^2,$$

whence a^2 is at once found.

Similar propositions are given for the hyperbola (see III. 24–26, 28, 29 [Props. 60 and 61 (2)]). The construction in the case of the hyperbola is also facilitated by means of the asymptote properties. In this case, if the letters have the same significations as in the figure for the ellipse, we find the centre by means of the chord BE or by using the auxiliary similar figure. The asymptotes are then determined by the ratio $\frac{a}{b}$. If these cut the chord AD in K, L, then

$$AK.AL=\tfrac{1}{4}b^2,$$

or

$$AK.KD=\tfrac{1}{4}b^2.$$

If the required curve is a *parabola*, the determination of the auxiliary similar figure after the manner of the first of the four cases detailed above would show that P', the end of the diameter, is at the middle point of the intercept between the intersection of the diameter with the tangent at Q' and with the ordinate from Q' respectively. The curve can then be determined by the simple use of the ordinary equation of the parabola.

So far the determination of the four-line locus has only been considered in the particular case where two opposite sides of the inscribed quadrilateral are parallel. It remains to consider the possible means by which the determination of the locus with reference to a quadrilateral of any form whatever might have been reduced to the problem of finding the locus with reference to a trapezium. As Apollonius' third Book contains no propositions which can well be used for effecting the transition, it must be concluded that the transition itself was not affected by Apollonius' completion of the theory of the locus, but that the key must be looked for elsewhere. Zeuthen (Chapter 8) finds the key in the *Porisms* of Euclid*. He notes first that Archimedes' proposition (given on p. lix, lx above) respecting the parabola exhibits the curve as a four-line locus with respect to two quadrilaterals, of which one is obtained from the other by turning two adjacent sides about the points on the parabola in which they meet the two other sides. (Thus PQ is turned about Q and takes the position QT, while PV is turned about its intersection with the parabola at infinity and takes the position of the diameter through Q.) This suggests the inquiry whether the same means which are used to effect the transition in this very special case cannot also be employed in the more general case now under consideration.

As the *Porisms* of Euclid are themselves lost, it is necessary to resort to the account which Pappus gives of their contents; and the only one of the *Porisms* which is there preserved in its original form is as follows†:

If from two given points there be drawn straight lines which intersect one another on a straight line given in position, and if one of the straight lines so drawn cuts off from a straight line given in position a certain length measured from a given point on it, then the other straight line also will cut off a portion from another straight line bearing a given ratio [*to the former intercept*].

The same proposition is true also when a four-line locus is substituted for the first-mentioned given straight line and the two fixed points are any two fixed points on the locus. Suppose that we take as the two fixed points the points A and C, being two opposite corners of the quadrilateral $ABCD$ to which the locus is referred, and suppose the lines from which the intercepts are cut off to be CE, AE drawn respectively parallel to the sides BA, BC of the quadrilateral.

Let M be a point on the required locus, and let AD, AM meet

* That the *Porisms* of Euclid were a very important contribution to geometry is indicated by the description of them in Pappus (p. 648) as a collection most ingeniously adapted for the solution of the more weighty problems (*ἄθροισμα φιλοτεχνότατον εἰς τὴν ἀνάλυσιν τῶν ἐμβριθεστέρων προβλημάτων*).

† Pappus, p. 656.

CE in D', M' respectively, while CD, CM meet AE in D'', M'' respectively.

For the purpose of determining the geometrical locus, let the distances of M from AB, CD be measured parallel to BC, and its distances from BC, AD parallel to BA.

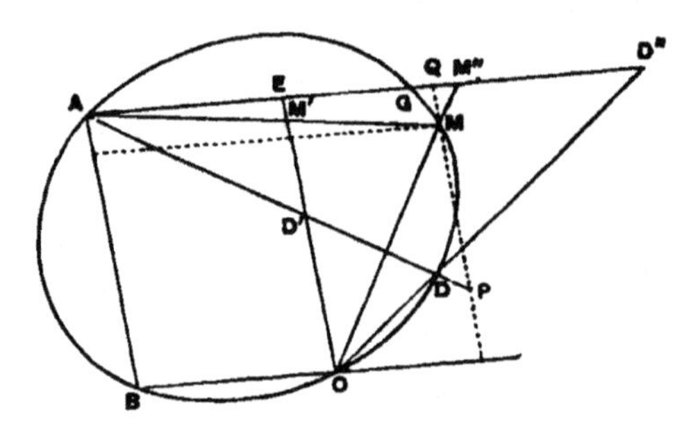

Then the ratio of the distances of M from CD, BC respectively will be equal to $\frac{D''M''}{CE}$, and the ratio of the distances of M from AB, DA will be equal to $\frac{AE}{D'M'}$.

Therefore the fact that the ratio of the rectangles under the distances of M from each pair of opposite sides of the quadrilateral $ABCD$ is constant may be expressed by the equation

$$\frac{D''M''}{D'M'} = \lambda \cdot \frac{CE}{AE} = \mu, \text{ say} \quad\text{.....................(1)},$$

where μ is a new constant independent of the position of M.

If now λ be determined by means of the position of a point F of the locus, we have

$$\frac{D''M''}{D'M'} = \frac{D''F''}{D'F'} = \frac{F''M''}{F'M'} \quad\text{.....................(2)},$$

where F', F'' are the intersections of AF, CE and of CF, AE respectively.

And, since the last ratio in (2), which is derived from the other two, remains constant while M moves along the required locus, it follows that that locus is also a four-line locus with reference to the four sides of the quadrilateral $ABCF$.

Thus, in order to extend the proposition about an inscribed

trapezium to a quadrilateral of any form, or, conversely, to reduce the determination of a four-line locus with reference to any quadrilateral to a similar locus with reference to a trapezium, it was only necessary to consider the case in which one of the lines AD or AF coincides with AE. It follows that the four-line locus with reference to any quadrilateral is, like the four-line locus with reference to a trapezium, a conic section.

The actual determination of the locus in the general form can be effected by expressing it in the more particular form.

Suppose that the distances of M from AB, CD (reckoned parallel to BC) are denoted by x, z, and the distances of M from BC, AD (reckoned parallel to BA) are y, u respectively. Then the locus is determined by an equation of the form

$$xz = \lambda . yu \text{ ....(1)},$$

where λ is a constant, and x, y are the coordinates of the point M with reference to BC, BA as axes.

If P, Q are the points in which the ordinate (y) of M meets AD, AE respectively,

$$u = PM$$

$$= PQ - MQ \text{ ....(2)}.$$

Since ($-MQ$) is the distance of M from AE measured parallel to BA, let it be denoted by u_1.

Then, from the figure,

$$PQ = \frac{D'E}{AE} \cdot x.$$

Therefore, from (1),

$$x\left(z - \lambda \frac{D'E}{AE} y\right) = \lambda y u_1.$$

In order to substitute a single term for $\left(z - \lambda \frac{D'E}{AE} y\right)$, we derive from the figure

$$z = \frac{D''M''}{CE} \cdot y,$$

and we have then to take a point G on AE such that

$$\lambda \frac{D'E}{AE} = \frac{D''G}{CE}.$$

(The point G is thus seen to be a point on the locus.)

Hence
$$z - \lambda \frac{D'E}{AE} y = \frac{D'M''}{CE} \cdot y - \frac{D'G}{CE} y$$
$$= \frac{GM''}{CE} \cdot y$$
$$= z_1,$$

where z_1 is the distance of the point M from the line CG measured parallel to BC.

The equation representing the locus is accordingly transformed into the equation

$$xz_1 = \lambda \cdot yu_1,$$

and the locus is expressed as a four-line locus with reference to the *trapezium* $ABCG$.

The method here given contains nothing which would be beyond the means at the disposal of the Greek geometers except the mere notation and the single use of the negative sign in $(-MQ)$, which however is not an essential difference, but only means that, whereas by the use of the negative sign we can combine several cases into one, the Greeks would be compelled to treat each separately.

Lastly, it should be observed that the four-line locus with reference to a trapezium corresponds to the equation

$$\alpha x^2 + \beta xy + \gamma y^2 + dx + ey = 0,$$

which may be written in the form

$$x(\alpha x + \beta y + d) = -y(\gamma y + e).$$

Thus the exact determination of the four-line locus with reference to a trapezium is the problem corresponding to that of tracing a conic from the general equation of the second degree wanting only the constant term.

CHAPTER VI.

THE CONSTRUCTION OF A CONIC THROUGH FIVE POINTS.

SINCE Apollonius was in possession of a complete solution of the problem of constructing the four-line locus referred to the sides of a quadrilateral of any form, it is clear that he had in fact solved the problem of constructing a conic through five points. For, given the quadrilateral to which the four-line locus is referred, and given a fifth point, the ratio (λ) between the rectangles contained by the distances of any point on the locus from each pair of opposite sides of the quadrilateral measured in any fixed directions is also given. Hence the construction of the conic through the five points is reduced to the construction of the four-line locus where the constant ratio λ is given.

The problem of the construction of a conic through five points is, however, not found in the work of Apollonius any more than the actual determination of the four-line locus. The omission of the latter is easily explained by the fact that, according to the author's own words, he only professed to give the theorems which were necessary for the solution, no doubt regarding the actual construction as outside the scope of his treatise. But, as in Euclid we find the problem of describing a circle about a triangle, it would have been natural to give in a treatise on conics the construction of a conic through five points. The explanation of the omission may be that it was not found possible to present the general problem in a form sufficiently concise to be included in a treatise embracing the whole subject of conics. This may be easily understood when it is remembered that, in the first place, a Greek geometer would regard the problem as being in reality *three* problems and involving a separate construction for each of the three conics, the parabola, the ellipse, and the hyperbola. He would then discover that the construction was not always possible for a parabola, since four points are sufficient to determine a parabola; and the construction of a parabola through four points would be a completely different problem not solved along with the construction of the four-line locus. Further, if the curve were an ellipse or a hyperbola, it would be necessary to find a διορισμός expressing the conditions which must be satisfied by the particular points in order that the conic might be the one or the other. If it were an ellipse, it might have been considered necessary to provide against its degeneration into a circle. Again, at all events until the time of Apollonius, it would have been regarded as necessary to find a διορισμός expressing the conditions for securing that the five points should not be distributed over both branches of the hyperbola. Thus it would follow that the complete treatment of the problem by the methods then in use must have involved a discussion of considerable length which would have been disproportionate in such a work as that of Apollonius.

It is interesting to note how far what we actually find in Apollonius can be employed for the direct construction of a conic through five points independently of the theory of the four-line locus. The methods of Book IV. on the number of points in which two conics may intersect are instructive in this connexion. These methods depend (1) on the harmonic polar property and (2) on the relation between the rectangles under the segments of intersecting chords drawn in fixed directions. The former property gives a method, when five points are given, of determining a sixth; and by repeating the process over and over again we may obtain as many separate points on the curve as we please. The latter proposition has the additional advantage that it allows us to choose more freely the particular points to be determined; and by this method we can find conjugate diameters and thence the axes. This is the method employed by Pappus in determining an ellipse passing through five points respecting which it is known beforehand that an ellipse can be drawn through them*. It is to be noted that Pappus' solution is not given as an independent problem in conic sections, but it is an intermediate step in another problem, that of finding the dimensions of a cylinder of which only a broken fragment is given such that no portion of the circumference of either of its bases is left whole. Further, the solution is made to depend on what is to be

* Pappus (ed. Hultsch), p. 1076 seqq.

found in Apollonius, and no claim is advanced that it contains anything more than any capable geometer could readily deduce for himself from the materials available in the *Conics*.

Pappus' construction is substantially as follows. If the given points are A, B, C, D, E, and are such that no two of the lines connecting the different pairs are parallel, we can reduce the problem to the construction of a conic through A, B, D, E, F, where EF is parallel to AB.

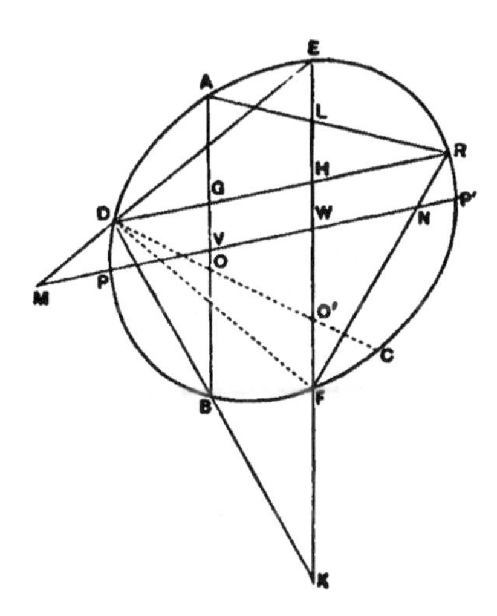

For, if EF be drawn through E parallel to AB, and if CD meet AB in O and EF in O', we have, by the proposition relating to intersecting chords,

$$CO.OD : AO.OB = CO'.O'D : EO'.O'F,$$

whence $O'F$ is known, and therefore F is determined.

We have therefore to construct an ellipse through A, B, D, E, F, where EF is parallel to AB.

And, if V, W be the middle points of AB, EF respectively, the line joining V and W is a diameter.

Suppose DR to be the chord through D parallel to the diameter, and let it meet AB, EF in G, H respectively. Then R is determined by means of the relation

$$RG.GD : BG.GA = RH.HD : FH.HE \text{ ....(1).}$$

In order to determine R, let DB, RA be joined meeting EF in K, L respectively.

Then

$$RG.GD : BG.GA = (RH : HL).(DH : HK), \text{ by similar triangles,}$$
$$= RH.HD : KH.HL.$$

Therefore, from (1), we have

$$FH.HE = KH.HL,$$

whence HL is found, and therefore L is determined. And the intersection of AL, DH determines R.

In order to find the extremities of the diameter (PP'), we draw ED, RF meeting the diameter in M, N respectively. And, by the same procedure as before, we obtain

$$FH.HE : RH.HD = FW.WE : P'W.WP,$$

by the property of the ellipse.

Also $\quad FH.HE : RH.HD = FW.WE : NW.WM,$

by similar triangles.

Hence $\quad P'W.WP = NW.WM;$

and similarly we can find the value of $P'V.VP$.

Pappus' method of determining P, P' by means of the given values of $P'V.VP$ and $P'W.WP$ amounts to an elimination of one of the unknown points and the determination of the other by an equation of the second degree.

Take two points Q, Q' on the diameter such that

$$P'V.VP = WV.VQ \quad\ldots\ldots\ldots\ldots(\alpha),$$
$$P'W.WP = VW.WQ' \quad\ldots\ldots\ldots\ldots(\beta),$$

and V, W, Q, Q' are thus known, while P, P' remain to be found.

It follows from (α) that

$$P'V : VW = QV : VP,$$

whence $\quad P'W : VW = PQ : PV.$

From this we obtain, by means of (β),

$$PQ : PV = Q'W : WP,$$

so that $\quad PQ : QV = Q'W : PQ',$

or $\quad PQ.PQ' = QV.Q'W.$

Thus P can be found, and similarly P'.

It is noteworthy that Pappus' method of determining the extremities of the diameter PP' (which is the principal object of his construction) can be applied to the direct construction of the points of intersection of a conic determined by five points with any straight line whatever, and there is no reason to doubt that this construction could have been effected by Apollonius. But there is a simpler expedient which we know from other sources that Apollonius was acquainted with, and which can be employed for the same purpose when once it is known that the four-line locus is a conic.

The auxiliary construction referred to formed the subject of a whole separate treatise of Apollonius *On determinate section* (περὶ διωρισμένης τομῆς). The problem is as follows:

Given four points A, B, C, D on a straight line, to determine another point P on the same straight line so that the ratio

$$AP \cdot CP : BP \cdot DP$$

has a given value.

The determination of the points of intersection of the given straight line and a four-line locus can be immediately transformed into this problem, A, B, C, D being in fact the points of intersection of the given straight line with the four lines to which the locus has reference.

Hence it is important to examine all the evidence which we possess about the separate treatise referred to. This is contained in the seventh Book of Pappus, who gives a short account of the contents of the work* as well as a number of lemmas to the different propositions in it. It is clear that the question was very exhaustively discussed, and in fact at much greater length than would have been likely had the investigation not been intended as a means of solving other important problems. The conclusion is therefore irresistible that, like the Books λόγου ἀποτομῆς and χωρίου ἀποτομῆς above mentioned, that *On determinate section* also was meant to be used for solving problems in conic sections.

To determine P by means of the equation

$$AP \cdot CP = \lambda \cdot BP \cdot DP,$$

where A, B, C, D, λ are given, is now an easy matter because the problem can at once be put into the form of a quadratic equation, and the Greeks also would have no difficulty in reducing it to the usual *application of areas*. But, if it was intended for application

* Pappus, pp. 642—644.

in further investigations, the complete discussion of it would naturally include, not only the finding of a solution, but also the determination of the limits of possibility and the number of possible solutions for different positions of the given pairs of points A, C and B, D, for the cases where the points in either pair coincide, where one is infinitely distant, and so forth: so that we should expect the subject to occupy considerable space. And this agrees with what we find in Pappus, who further makes it clear that, though we do not meet with any express mention of *series* of point-pairs determined by the equation for different values of λ, yet the treatise contained what amounts to a complete *theory of Involution*. Thus Pappus says that the separate cases were dealt with in which the given ratio was that of either (1) the square of one abscissa measured from the required point or (2) the rectangle contained by two such abscissae to any one of the following: (1) the square of one abscissa, (2) the rectangle contained by one abscissa and another separate line of given length independent of the position of the required point, (3) the rectangle contained by two abscissae. We also learn that maxima and minima were investigated. From the lemmas too we may draw other conclusions, e.g.

(1) that, in the case where $\lambda = 1$, and therefore P has to be determined by the equation

$$AP . CP = BP . DP,$$

Apollonius used the relation*

$$BP : DP = AB . BC : AD . DC;$$

(2) that Apollonius probably obtained a double point E of the involution determined by the point-pairs A, C and B, D by means of the relation†

$$AB . BC : AD . DC = BE^2 : DE^2.$$

Assuming then that the results of the work *On determinate section* were used for finding the points of intersection of a straight line with a conic section represented as a four-line locus, or a conic determined by five points on it, the special cases and the various διορισμοί would lead to the same number of properties of the conics under consideration. There is therefore nothing violent in the supposition that Apollonius had already set up many landmarks in the field explored eighteen centuries later by Desargues.

* This appears in the first lemma (p. 704) and is proved by Pappus for several different cases.

† Cf. Pappus' prop. 40 (p. 732).

APPENDIX TO INTRODUCTION.

NOTES ON THE TERMINOLOGY OF GREEK GEOMETRY.

THE propositions from the *Conics* of Apollonius which are given at length in Chapter II. above will have served to convey some idea of the phraseology of the Greek geometers; and the object of the following notes is to supplement what may be learnt from those propositions by setting out in detail the principal technical terms and expressions, with special reference to those which are found in Apollonius. It will be convenient to group them under different headings.

1. Points and lines.

A *point* is σημεῖον, *the point* A τὸ Α σημεῖον or τὸ Α simply; a fuller expression commonly used by the earlier geometers was τὸ (σημεῖον) ἐφ' οὗ Α, "the point on which (is put the letter) A*." *Any point* is τυχὸν σημεῖον, *the point* (*so*) *arising* τὸ γενόμενον σημεῖον, *the point* (*so*) *taken* τὸ ληφθὲν σημεῖον, *a point not within the section* σημεῖον μὴ ἐντὸς τῆς τομῆς, *any point within the surface* σημεῖόν τι τῶν ἐντὸς τῆς ἐπιφανείας; *in one point only* καθ' ἓν μόνον σημεῖον, *in two points* κατὰ δύο, and so on.

The following are names for particular points: *apex* or *vertex* κορυφή, *centre* κέντρον, *point of division* διαίρεσις, *point of bisection* διχοτομία, *extremity* πέρας.

A *line* is γραμμή, a *straight line* εὐθεῖα γραμμή or εὐθεῖα alone, a *finite straight line* εὐθεῖα πεπερασμένη; a *curved line* is καμπύλη

* A similar expression was ἡ (εὐθεῖα) ἐφ' ᾗ AB *the straight line* (*on which are the letters*) AB. The same phrases, with the same variation of case after ἐπί, are found frequently in Aristotle, particularly in the logical treatises and the *Physics*.

γραμμή, but γραμμή alone is often used of a curve, e.g. a circle or a conic; thus τὸ πέρας τῆς εὐθείας τὸ πρὸς τῇ γραμμῇ is *that extremity of the straight line which is on the curve*. A *segment* (of a line as well as a curve) is τμῆμα.

Of lines in relation to other lines we find the terms *parallel* παράλληλος, *a perpendicular to* κάθετος ἐπί (with acc.); a *straight line produced* is ἡ ἐπ' εὐθείας αὐτῇ.

For a line passing through particular points we have the following expressions used with διά and the genitive, ἥξει, ἔρχεται, ἐλεύσεται, πορεύεται; likewise πίπτω διά, or κατά (with acc.).

Of a line *meeting* another line πίπτειν ἐπί (with acc.), συμπίπτειν, συμβάλλειν, ἅπτομαι are used; *until it meets* is ἕως οὗ συμπέσῃ or ἄχρις ἂν συμπέσῃ, *point of meeting* σύμπτωσις; *the line from the point of concourse to* Δ, ἡ ἀπὸ τῆς συμπτώσεως ἐπὶ τὸ Δ; *the straight line joining* H, Θ, ἡ ἐπὶ τὰ H, Θ ἐπιζευγνυμένη εὐθεῖα; BA *passes through the points of contact*, ἐπὶ τὰς ἁφάς ἐστιν ἡ BA.

The line ZΘ *is bisected in* M, δίχα τέτμηται ἡ ZΘ κατὰ τὸ M; *bisecting one another* δίχα τέμνουσαι ἀλλήλας, *the line joining their middle points* ἡ τὰς διχοτομίας αὐτῶν ἐπιζευγνύουσα, *is cut into equal and unequal parts* εἰς μὲν ἴσα, εἰς δὲ ἄνισα τέτμηται.

Straight lines *cut off* or *intercepted* are ἀποτεμνόμεναι or ἀπολαμβανόμεναι, *the part cut off without* (the curve) ἡ ἐκτὸς ἀπολαμβανομένη, *will cut off an equal length* ἴσην ἀπολήψεται, *the lengths intercepted on it by the* (conic) *section towards the asymptotes* αἱ ἀπολαμβανόμεναι ἀπ' αὐτῆς πρὸς ταῖς ἀσυμπτώτοις.

A point on a line is often elegantly denoted by an adjective agreeing with it: thus ἀπ' ἄκρας αὐτῆς *from its extremity*, ἀπ' ἄκρου τοῦ ἄξονος *from the extremity of the axis*, ἡ ἐπ' ἄκραν τὴν ἀποληφθεῖσαν ἀγομένη *the line drawn to the extremity of the intercept*, αἱ πρὸς μέσην τὴν τομὴν κλώμεναι εὐθεῖαι *the straight lines drawn so as to meet at the middle point of the section*.

2. Angles.

An *angle* is γωνία, an *acute angle* ὀξεῖα γωνία, *obtuse* ἀμβλεῖα, *right* ὀρθή; *at right angles to* πρὸς ὀρθάς (with dative) or ὀρθὸς πρὸς (with acc.); *the line* ΔA (*drawn*) *from* Δ *at right angles to* EΔ, ἀπὸ τοῦ Δ τῇ EΔ ὀρθὴ ἡ ΔA; *to cut at right angles* πρὸς ὀρθὰς τέμνειν, *will not in general be at right angles but only when*... οὐκ αἰεὶ πρὸς ὀρθὰς ἔσται, ἀλλ' ὅταν...

At any angles ἐν τυχούσαις γωνίαις, *at a given angle* ἐν δοθείσῃ γωνίᾳ.

Vertically opposite (angles) κατὰ κορυφὴν ἀλλήλαις κείμεναι; *the angle vertically opposite to the angle* ΖΘΕ, ἡ κατὰ κορυφὴν τῆς ὑπὸ ΖΘΕ γωνίας; the same expression is also used of triangles (e.g. in τὰ γινόμενα κατὰ κορυφὴν τρίγωνα), and of the two halves of a double cone, which are called *vertically opposite surfaces* αἱ κατὰ κορυφὴν ἐπιφάνειαι.

The exterior angle of the triangle is ἡ ἐκτὸς τοῦ τριγώνου γωνία.

For *the angle* ΔΓΕ we find the full expression ἡ περιεχομένη γωνία ὑπὸ τῶν ΔΓΕ or "the angle contained by the lines ΔΓ, ΓΕ," but more usually ἡ ὑπὸ τῶν ΔΓΕ or ἡ ὑπὸ ΔΓΕ. *The angles* ΑΓΖ, ΑΖΓ *are (together) equal to a right angle* αἱ ὑπὸ ΑΓΖ, ΑΖΓ μιᾷ ὀρθῇ ἴσαι εἰσίν.

The *adjacent angle*, or the *supplement* of an angle, is ἡ ἐφεξῆς γωνία.

To *subtend* (an angle) is ὑποτείνειν either with a simple accusative, or with ὑπὸ and acc. (*extend under*) as in αἱ γωνίαι, ὑφ' ἃς αἱ ὁμόλογοι πλευραὶ ὑποτείνουσιν *the angles which the homologous sides subtend.*

3. Planes and plane figures.

A *plane* is ἐπίπεδον, a *figure* σχῆμα or εἶδος, a *figure* in the sense of a *diagram* καταγραφή or σχῆμα.

(A circle) *which is not in the same plane with the point* ὃς οὐκ ἔστιν ἐν τῷ αὐτῷ ἐπιπέδῳ τῷ σημείῳ.

The *line of intersection* of two planes is their κοινὴ τομή.

A *rectilineal figure* is σχῆμα εὐθύγραμμον (Euclid), and among the figures of this kind are *triangle* τρίγωνον, *quadrilateral* τετράπλευρον, *a five-sided figure* πεντάπλευρον etc., πλευρά being a *side*.

A *circle* is κύκλος, its *circumference* περιφέρεια, a *semicircle* ἡμικύκλιον, a *segment* of a circle τμῆμα κύκλου, a *segment greater*, or *less, than a semicircle* τμῆμα μεῖζον, or ἔλασσον, ἡμικυκλίου; a *segment of a circle containing an angle equal to the angle* ΑΓΒ is κύκλου τμῆμα δεχόμενον γωνίαν ἴσην τῇ ὑπὸ ΑΓΒ.

Of quadrilaterals, a *parallelogram* is παραλληλόγραμμον, a *square* τετράγωνον, a *rectangle* ὀρθογώνιον or frequently χωρίον with or without ὀρθογώνιον. *Diagonal* is διάμετρος.

To *describe* a figure *upon* a given line (as base) is ἀναγράφειν ἀπό. Thus *the figure* ΘΙΗ *has been described upon the radius* ΘΗ is ἀναγέγραπται ἀπὸ τῆς ἐκ τοῦ κέντρου τῆς ΘΗ εἶδος τὸ ΘΙΗ, *the square on* ΖΘ is τὸ ἀπὸ τῆς ΖΘ (τετράγωνον), *the figures on* ΚΛ, ΛΖ is τὰ ἀπὸ ΚΛΖ εἴδη. But ἐπί with the genitive is used of *describing a semicircle, or a segment of a circle, on a* given straight line, e.g. ἐπὶ τῆς ΑΔ γεγράφθω ἡμικύκλιον, τμῆμα κύκλου. Similarly *quadrilaterals standing on the diameters as bases* are βεβηκότα ἐπὶ τῶν διαμέτρων τετράπλευρα.

A rectangle *applied to* a given straight line is παρακείμενον παρά (with acc.), and its *breadth* is πλάτος. The *rectangle contained by* ΔΖ, ΖΕ is τὸ ὑπὸ τῶν ΔΖ, ΖΕ or τὸ ὑπὸ (τῶν) ΔΖΕ; *will contain* (with another straight line) *a rectangle equal to the square on* is ἴσον περιέξει τῷ ἀπό.

With reference to *squares* the most important point to notice is the use of the word δύναμις and the various parts of the verb δύναμαι. δύναμις expresses a *square* (literally a *power*); thus in Diophantus it is used throughout as the technical term for the square of the unknown in an algebraical equation, i.e. for x^2. In geometrical language it is used most commonly in the dative singular, δυνάμει, in such expressions as the following: λόγος ὃν ἔχει τὸ ἐντὸς τμῆμα πρὸς τὸ λοιπὸν δυνάμει, "the ratio which" (as one might say) "the inner segment has to the remaining segment *potentially*," meaning the *ratio of the square of the inner segment to that of the other*. (Similarly Archimedes speaks of the radius of a circle as being δυνάμει ἴσα to the sum of two areas, meaning that the *square of the radius is equal* etc.) In like manner, when δύναται is used of a straight line, it means literally that the line *is* (if squared) *capable* of producing an area equal to another. ἴσον δυνάμεναι τῷ ὑπὸ is in Apollonius (*straight lines*) *the squares on which are equal to the rectangle contained by*; δύναται τὸ περιεχόμενον ὑπὸ *the square on it is equal to the rectangle contained by*; ΜΝ δύναται τὸ ΖΞ, *the square on* ΜΝ *is equal to the rectangle* ΖΞ; δυνήσεται τὸ παρακείμενον ὀρθογώνιον πρὸς τὴν προσπορισθεῖσαν *the square on it will be equal to the rectangle applied to the straight line so taken in addition* (to the figure); and so on.

To construct a triangle out of three straight lines is in Euclid ἐκ τριῶν εὐθειῶν τρίγωνον συστήσασθαι, and similarly Apollonius speaks of its being possible συστήσασθαι τρίγωνον ἐκ τῆς Θ καὶ δύο τῶν ΕΑ, *to construct a triangle from the straight line* Θ *and two straight lines* (*equal to*) ΕΑ. *The triangle formed by* three straight lines is τὸ γινόμενον ὑπ' αὐτῶν τρίγωνον.

Equiangular is ἰσογώνιος, *similar* ὅμοιος, *similar and similarly situated* ὅμοιος καὶ ὁμοίως κείμενος; *because of the similarity of the triangles* ΘΕΝ, ΚΕΟ is διὰ τὴν ὁμοιότητα τῶν ΘΕΝ, ΚΕΟ τριγώνων.

4. Cones and sections of cones.

A *cone* is κῶνος, a *right cone* ὀρθὸς κῶνος, an *oblique* or *scalene cone* σκαληνὸς κῶνος, the *surface of a cone* is κωνικὴ ἐπιφάνεια, the *straight line generating* the surface by its motion about the circumference of a circle is ἡ γράφουσα εὐθεῖα, the *fixed point* through which the straight line always passes is τὸ μεμενηκὸς σημεῖον, the surface of the double cone is *that which consists of two surfaces lying vertically opposite to one another* ἣ συγκεῖται ἐκ δύο ἐπιφανειῶν κατὰ κορυφὴν ἀλλήλαις κειμένων, the circular *base* is βάσις, the *apex* κορυφή, the *axis* ἄξων.

A circular section *subcontrary* to the base is ὑπεναντία τομή.

In addition to the names *parabola*, *ellipse*, and *hyperbola* (which last means only one branch of a hyperbola), Apollonius uses the expression τομαὶ ἀντικείμεναι or αἱ ἀντικείμεναι denoting the *opposite branches* of a hyperbola; also αἱ κατ' ἐναντίον τομαί has the same meaning, and we even find the expression διάμετρος τῶν δύο συζυγῶν for a *diameter of two pairs of opposite branches*, so that *conjugate* here means opposite branches. (Cf. too ἐν μὲν τῇ ἑτέρᾳ συζυγίᾳ *in the one pair of opposites*.) Generally, however, the expression τομαὶ συζυγεῖς is used of *conjugate hyperbolas*, which are also called αἱ κατὰ συζυγίαν ἀντικείμεναι or συζυγεῖς ἀντικείμεναι *conjugate opposites*. Of the four branches of two conjugate hyperbolas any two *adjoining branches* are αἱ ἐφεξῆς τομαί.

In the middle of a proposition, where we should generally use the word *curve* to denote the conic, Apollonius generally uses τομή *section*, sometimes γραμμή.

5. Diameters and chords of conics.

Diameter is ἡ διάμετρος, *conjugate diameters* συζυγεῖς διάμετροι, of which the *transverse* is ἡ πλαγία, the other ἡ ὀρθία (*erect*) or δευτέρα (*secondary*).

The *original diameter* (i.e. that first arising out of the cutting of a cone in a certain manner) is ἡ ἐκ τῆς γενέσεως διάμετρος or ἡ προϋπάρχουσα διάμετρος, and (in the plural) αἱ ἀρχικαὶ διάμετροι. *The bisecting diameter* is ἡ διχοτομοῦσα διάμετρος. A *radius* of a central conic is simply ἐκ τοῦ κέντρου (with or without the definite article).

Chords are simply αἱ ἀγόμεναι ἐν τῇ τομῇ.

6. Ordinates.

The word used is the adverb τεταγμένως *ordinate-wise*, and the advantage of this is that it can be used with any part of the verb signifying to *draw*. This verb is either κατάγειν or ἀνάγειν, the former being used when the ordinate is drawn *down* to the diameter from a point on the curve, and the latter when it is drawn *upwards* from a point on a diameter. Thus τεταγμένως κατήχθω ἐπὶ τὴν διάμετρον means *suppose an ordinate drawn to the diameter*, which diameter is then sometimes called ἡ ἐφ' ἣν ἄγονται or κατῆκται. *An ordinate* is τεταγμένως καταγομένη or κατηγμένη, and sometimes τεταγμένως alone or κατηγμένη alone, the other word being understood; similarly κατῆκται and ἀνῆκται are used alone for *is an ordinate* or *has been drawn ordinate-wise*. τεταγμένως is also used of the tangent at the extremity of a diameter.

Parallel to an ordinate is παρὰ τεταγμένως κατηγμένην or παρατεταγμένως in one word.

7. Abscissa.

The *abscissa* of an ordinate is ἡ ἀπολαμβανομένη ὑπ' αὐτῆς ἀπὸ τῆς διαμέτρου πρὸς τῇ κορυφῇ *the* (*portion*) *cut off by it from the diameter towards the vertex*. Similarly we find the expressions αἱ ἀποτεμνόμεναι ὑπὸ τῆς κατηγμένης, or αἱ ἀπολαμβανόμεναι ὑπ' αὐτῶν, πρὸς τοῖς πέρασι τῆς πλαγίας πλευρᾶς τοῦ εἴδους *the* (*portions*) *cut off by the ordinate*, or *by them*, *towards the extremities of the transverse side of the figure* (as to which last expression see paragraph 9 following).

8. Parameter.

The full phrase is the *parameter of the ordinates*, which is ἡ παρ' ἣν δύνανται αἱ καταγόμεναι τεταγμένως, i.e. the straight line *to which are applied* the rectangles which in each conic are equal to the squares on the ordinates, or (perhaps) to which the said squares *are related* (by comparison).

9. The "figure" of a central conic.

The *figure* (τὸ εἶδος) is the technical term for a rectangle supposed to be described on the transverse diameter as base and with altitude equal to the parameter or latus rectum. Its area is therefore equal to the square on the conjugate diameter, and, with reference to the rectangle, the transverse diameter is called the *transverse side* (πλαγία πλευρά) and the parameter is the *erect side* (ὀρθία πλευρά) of the *figure* (εἶδος). We find the following different expressions, τὸ πρὸς τῇ ΒΔ εἶδος *the figure on* (*the diameter*) ΒΔ; τὸ παρὰ τὴν ΑΒ εἶδος *the figure applied to* (*the diameter*) ΑΒ; τὸ ὑπὸ ΔΕ, Η εἶδος *the figure contained by* (*the diameter*) ΔΕ *and* (*the parameter*)

H. Similarly τὸ γινόμενον εἶδος πρὸς τῇ διὰ τῆς ἁφῆς ἀγομένῃ διαμέτρῳ is *the figure formed on the diameter drawn through the point of contact* and τὸ πρὸς τῇ τὰς ἁφὰς ἐπιζευγνυούσῃ εἶδος is *the figure on* (*the diameter which is*) *the chord joining the points of contact* (of two parallel tangents).

τὸ τέταρτον τοῦ εἴδους *one-fourth of the figure* is, with reference to a diameter PP', one-fourth of the square of the conjugate diameter DD', i.e. CD^2.

10. Tangents etc.

To *touch* is most commonly ἐφάπτεσθαι, whether used of straight lines touching curves or of curves touching each other, a *tangent* being of course ἐφαπτομένη; *the tangent at* Δ, ἡ κατὰ τὸ Δ ἐφαπτομένη. (The simple verb ἅπτεσθαι is not generally used in this sense but as a rule means to *meet*, or is used of points *lying on* a locus. Cf. Pappus, p. 664, 28, ἅψεται τὸ σημεῖον θέσει δεδομένης εὐθείας *the point will lie on a straight line given in position*; p. 664, 2, ἐὰν ἅπτηται ἐπιπέδου τόπου θέσει δεδομένου *if it lies on a plane locus given in position*). The word ἐπιψαύειν is also commonly used of *touching*, e.g. καθ' ἓν ἐπιψαύουσα τῆς τομῆς is *touching the section in one point*, ἧς ἔτυχε τῶν τομῶν ἐπιψαύουσα *touching any one of the sections at random*.

Point of contact is ἁφή, *chord of contact* ἡ τὰς ἁφὰς ἐπιζευγνύουσα.

The point of intersection of two tangents is ἡ σύμπτωσις τῶν ἐφαπτομένων.

The following elliptical expressions are found in Apollonius: ἀπ' αὐτοῦ ἡ ΔΒ ἐφαπτέσθω *let* ΔΒ *be the tangent* (*drawn*) *from* Δ (outside the curve); ἐὰν ἀπ' αὐτοῦ ἡ μὲν ἐφάπτηται, ἡ δὲ τέμνῃ *if* (*there be drawn*) *from it* (*two straight lines of which*) *one touches, and the other cuts* (*the curve*).

11. Asymptotes.

Though the technical term used by Apollonius for the *asymptotes* is ἀσύμπτωτος, it is to be observed that the Greek word has a wider meaning and was used of any lines which do not meet, in whatever direction they are produced. Thus Proclus*, quoting from Geminus, distinguishes between (*a*) ἀσύμπτωτοι which are in one plane and (*b*) those which are not. He adds that of ἀσύμπτωτοι which are in one plane "some are always at the same distance from one another (i.e. parallel), while others continually diminish the distance, as a hyperbola approaches the straight line and the

* *Comment. in Eucl.* I. p. 177.

conchoid the straight line." The same use of ἀσύμπτωτος in its general sense is found even in Apollonius, who says (II. 14) πασῶν τῶν ἀσυμπτώτων τῇ τομῇ ἐγγιόν εἰσιν αἱ ΑΒ, ΑΓ, *the lines* AB, AΓ (the asymptotes proper) *are nearer than any of the lines which do not meet the section.*

The original enunciation of II. 14 [Prop. 36] is interesting: αἱ ἀσύμπτωτοι καὶ ἡ τομὴ εἰς ἄπειρον ἐκβαλλόμεναι ἐγγιόν τε προσάγουσιν ἑαυταῖς καὶ παντὸς τοῦ δοθέντος διαστήματος εἰς ἔλαττον ἀφικνοῦνται διάστημα, *the asymptotes and the section, if produced to infinity, approach nearer to one another and come within a distance less than any given distance.*

One of the angles formed by the asymptotes is ἡ περιέχουσα τὴν ὑπερβολήν *the angle containing* (or *including*) *the hyperbola*, and similarly we find the expression ἐπὶ μιᾶς τῶν ἀσυμπτώτων τῶν περιεχουσῶν τὴν τομήν *on one of the asymptotes containing the section.*

The space between the asymptotes and the curve is ὁ ἀφοριζόμενος τόπος ὑπὸ τῶν ἀσυμπτώτων καὶ τῆς τομῆς.

12. Data and hypotheses.

Given is δοθείς or δεδομένος; *given in position* θέσει δεδομένη, *given in magnitude* τῷ μεγέθει δεδομένη (of straight lines). For *is* or *will be, given in position* we frequently find θέσει ἐστίν, ἔσται without δεδομένος, or even θέσει alone, as in θέσει ἄρα ἡ ΑΕ. A more remarkable ellipse is that commonly found in such expressions as παρὰ θέσει τὴν ΑΒ, *parallel to* AB (*given*) *in position*, and πρὸς θέσει τῇ ΑΒ, used of an angle made *with* AB (*given*) *in position.*

Of *hypotheses* ὑπόκειται and the other parts of the same verb are used, either alone, as in ὑποκείσθω τὰ μὲν ἄλλα τὰ αὐτά *let all the other suppositions be the same*, τῶν αὐτῶν ὑποκειμένων *with the same suppositions*, or with substantives or adjectives following, e.g. κύκλος ὑπόκειται ἡ ΔΚΕΛ γραμμή *the line* ΔKEΛ *is by hypothesis a circle*, ὑπόκειται ἴση *is by hypothesis equal*, ὑπόκεινται συμπίπτουσαι *they meet by hypothesis*. In accordance with the well-known Greek idiom ὅπερ οὐχ ὑπόκειται means *which is contrary to the hypothesis.*

13. Theorems and problems.

In a theorem *what is required to be proved* is sometimes denoted by τὸ προτεθέν, and the requirement in a problem is τὸ ἐπιταχθέν. Thus εἰ μὲν οὖν ἡ ΑΒ ἄξων ἐστί, γεγονὸς ἂν εἴη τὸ ἐπιταχθέν *if then* AB *is an axis, that which was required would have been done. To draw in the manner required* is ἀγαγεῖν ὡς πρόκειται. When the solution of a problem has been arrived at, e.g. when a required tangent has been drawn, the tangent is said ποιεῖν τὸ πρόβλημα.

In the ἔκθεσις or *setting out* of a theorem the re-statement of what it is required to prove is generally introduced by Apollonius as well as Euclid by the words λέγω, ὅτι; and in one case Apollonius abbreviates the re-statement by saying simply λέγω, ὅτι ἔσται τὰ τῆς προτάσεως *I say that the property stated in the enunciation will be true*; *it is to be proved* is δεικτέον, *it remains to be proved* λοιπὸν ἄρα δεικτέον, *let it be required to draw* δέον ἔστω ἀγαγεῖν.

The *synthesis* of a problem regularly begins with the words συντεθήσεται δὴ (τὸ πρόβλημα) οὕτως.

14. Constructions.

These are nearly always expressed by the use of the perfect imperative passive (with which may be classified such perfect imperatives as γεγονέτω from γίνεσθαι, συνεστάτω from συνιστάναι, and the imperative κείσθω from κεῖμαι). The instances in Apollonius where active forms of transitive verbs appear in constructions are rare; but we find the following, ἐὰν ποιήσωμεν *if we make* (one line in a certain ratio to another), ὁμοίως γὰρ τῷ προειρημένῳ ἀγαγὼν τὴν ΑΒ ἐφαπτομένην λέγω, ὅτι *for in the same manner as before, after drawing the tangent* AB, *I say that...*, ἐπιζεύξαντες τὴν ΑΒ ἐροῦμεν *having joined* AB *we shall prove*; while in ἀγαγόντες γὰρ ἐπιψαύουσαν τὴν ΘΕ ἐφάπτεται αὕτη we have a somewhat violent anacoluthon, *for, having drawn the tangent* ΘE, *this touches*.

Of the words used in constructions the following are the most common: *to draw* ἄγειν, διάγειν and other compounds, *to join* ἐπιζευγνύναι, *to produce* ἐκβάλλειν, προσεκβάλλειν, *to take* or *supply* πορίζειν, *to cut off* ἀπολαμβάνειν, ἀποτέμνειν, ἀφαιρεῖν, *to construct* συνίστασθαι, κατασκευάζειν, *to describe* γράφω and its compounds, *to apply* παραβάλλειν, *to erect* ἀνιστάναι, *to divide* διαιρεῖν, *to bisect* διχοτομεῖν.

Typical expressions are the following: τῇ ὑπὸ τῶν ΗΘΕ γωνίᾳ ἴση συνεστάτω ἡ ὑπὸ τῶν ΒΑΓ *let the angle* ΒΑΓ *be constructed equal to the angle* ΗΘΕ; ὁ κέντρῳ τῷ Κ διαστήματι δὲ τῷ ΚΓ κύκλος γραφόμενος *the circle described with* K *as centre and at a distance* ΚΓ; ἀνεστάτω ἀπὸ τῆς ΑΒ ἐπίπεδον ὀρθὸν πρὸς τὸ ὑποκείμενον ἐπίπεδον *let a plane be erected on* AB *at right angles to the supposed plane*; κείσθω αὐτῇ ἴση *let* (*an angle*) *be made equal to it*, ἐκκείσθω *let* (a line, circle etc.) *be set out*, ἀφῃρήσθω ἀπ᾽ αὐτοῦ τμῆμα *let a segment be cut off from it*, τῶν αὐτῶν κατασκευασθέντων *with the same construction*.

No detailed enumeration of the various perfect imperatives is necessary; but γεγονέτω for *suppose it done* deserves mention for its elegance.

Let it be conceived is νοείσθω: thus νοείσθω κῶνος, οὗ κορυφὴ τὸ Ζ σημεῖον *let a cone be conceived whose apex is the point* Ζ.

A curious word is κλάω, meaning literally to *break off* and generally used of two straight lines meeting and forming an angle, e.g. of two straight lines drawn from the foci of a central conic to one and the same point on the curve, ἀπὸ τῶν Ε, Δ σημείων κεκλάσθωσαν πρὸς τὴν γραμμὴν αἱ ΕΖ, ΖΔ, (literally) *from the points* Ε, Δ *let* ΕΖ, ΖΔ *be broken short off against the curve.* Similarly, in a proposition of Apollonius quoted by Eutocius from the Ἀναλυόμενος τόπος, *the straight lines drawn from the given points to meet on the circumference of the circle* are αἱ ἀπὸ τῶν δοθέντων σημείων ἐπὶ τὴν περιφέρειαν τοῦ κύκλου κλώμεναι εὐθεῖαι.

15. Operations (Addition, Subtraction etc.).

The usual word for *being added* is πρόσκειμαι: thus δίχα τέτμηται ἡ ΖΘ κατὰ τὸ Μ προσκειμένην ἔχουσα τὴν ΔΖ, or ΖΘ *is bisected in* Μ *and has* ΔΖ *added.* Of a magnitude *having another added* to it the participle of προσλαμβάνειν is used in the same way as λιπών for *having something subtracted.* Thus τὸ ΚΡ λιπὸν ἢ προσλαβὸν τὸ ΒΟ ἴσον ἐστὶ τῷ ΜΠ means ΚΡ *minus or plus* ΒΟ *is equal to* ΜΠ. μετά (with gen.) is also used for *plus*, e.g. τὸ ὑπὸ ΑΕΒ μετὰ τοῦ ἀπὸ ΖΕ is equivalent to $AE \cdot EB + ZE^2$.

A curious expression is συναμφότερος ἡ ΑΔ, ΔΒ, or συναμφότερος ἡ ΓΖΔ meaning *the sum of* ΑΔ, ΔΒ, or *of* ΓΖ, ΖΔ.

Of adding or subtracting a *common* magnitude κοινός is used: thus κοινὸν προσκείσθω or ἀφῃρήσθω is *let the common (magnitude) be added*, or *taken away*, the adjective λοιπός being applied to the remainder in the latter case.

To *exceed* is ὑπερβάλλειν or ὑπερέχειν, *the excess* is often ἡ ὑπεροχή, ᾗν ὑπερέχει κ.τ.λ., ΠΑ *exceeds* ΑΟ *by* ΟΠ is τὸ ΠΑ τοῦ ΑΟ ὑπερέχει τῷ ΟΠ, to *differ from* is διαφέρειν with gen., to *differ by* is expressed by the dative, e.g. (a certain triangle) *differs from* ΓΔΘ *by the triangle on* ΑΘ *as base similar to* ΓΔΑ, διαφέρει τοῦ ΓΔΘ τῷ ἀπὸ τῆς ΑΘ τριγώνῳ ὁμοίῳ τῷ ΓΔΑ; (the area) *by which the square on* ΓΡ *differs from the square on* ΑΣ, ᾧ διαφέρει τὸ ἀπὸ ΓΡ τοῦ ἀπὸ ΑΣ.

For multiplications and divisions the geometrical equivalents are the methods of proportions and the application of areas; but of *numerical* multiples or fractions of magnitudes the following are typical instances: *the half of* AB, ἡ ἡμίσεια τῆς AB; the fourth part of the figure, τὸ τέταρτον τοῦ εἴδους; *four times the rectangle* AE . EΔ, τὸ τετράκις ὑπὸ AEΔ.

16. Proportions.

Ratio is λόγος, *will be cut in the same ratio* εἰς τὸν αὐτὸν λόγον τμηθήσονται, *the three proportionals* αἱ τρεῖς ἀνάλογον; *being a mean proportional between* ΞΘ, ΘΑ, μέσον λόγον ἔχουσα, or μέση ἀνάλογον, τῶν ΞΘΑ. *The sides about the right angles (are) proportional* περὶ ὀρθὰς γωνίας αἱ πλευραὶ ἀνάλογον.

The ratio of A *to* B is ὁ λόγος, ὃν ἔχει τὸ A πρὸς τὸ B, or ὁ τοῦ A πρὸς τὸ B λόγος; *suppose the ratio of* ΓΔ *to* ΔB *made the same as the ratio of* ΓΗ *to* ΗB, τῷ τῆς ΓΗ πρὸς ΗB λόγῳ ὁ αὐτὸς πεποιήσθω ὁ τῆς ΓΔ πρὸς ΔB; A *has to* B *a greater* (or *less*) *ratio than* Γ *has to* Δ, τὸ A πρὸς τὸ B μείζονα (or ἐλάσσονα) λόγον ἔχει ἤπερ τὸ Γ πρὸς τὸ Δ, or τοῦ, ὃν ἔχει τὸ Γ πρὸς τὸ Δ; *the ratio of the square of the inner segment to the square of the remaining segment*, λόγος, ὃν ἔχει τὸ ἐντὸς τμῆμα πρὸς τὸ λοιπὸν δυνάμει.

The following is the ordinary form of a proportion: *as the square on* AΣ *is to the rectangle under* BΣ, ΣΓ *so is* EΘ *to* EΠ, ὡς τὸ ἀπὸ AΣ πρὸς τὸ ὑπὸ BΣΓ, οὕτως ἡ EΘ πρὸς EΠ. In a proportion the *antecedents* are τὰ ἡγούμενα, i.e. the *leading* terms, the *consequents* τὰ ἑπόμενα; *as one of the antecedents is to one of the consequents so are all the antecedents (taken together) to all the consequents (taken together)* ὡς ἓν τῶν ἡγουμένων πρὸς ἓν τῶν ἑπομένων, οὕτως ἅπαντα τὰ ἡγούμενα πρὸς ἅπαντα τὰ ἑπόμενα.

A very neat and characteristic sentence is that which forms the enunciation of Euclid v. 19: ἐὰν ᾖ ὡς ὅλον πρὸς ὅλον, οὕτως ἀφαιρεθὲν πρὸς ἀφαιρεθέν, καὶ τὸ λοιπὸν πρὸς τὸ λοιπὸν ἔσται ὡς ὅλον πρὸς ὅλον. *If as a whole is to a whole so is (a part) taken away to (a part) taken away, the remainder also will be to the remainder as the whole to the whole.* Similarly in Apollonius we have e.g. ἐπεὶ οὖν ὡς ὅλον ἐστὶ τὸ ἀπὸ AE πρὸς ὅλον τὸ AZ, οὕτως ἀφαιρεθὲν τὸ ὑπὸ AΔB πρὸς ἀφαιρεθὲν τὸ ΔΗ, καὶ λοιπόν ἐστι πρὸς λοιπόν, ὡς ὅλον πρὸς ὅλον, *since then, as the whole the square on* AE *is to the whole the (parallelogram)* AZ, *so is (the part) taken away the rectangle under* AΔ, ΔB *to (the part) taken away the (parallelogram)* ΔΗ, *remainder is also to remainder as whole to whole.*

To be compounded of is συγκεῖσθαι, *the ratio compounded of* ὁ συγκείμενος (or συνημμένος, from συνάπτειν) λόγος (ἔκ τε τοῦ, ὃν ἔχει κ.τ.λ.), *the ratio compounded (of the ratios) of the sides* ὁ συγκείμενος λόγος ἐκ τῶν πλευρῶν. συγκεῖσθαι is moreover used not only of *being* a compounded ratio, but also of *being equal to* a ratio compounded of two others, even when none of the terms in the two latter ratios are the same as either term of the first ratio.

Another way of describing the ratio compounded of two others is to use μετά (with gen.) which here implies multiplication and not addition. Thus ὁ τῆς ΑΣ πρὸς ΣΓ λόγος μετὰ τοῦ τῆς ΑΣ πρὸς ΣΒ is *the ratio compounded of the ratio of* ΑΣ *to* ΣΓ *and that of* ΑΣ *to* ΣΒ. Similarly κοινὸς ἀφῃρήσθω ὁ τῆς ΓΔ πρὸς ΓΘ means *let the common ratio of* ΓΔ *to* ΓΘ *be divided out* (and not, as usual, subtracted), κοινοῦ ἀφαιρεθέντος τούτου τοῦ λόγου *dividing out by this common ratio.*

Taking the rectangle contained by ΘΕ, ΕΖ *as a middle term* is τοῦ ὑπὸ ΘΕΖ μέσου λαμβανομένου, *taking* ΑΗ *as a common altitude* τῆς ΑΗ κοινοῦ ὕψους λαμβανομένης.

So that the corresponding terms are continuous ὥστε τὰς ὁμολόγους συνεχεῖς εἶναι; *so that the segments adjoining the vertex are corresponding terms* ὥστε ὁμόλογα εἶναι τὰ πρὸς τῇ κορυφῇ τμήματα.

There remain the technical terms for transforming such a proportion as $a : b = c : d$. These correspond with the definitions at the beginning of Eucl. Book v. Thus ἐναλλάξ *alternately* (usually called *permutando* or *alternando*) means transforming the proportion into $a : c = b : d$.

ἀνάπαλιν *reversely* (usually *invertendo*), $b : a = d : c$.

σύνθεσις λόγου is *composition of a ratio*, by which the ratio $a : b$ becomes $a + b : b$. The corresponding Greek term to *componendo* is συνθέντι which means no doubt, literally, "to one who has compounded," or "if we compound," the ratios. Thus συνθέντι is used of the inference that $a + b : b = c + d : d$.

διαίρεσις λόγου means *division of a ratio* in the sense of *separation* or *subtraction* in the same way as σύνθεσις signifies addition. Similarly διελόντι (the translation of which as *dividendo* or *dirimendo* is misleading) means really *separating* in the sense of *subtracting*: thus $a - b : b = c - d : d$.

ἀναστροφὴ λόγου *conversion of a ratio* and ἀναστρέψαντι *convertendo* correspond respectively to the ratio $a : a - b$ and to the inference that $a : a - b = c : c - d$.

δι' ἴσου, generally translated *ex aequali* (sc. *distantia*), is applied to the inference e.g. from the proportions

$$a : b : c : d \text{ etc.} = A : B : C : D \text{ etc.}$$

that $$a : d = A : D.$$

All the expressions above explained, ἐναλλάξ, ἀνάπαλιν, συνθέντι, διελόντι, ἀναστρέψαντι, δι' ἴσου are constantly used in Apollonius as in Euclid. In one place we find the variant διὰ δὲ τὸ ἀνάπαλιν.

Are in reciprocal proportion is ἀντιπεπόνθασιν.

17. Inferences.

The usual equivalent for *therefore* is ἄρα, e.g. ἐν τῇ ἐπιφανείᾳ ἄρα ἐστί *it is therefore on the surface*, εὐθεῖα ἄρα ἐστὶν ἡ AB *therefore* AB *is a straight line*; οὖν is generally used in a somewhat weaker sense, and in conjunction with some other word, in order to mark the starting point of an argument rather than to express a formal inference, so that we can usually translate it by *then*, e.g. ἐπεὶ οὖν *since, then*, ὅτι μὲν οὖν...φανερόν *it is, then, clear that*.... δή is somewhat similarly used in taking up an argument. *So that* is ὥστε, *that is* τουτέστιν. A corollary is often introduced by καὶ φανερόν, ὅτι, or by συναποδέδεικται *it is proved at the same time*.

It is at once clear φανερὸν αὐτόθεν, *from this it is clear* ἐκ δὴ τούτου φανερόν, *for this reason* διὰ τοῦτο, *for the same reason* διὰ τὰ αὐτά, *wherefore* διόπερ, *in the same way as above* or *before* κατὰ τὰ αὐτὰ τοῖς ἐπάνω or ἔμπροσθεν, *similarly it will be shown* ὁμοίως καὶ δειχθήσεται, *the same results as before will follow* τὰ αὐτὰ τοῖς πρότερον συμβήσεται, *the same proofs will apply* αἱ αὐταὶ ἀποδείξεις ἁρμόσουσι.

Conversely ἀντιστρόφως, *by the converse of the theorem* διὰ τὴν ἀντιστροφὴν τοῦ θεωρήματος, *by what was proved and its converse* διὰ τὰ εἰρημένα καὶ τὰ ἀντίστροφα αὐτῶν.

By what was before proved in the case of the hyperbola διὰ τὸ προδεδειγμένον ἐπὶ τῆς ὑπερβολῆς; *for the same (facts) have been proved in the case of the parallelograms which are their doubles* καὶ γὰρ ἐπὶ τῶν διπλασίων αὐτῶν παραλληλογράμμων τὰ αὐτὰ δέδεικται.

By the similarity of the triangles διὰ τὴν ὁμοιότητα τῶν τριγώνων, *by parallels* διὰ τὰς παραλλήλους, *by the (property of the) section, parabola, hyperbola* διὰ τὴν τομήν, παραβολήν, ὑπερβολήν.

The properties which have already been proved true of the sections when the original diameters are taken (as axes of reference) ὅσα προδέδεικται περὶ τὰς τομὰς συμβαίνοντα συμπαραβαλλομένων τῶν ἀρχικῶν διαμέτρων.

Much more πολλῷ μᾶλλον. Cf. πολὺ πρότερον τέμνει τὴν τομήν *much sooner does it cut the section*.

18. Conclusions.

Which it was required to do, to prove ὅπερ ἔδει ποιῆσαι, δεῖξαι; *which is absurd* ὅπερ ἄτοπον; *and this is impossible, so that the original supposition is so also* τοῦτο δὲ ἀδύνατον· ὥστε καὶ τὸ ἐξ ἀρχῆς. *And again the absurdity will be similarly inferred* καὶ πάλιν ὁμοίως συναχθήσεται τὸ ἄτοπον.

19. Distinctions of cases.

These properties are general, but for the hyperbola only etc. ταῦτα μὲν κοινῶς, ἐπὶ δὲ τῆς ὑπερβολῆς μόνης κ.τ.λ., *in the third figure* ἐπὶ τῆς τρίτης καταγραφῆς or τοῦ τρίτου σχήματος, *in all the possible cases* κατὰ πάσας τὰς ἐνδεχομένας διαστολάς.

20. Direction, concavity, convexity.

In both directions ἐφ' ἑκάτερα, *towards the same parts as the section* ἐπὶ ταὐτὰ τῇ τομῇ; *towards the direction of the point E,* ἐπὶ τὰ μέρη, ἐφ' ἅ ἐστι τὸ E; *on the same side of the centre as AB,* ἐπὶ τὰ αὐτὰ μέρη τοῦ κέντρου, ἐν οἷς ἐστιν ἡ AB. There is also the expression κατὰ τὰ ἑπόμενα μέρη τῆς τομῆς, meaning literally *in the succeeding parts of the section*, and used of a line cutting a branch of a hyperbola and passing inside.

The concave parts τὰ κοῖλα, *the convexities* τὰ κυρτά, *not having its concavity (convexity) towards the same parts* μὴ ἐπὶ τὰ αὐτὰ μέρη τὰ κοῖλα (τὰ κυρτὰ) ἔχουσα, *towards the same parts as the concavity of the curve* ἐπὶ τὰ αὐτὰ τοῖς κοίλοις τῆς γραμμῆς, *if it touches with its concave side* ἐὰν ἐφάπτηται τοῖς κοίλοις αὐτῆς, *will touch on its concave side* ἐφάψεται κατὰ τὰ κοῖλα.

Having its convexity turned the opposite way ἀνεστραμμένα τὰ κυρτὰ ἔχουσα.

21. Infinite, Infinity.

Unlimited or *infinite* ἄπειρος, *to increase without limit* or *indefinitely* εἰς ἄπειρον αὐξάνεσθαι.

ἄπειρος is also used in a numerical sense; thus *in the same way we shall find an infinite number of diameters* τῷ δὲ αὐτῷ τρόπῳ καὶ ἀπείρους εὑρήσομεν διαμέτρους.

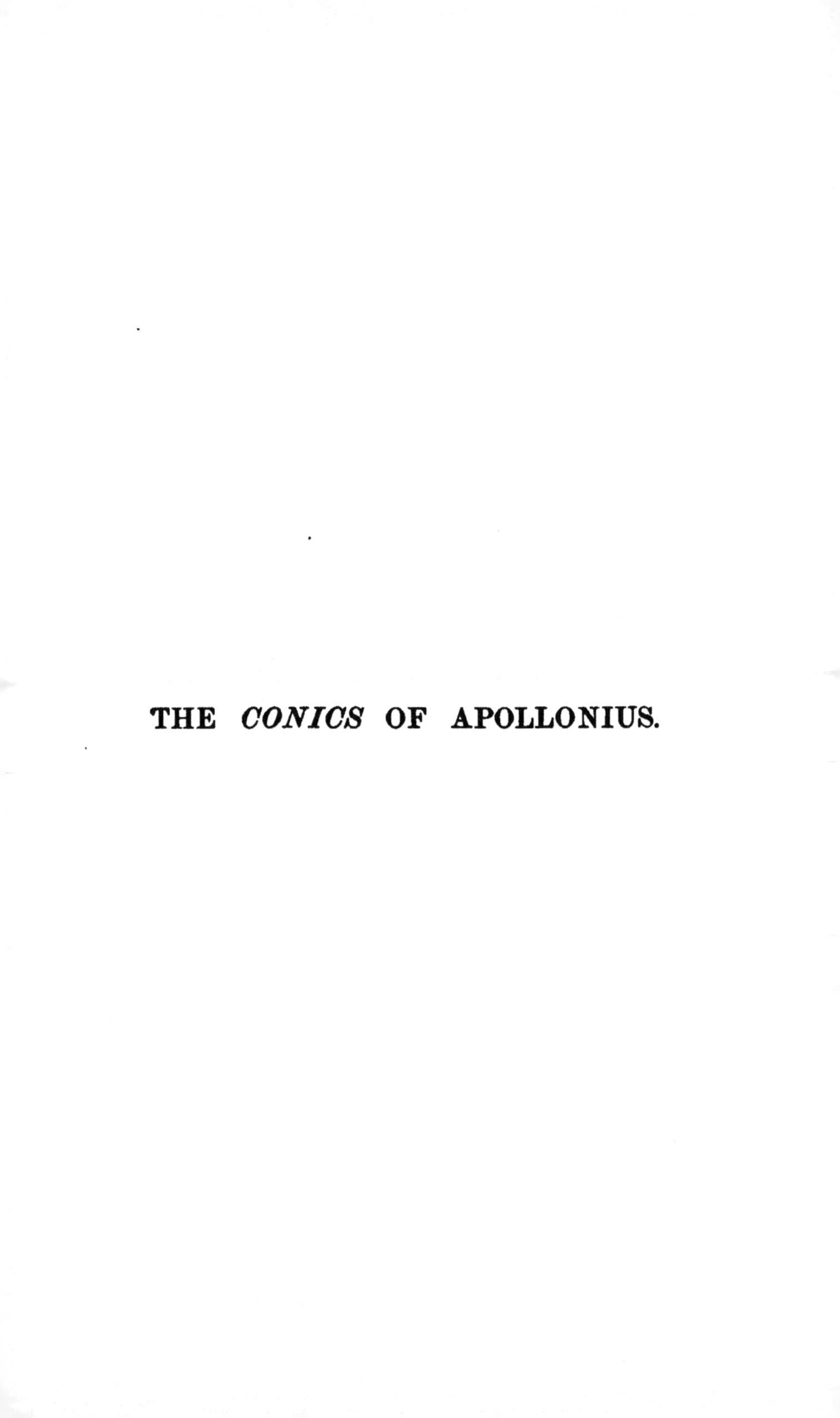

THE *CONICS* OF APOLLONIUS.

THE CONE.

IF a straight line indefinite in length, and passing always through a fixed point, be made to move round the circumference of a circle which is not in the same plane with the point, so as to pass successively through every point of that circumference, the moving straight line will trace out the surface of a **double cone**, or two similar cones lying in opposite directions and meeting in the fixed point, which is the **apex** of each cone.

The circle about which the straight line moves is called the **base** of the cone lying between the said circle and the fixed point, and the **axis** is defined as the straight line drawn from the fixed point or the apex to the centre of the circle forming the base.

The cone so described is a **scalene** or **oblique** cone except in the particular case where the axis is perpendicular to the base. In this latter case the cone is a **right** cone.

If a cone be cut by a plane passing through the apex, the resulting section is a triangle, two sides being straight lines lying on the surface of the cone and the third side being the straight line which is the intersection of the cutting plane and the plane of the base.

Let there be a cone whose apex is A and whose base is the circle BC, and let O be the centre of the circle, so that AO is the axis of the cone. Suppose now that the cone is cut by any plane parallel to the plane of the base BC, as DE, and let ıe axis AO meet the plane DE in o. Let p be any point on ıe intersection of the plane DE and the surface of the cone. ›in Ap and produce it to meet the circumference of the circle C in P. Join OP, op.

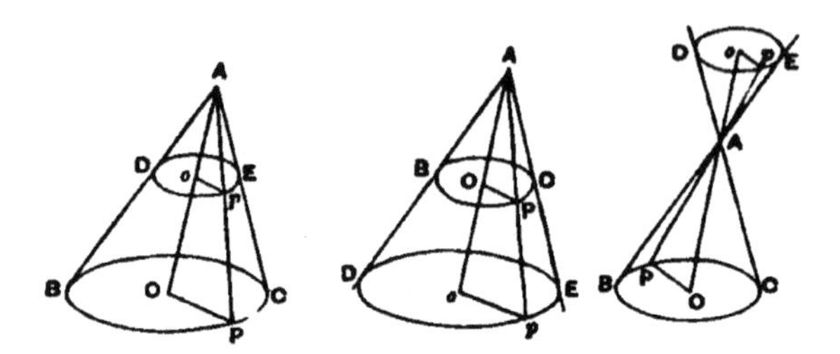

Then, since the plane passing through the straight lines O, AP cuts the two parallel planes BC, DE in the straight ıes OP, op respectively, OP, op are parallel.

$$\therefore op : OP = Ao : AO.$$

nd, BPC being a circle, OP remains constant for all positions ' p on the curve DpE, and the ratio $Ao : AO$ is also constant.

Therefore op is constant for all points on the section of the ırface by the plane DE. In other words, that section is circle.

Hence *all sections of the cone which are parallel to the rcular base are circles.* [I. 4.]*

Next, let the cone be cut by a plane passing through the ɩis and perpendicular to the plane of the base BC, and let the ɩction be the triangle ABC. Conceive another plane HK ·awn at right angles to the plane of the triangle ABC ıd cutting off from it the triangle AHK such that AHK is milar to the triangle ABC but lies in the contrary sense, ɜ. such that the angle AKH is equal to the angle ABC. hen the section of the cone by the plane HK is called a **ıbcontrary** section (ὑπεναντία τομή).

* The references in this form, here and throughout the book, are to the ɩginal propositions of Apollonius.

Let P be any point on the intersection of the plane HK with the surface, and F any point on the circumference of the circle BC. Draw PM, FL each perpendicular to the plane of the triangle ABC, meeting the straight lines HK, BC respectively in M, L. Then PM, FL are parallel.

Draw through M the straight line DE parallel to BC, and it follows that the plane through DME, PM is parallel to the base BC of the cone.

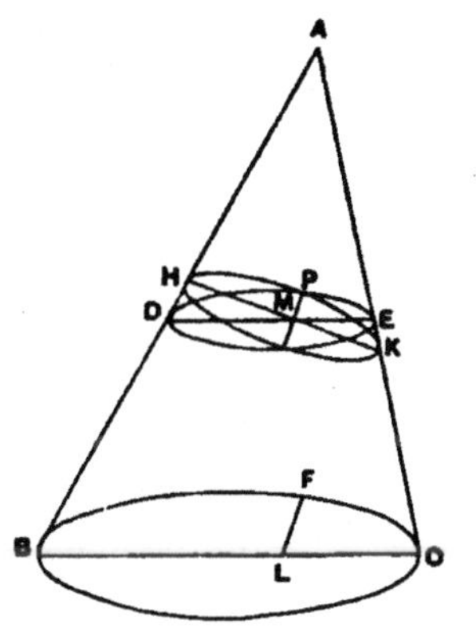

Thus the section DPE is a circle, and $DM . ME = PM^2$.

But, since DE is parallel to BC, the angle ADE is equal to the angle ABC which is by hypothesis equal to the angle AKH.

Therefore in the triangles HDM, EKM the angles HDM, EKM are equal, as also are the vertical angles at M.

Therefore the triangles HDM, EKM are similar.

Hence $$HM : MD = EM : MK.$$

$$\therefore HM . MK = DM . ME = PM^2.$$

And P is *any* point on the intersection of the plane HK with the surface. Therefore the section made by the plane HK is a circle.

Thus *there are two series of circular sections of an oblique cone, one series being parallel to the base, and the other consisting of the sections subcontrary to the first series.* [I. 5.]

Suppose a cone to be cut by any plane through the axis making the triangular section ABC, so that BC is a diameter of the circular base. Let H be any point on the circumference of the base, let HK be perpendicular to the diameter BC, and let a parallel to HK be drawn from any point Q on the surface of the cone but not lying in the plane of the axial triangle. Further, let AQ be joined and produced, if necessary, to meet the circumference of the base in F, and let FLF' be the chord perpendicular to BC. Join AL, AF'. Then the straight line through Q parallel to HK is also parallel to FLF'; it follows therefore that the parallel through Q will meet both AL and AF'. And AL is in the plane of the axial triangle ABC. Therefore the parallel through Q will meet both the plane of the axial triangle and the other side of the surface of the cone, since AF' lies on the cone.

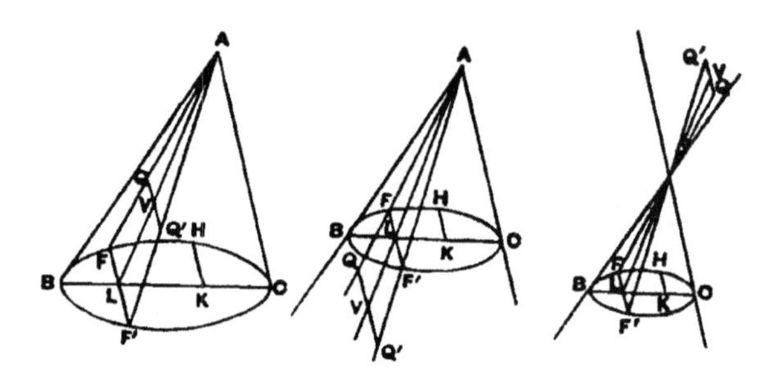

Let the points of intersection be V, Q' respectively.

Then $QV : VQ' = FL : LF'$, and $FL = LF'$.

$$\therefore QV = VQ',$$

or QQ' is bisected by the plane of the axial triangle. [I. 6.]

Again, let the cone be cut by another plane not passing through the apex but intersecting the plane of the base in a straight line DME perpendicular to BC, the base of any axial triangle, and let the resulting section of the surface of the cone be DPE, the point P lying on either of the sides AB, AC of the axial triangle. The plane of the section will then cut the plane of the axial triangle in the straight line PM joining P to the middle point of DE.

Now let Q be any point on the curve of section, and through Q draw a straight line parallel to DE.

Then this parallel will, if produced to meet the other side of the surface in Q', meet, and be bisected by, the axial triangle. But it lies also in the plane of the section *DPE*; it will therefore meet, and be bisected by, *PM*.

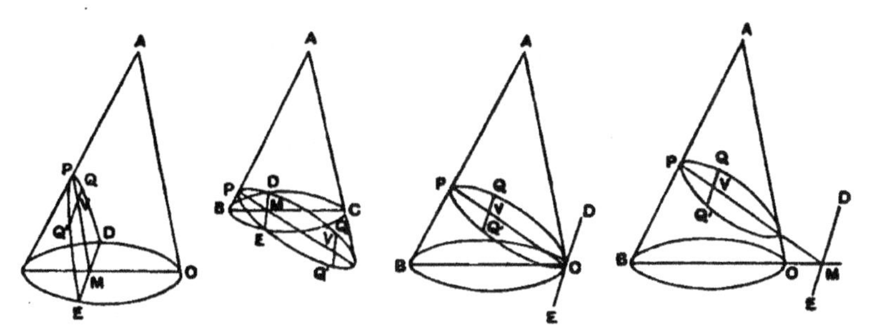

Therefore *PM* bisects any chord of the section which is parallel to *DE*.

Now a straight line bisecting each of a series of parallel chords of a section of a cone is called a **diameter**.

Hence, *if a cone be cut by a plane which intersects the circular base in a straight line perpendicular to the base of any axial triangle, the intersection of the cutting plane and the plane of the axial triangle will be a diameter of the resulting section of the cone.* [I. 7.]

If the cone be a *right* cone it is clear that the diameter so found will, for all sections, be at right angles to the chords which it bisects.

If the cone be *oblique*, the angle between the diameter so found and the parallel chords which it bisects will in general not be a right angle, but will be a right angle in the particular case only where the plane of the axial triangle *ABC* is at right angles to the plane of the base.

Again, if *PM* be the diameter of a section made by a plane cutting the circular base in the straight line *DME* perpendicular to *BC*, and if *PM* be in such a direction that it does not meet *AC* though produced to infinity, i.e. if *PM* be either parallel to *AC*, or makes with *PB* an angle less than the angle *BAC* and therefore meets *CA* produced beyond the apex of the cone, the section made by the said plane extends to infinity.

For, if we take any point V on PM produced and draw through it HK parallel to BC, and QQ' parallel to DE, the plane through HK, QQ' is parallel to that through DE, BC, i.e. to the base. Therefore the section $HQKQ'$ is a circle. And D, E, Q, Q' are all on the surface of the cone and are also on the cutting plane. Therefore the section DPE extends to the circle HQK, and in like manner to the circular section through any point on PM produced, and therefore to any distance from P. [I. 8.]

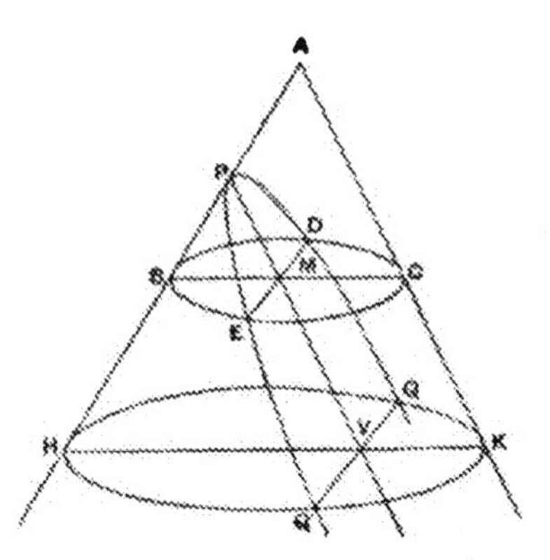

[It is also clear that $DM^2 = BM . MC$, and $QV^2 = HV . VK$; and $HV . VK$ becomes greater as V is taken more distant from P. For, in the case where PM is parallel to AC, VK remains constant while HV increases; and in the case where the diameter PM meets CA produced beyond the apex of the cone, both HV, VK increase together as V moves away from P. Thus QV increases indefinitely as the section extends to infinity.]

If on the other hand PM meets AC, the section does not extend to infinity. In that case the section will be a circle if its plane is parallel to the base or subcontrary. But, if the section is neither parallel to the base nor subcontrary, it will not be a circle. [I. 9.]

For let the plane of the section meet the plane of the base in DME, a straight line perpendicular to BC, a diameter of the circular base. Take the axial triangle through BC meeting the plane of section in the straight line PP'. Then P, P', M are all points in the plane of the axial triangle and in the plane of section. Therefore $PP'M$ is a straight line.

If possible, let the section PP' be a circle. Take any point Q on it and draw QQ' parallel to DME. Then if QQ' meets the axial triangle in V, $QV = VQ'$. Therefore PP' is the diameter of the supposed circle.

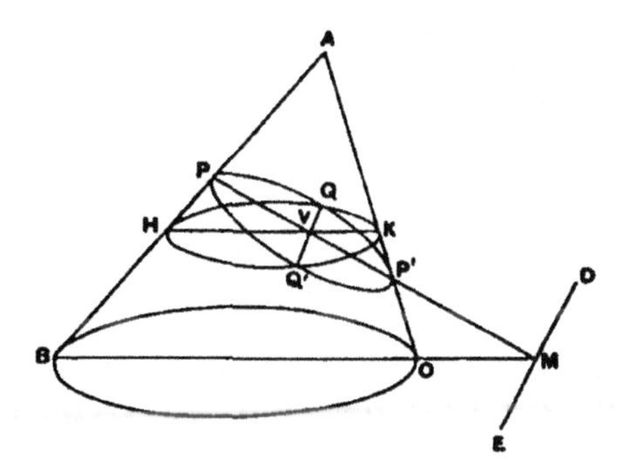

Let $HQKQ'$ be the circular section through QQ' parallel to the base.

Then, from the circles, $QV^2 = HV \,.\, VK$,

$$QV^2 = PV \,.\, VP'.$$

$$\therefore HV \,.\, VK = PV \,.\, VP',$$

so that $$HV : VP = P'V : VK.$$

$\therefore$ the triangles VPH, VKP' are similar, and

$$\angle PHV = \angle KP'V;$$

$\therefore \angle KP'V = \angle ABC$, and the section PP' is subcontrary: which contradicts the hypothesis.

$$\therefore PQP' \text{ is not a circle.}$$

It remains to investigate the character of the sections mentioned on the preceding page, viz. (*a*) those which extend to infinity, (*b*) those which are finite but are not circles.

Suppose, as usual, that the plane of section cuts the circular base in a straight line DME and that ABC is the axial triangle whose base BC is that diameter of the base of the cone which bisects DME at right angles at the point M. Then, if the plane of the section and the plane of the axial triangle intersect in the straight line PM, PM is a diameter of the section bisecting all chords of the section, as QQ', which are drawn parallel to DE.

If QQ' is so bisected in V, QV is said to be an **ordinate**, or a straight line **drawn ordinate-wise** (τεταγμένως κατηγμένη), to the diameter PM; and the length PV cut off from the diameter by any ordinate QV will be called the **abscissa** of QV.

Proposition 1.

[I. 11.]

First let the diameter PM of the section be parallel to one of the sides of the axial triangle as AC, and let QV be any ordinate to the diameter PM. Then, if a straight line PL (supposed to be drawn perpendicular to PM in the plane of the section) be taken of such a length that $PL : PA = BC^2 : BA . AC$, it is to be proved that

$$QV^2 = PL . PV.$$

Let HK be drawn through V parallel to BC. Then, since QV is also parallel to DE, it follows that the plane through H, Q, K is parallel to the base of the cone and therefore

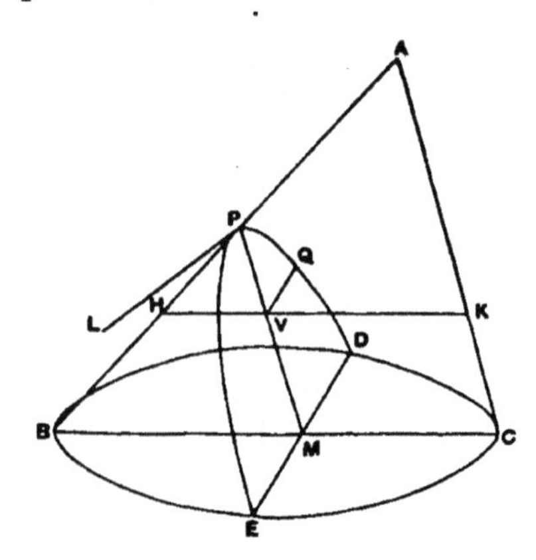

produces a circular section whose diameter is HK. Also QV is at right angles to HK.

$$\therefore HV.VK = QV^2.$$

Now, by similar triangles and by parallels,

$$HV : PV = BC : AC$$

and $$VK : PA = BC : BA.$$

$$\therefore HV.VK : PV.PA = BC^2 : BA.AC.$$

Hence $$QV^2 : PV.PA = PL : PA$$
$$= PL.PV : PV.PA.$$
$$\therefore QV^2 = PL.PV.$$

It follows that the square on any ordinate to the fixed diameter PM is equal to a rectangle applied (παραβάλλειν) to the fixed straight line PL drawn at right angles to PM with altitude equal to the corresponding abscissa PV. Hence the section is called a PARABOLA.

The fixed straight line PL is called the **latus rectum** (ὀρθία) or the **parameter of the ordinates** (παρ' ἣν δύνανται αἱ καταγόμεναι τεταγμένως).

This parameter, corresponding to the diameter PM, will for the future be denoted by the symbol p.

Thus $$QV^2 = p.PV,$$

or $$QV^2 \propto PV.$$

Proposition 2.

[I. 12.]

Next let PM not be parallel to AC but let it meet CA produced beyond the apex of the cone in P'. Draw PL at right angles to PM in the plane of the section and of such a length that $PL : PP' = BF.FC : AF^2$, where AF is a straight line through A parallel to PM and meeting BC in F. Then, if VR be drawn parallel to PL and $P'L$ be joined and produced to meet VR in R, it is to be proved that

$$QV^2 = PV.VR.$$

As before, let HK be drawn through V parallel to BC, so that $$QV^2 = HV.VK.$$

Then, by similar triangles,

$$HV : PV = BF : AF,$$
$$VK : P'V = FC : AF.$$

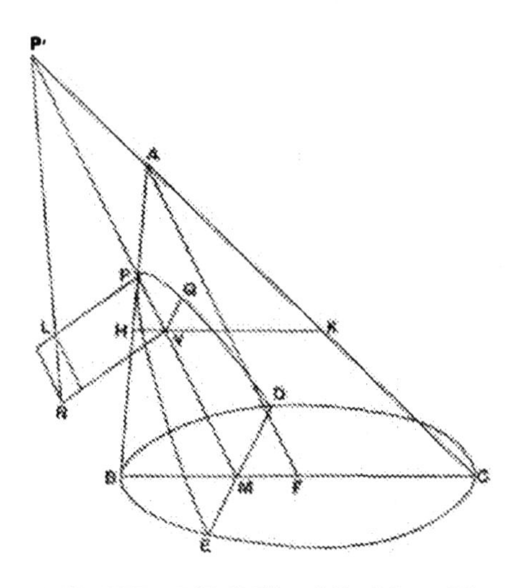

$$\therefore\ HV.VK : PV.P'V = BF.FC : AF^2.$$

Hence
$$QV^2 : PV.P'V = PL : PP'$$
$$= VR : P'V$$
$$= PV.VR : PV.P'V.$$
$$\therefore\ QV^2 = PV.VR.$$

It follows that the square on the ordinate is equal to a rectangle whose height is equal to the abscissa and whose base lies along the fixed straight line PL but overlaps (ὑπερβάλλει) it by a length equal to the difference between VR and PL*. Hence the section is called a HYPERBOLA.

* Apollonius describes the rectangle PR as *applied to the latus rectum but exceeding by a figure similar and similarly situated to that contained by PP' and PL*, i.e. exceeding the rectangle VL by the rectangle LR. Thus, if $QV=y$, $PV=x$, $PL=p$, and $PP'=d$,

$$y^2 = px + \frac{p}{d} . x^2,$$

which is simply the Cartesian equation of the hyperbola referred to oblique axes consisting of a diameter and the tangent at its extremity.

PL is called the **latus rectum** or the **parameter of the ordinates** as before, and PP' is called the **transverse** (ἡ πλαγία). The fuller expression **transverse diameter** (ἡ πλαγία διάμετρος) is also used; and, even more commonly, Apollonius speaks of the diameter and the corresponding parameter together, calling the latter the **latus rectum** (i.e. the *erect side*, ἡ ὀρθία πλευρά), and the former the **transverse side** (ἡ πλαγία πλευρά), of the **figure** (εἶδος) **on, or applied to, the diameter** (πρὸς τῇ διαμέτρῳ), i.e. of the rectangle contained by PL, PP' as drawn.

The parameter PL will in future be denoted by p.

[COR. It follows from the proportion

$$QV^2 : PV . P'V = PL : PP'$$

that, for any fixed diameter PP',

$$QV^2 : PV . P'V \text{ is a constant ratio,}$$

or QV^2 varies as $PV . P'V$.]

Proposition 3.

[I. 13.]

If PM meets AC in P' and BC in M, draw AF parallel to PM meeting BC produced in F, and draw PL at right angles to PM in the plane of the section and of such a length that $PL : PP' = BF . FC : AF^2$. Join $P'L$ and draw VR parallel to PL meeting $P'L$ in R. It will be proved that

$$QV^2 = PV . VR.$$

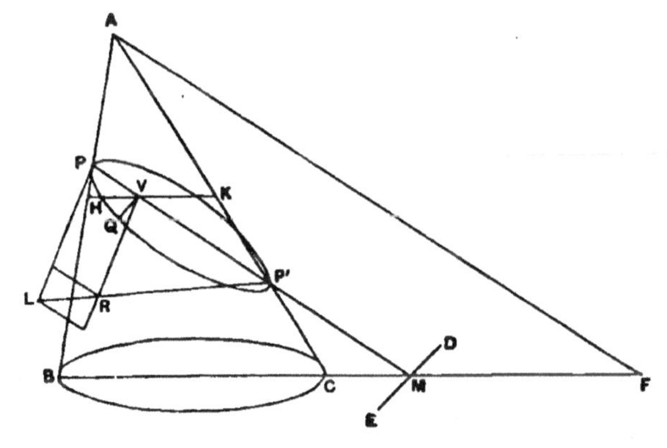

Draw HK through V parallel to BC. Then, as before,

$$QV^2 = HV . VK.$$

Now, by similar triangles,

$$HV : PV = BF : AF,$$
$$VK : P'V = FC : AF.$$
$$\therefore HV . VK : PV . P'V = BF . FC : AF^2.$$

Hence
$$QV^2 : PV . P'V = PL : PP'$$
$$= VR : P'V$$
$$= PV . VR : PV . P'V.$$
$$\therefore QV^2 = PV . VR.$$

Thus the square on the ordinate is equal to a rectangle whose height is equal to the abscissa and whose base lies along the fixed straight line PL but falls short of it (ἐλλείπει) by a length equal to the difference between VR and PL*. The section is therefore called an ELLIPSE.

As before, PL is called the **latus rectum**, or the **parameter** of the ordinates to the diameter PP', and PP' itself is called the **transverse** (with or without the addition of **diameter** or **side** of the **figure**, as explained in the last proposition).

PL will henceforth be denoted by p.

[COR. It follows from the proportion

$$QV^2 : PV . P'V = PL : PP'$$

that, for any fixed diameter PP',

$$QV^2 : PV . P'V \text{ is a constant ratio,}$$

or QV^2 varies as $PV . P'V$.]

* Apollonius describes the rectangle PR as *applied to the latus rectum but falling short by a figure similar and similarly situated to that contained by PP' and PL*, i.e. falling short of the rectangle VL by the rectangle LR.

If $QV = y$, $PV = x$, $PL = p$, and $PP' = d$,

$$y^2 = px - \frac{p}{d} . x^2.$$

Thus Apollonius' enunciation simply expresses the Cartesian equation referred to a diameter and the tangent at its extremity as (oblique) axes.

Proposition 4.

[I. 14.]

If a plane cuts both parts of a double cone and does not pass through the apex, the sections of the two parts of the cone will both be hyperbolas which will have the same diameter and equal latera recta corresponding thereto. And such sections are called OPPOSITE BRANCHES.

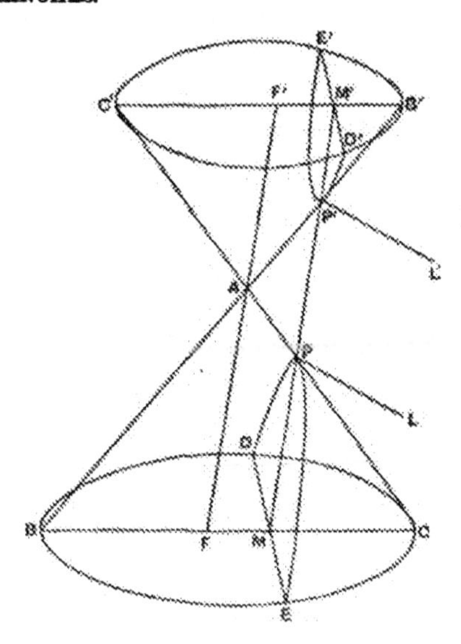

Let BC be the circle about which the straight line generating the cone revolves, and let $B'C'$ be any parallel section cutting the opposite half of the cone. Let a plane cut both halves of the cone, intersecting the base BC in the straight line DE and the plane $B'C'$ in $D'E'$. Then $D'E'$ must be parallel to DE.

Let BC be that diameter of the base which bisects DE at right angles, and let a plane pass through BC and the apex A cutting the circle $B'C'$ in $B'C'$, which will therefore be a diameter of that circle and will cut $D'E'$ at right angles, since $B'C'$ is parallel to BC, and $D'E'$ to DE.

Let FAF' be drawn through A parallel to MM', the straight line joining the middle points of DE, $D'E'$ and meeting CA, $B'A$ respectively in P, P'.

Draw perpendiculars PL, $P'L'$ to MM' in the plane of the section and of such length that

$$PL : PP' = BF . FC : AF^2,$$

$$P'L' : P'P = B'F' . F'C' : AF'^2.$$

Since now MP, the diameter of the section DPE, when produced, meets BA produced beyond the apex, the section DPE is a hyperbola.

Also, since $D'E'$ is bisected at right angles by the base of the axial triangle $AB'C'$, and $M'P$ in the plane of the axial triangle meets $C'A$ produced beyond the apex A, the section $D'P'E'$ is also a hyperbola.

And the two hyperbolas have the same diameter $MPP'M'$.

It remains to prove that $PL = P'L'$.

We have, by similar triangles,

$$BF : AF = B'F' : AF',$$

$$FC : AF = F'C' : AF'.$$

$$\therefore BF . FC : AF^2 = B'F' . F'C' : AF'^2.$$

Hence $$PL : PP' = P'L' : P'P.$$

$$\therefore PL = P'L'.$$

THE DIAMETER AND ITS CONJUGATE.

Proposition 5.

[I. 15.]

If through C, the middle point of the diameter PP' of an ellipse, a double ordinate DCD' be drawn to PP', DCD' will bisect all chords parallel to PP', and will therefore be a diameter the ordinates to which are parallel to PP'.

In other words, *if the diameter bisect all chords parallel to a second diameter, the second diameter will bisect all chords parallel to the first.*

Also the parameter of the ordinates to DCD' will be a third proportional to DD', PP'.

(1) Let QV be any ordinate to PP', and through Q draw QQ' parallel to PP' meeting DD' in v and the ellipse in Q'; and let $Q'V'$ be the ordinate drawn from Q' to PP'.

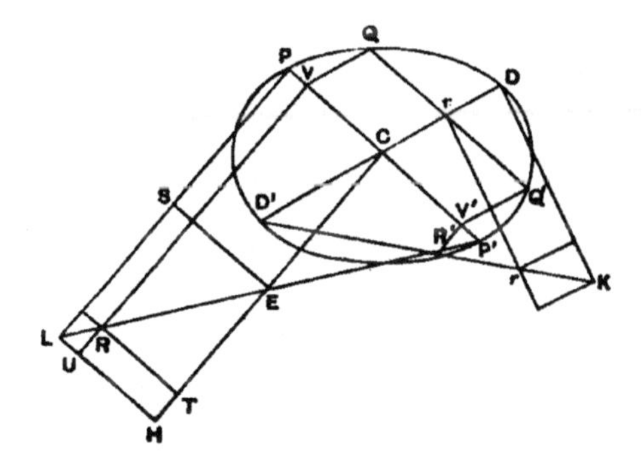

Then, if PL is the parameter of the ordinates, and if $P'L$ is joined and $VR, CE, V'R'$ drawn parallel to PL to meet $P'L$, we have [Prop. 3]

$$QV^2 = PV \cdot VR,$$

$$Q'V'^2 = PV' \cdot V'R';$$

and $QV = Q'V'$, because QV is parallel to $Q'V'$ and QQ' to PP'.

$$\therefore PV \cdot VR = PV' \cdot V'R'.$$

Hence $\quad PV : PV' = V'R' : VR = P'V' : P'V.$

$$\therefore PV : PV' \sim PV = P'V' : P'V \sim P'V',$$

or $\quad PV : VV' = P'V' : VV'.$

$$\therefore PV = P'V'.$$

Also $\quad CP = CP'.$

By subtraction, $\quad CV = CV',$

and $\therefore Qv = vQ'$, so that QQ' is bisected by DD'.

(2) Draw DK at right angles to DD' and of such a length that $DD' : PP' = PP' : DK$. Join $D'K$ and draw vr parallel to DK to meet $D'K$ in r.

Also draw TR, LUH and ES parallel to PP'.

Then, since $PC = CP'$, $PS = SL$ and $CE = EH$;

$\therefore$ the parallelogram $\quad (PE) = (SH).$

Also $\quad (PR) = (VS) + (SR) = (SU) + (RH).$

By subtraction, $\quad (PE) - (PR) = (RE);$

$$\therefore CD^2 - QV^2 = RT \cdot TE.$$

But $\quad CD^2 - QV^2 = CD^2 - Cv^2 = D'v \cdot vD.$

$$\therefore D'v \cdot vD = RT \cdot TE \quad \text{..................(A).}$$

Now $\quad DD' : PP' = PP' : DK$, by hypothesis.

$$\therefore DD' : DK = DD'^2 : PP'^2$$

$$= CD^2 : CP^2$$

$$= PC \cdot CE : CP^2$$

$$= RT \cdot TE : RT^2,$$

and $\quad DD' : DK = D'v : vr$

$$= D'v \cdot vD : vD \cdot vr;$$

$$\therefore D'v \cdot vD : Dv \cdot vr = RT \cdot TE : RT^2.$$

But $\quad D'v \cdot vD = RT \cdot TE$, from (A) above;

$$\therefore Dv \cdot vr = RT^2 = CV^2 = Qv^2.$$

Thus DK is the parameter of the ordinates to DD', such as Qv.

Therefore the parameter of the ordinates to DD' is a third proportional to DD', PP'.

COR. We have
$$CD^2 = PC . CE$$
$$= \tfrac{1}{2}PP' . \tfrac{1}{2}PL;$$
$$\therefore DD'^2 = PP' . PL,$$
or
$$PP' : DD' = DD' : PL,$$
and PL is a third proportional to PP', DD'.

Thus the relations of PP', DD' and the corresponding parameters are reciprocal.

DEF. Diameters such as PP', DD', each of which bisects all chords parallel to the other, are called **conjugate diameters**.

Proposition 6.

[I. 16.]

If from the middle point of the diameter of a hyperbola with two branches a line be drawn parallel to the ordinates to that diameter, the line so drawn will be a diameter conjugate to the former one.

If any straight line be drawn parallel to PP', the given diameter, and meeting the two branches of the hyperbola in Q, Q' respectively, and if from C, the middle point of PP', a straight line be drawn parallel to the ordinates to PP' meeting QQ' in v, we have to prove that QQ' is bisected in v.

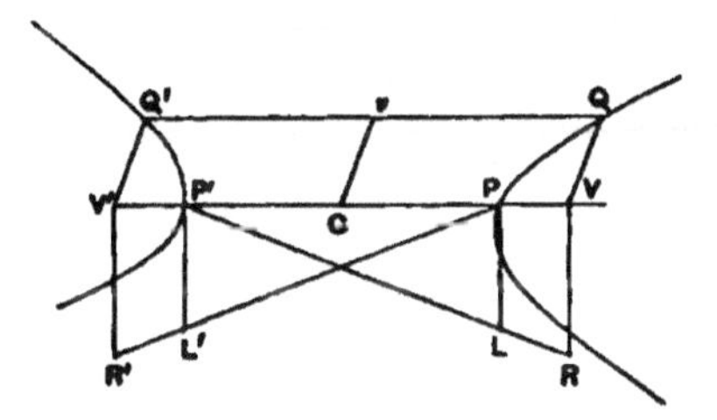

Let QV, $Q'V'$ be ordinates to PP', and let PL, $P'L'$ be the parameters of the ordinates in each branch so that [Prop. 4]

$PL = P'L'$. Draw VR, $V'R'$ parallel to PL, $P'L'$, and let PL', $P'L$ be joined and produced to meet $V'R'$, VR respectively in R', R.

Then we have $\quad QV^2 = PV \cdot VR,$

$$Q'V'^2 = P'V' \cdot V'R'.$$

$$\therefore PV \cdot VR = P'V' \cdot V'R', \text{ and } V'R' : VR = PV : P'V'.$$

Also $\quad PV' : V'R' = PP' : P'L' = P'P : PL = P'V : VR.$

$$\therefore PV' : P'V = V'R' : VR$$

$$= PV : P'V', \text{ from above;}$$

$$\therefore PV' : PV = P'V : P'V',$$

and $\quad PV' + PV : PV = P'V + P'V' : P'V',$

or $\quad VV' : PV = VV' : P'V';$

$$\therefore PV = P'V'.$$

But $\quad CP = CP';$

$$\therefore \text{ by addition, } CV = CV',$$

or $\quad Qv = Q'v.$

Hence Cv is a diameter conjugate to PP'.

[More shortly, we have, from the proof of Prop. 2,

$$QV^2 : PV \cdot P'V = PL : PP',$$

$$Q'V'^2 : P'V' \cdot PV' = P'L' : PP',$$

and $\quad QV = Q'V', \ PL = P'L';$

$$\therefore PV \cdot P'V = PV' \cdot P'V', \text{ or } PV : PV' = P'V' : P'V,$$

whence, as above, $\quad PV = P'V'$.]

Def. The middle point of the diameter of an ellipse or hyperbola is called the **centre**; and the straight line drawn parallel to the ordinates of the diameter, of a length equal to the mean proportional between the diameter and the parameter, and bisected at the centre, is called the **secondary diameter** (δευτέρα διάμετρος).

Proposition 7.

[I. 20.]

In a parabola the square on an ordinate to the diameter varies as the abscissa.

This is at once evident from Prop. 1.

Proposition 8.

[I. 21.]

In a hyperbola, an ellipse, or a circle, if QV be any ordinate to the diameter PP',

$$QV^2 \propto PV . P'V.$$

[This property is at once evident from the proportion

$$QV^2 : PV . P'V = PL : PP'$$

obtained in the course of Props. 2 and 3; but Apollonius gives a separate proof, starting from the property $QV^2 = PV . VR$ which forms the basis of the definition of the conic, as follows.]

Let QV, $Q'V'$ be two ordinates to the diameter PP'.

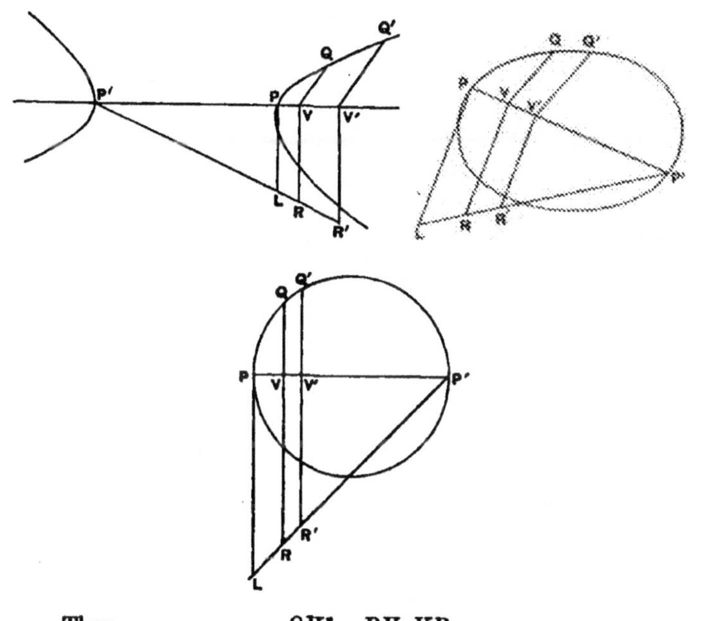

Then
$$QV^2 = PV . VR,$$
$$Q'V'^2 = PV' . V'R';$$
$$\therefore QV^2 : PV . P'V = PV . VR : PV . P'V$$
$$= VR : P'V = PL : PP'.$$

Similarly $Q'V'^2 : PV'.P'V' = PL : PP'$.

$$\therefore QV^2 : Q'V'^2 = PV.P'V : PV'.P'V';$$

and $QV^2 : PV.P'V$ is a constant ratio,

or $$QV^2 \propto PV.P'V.$$

Proposition 9.

[I. 29.]

If a straight line through the centre of a hyperbola with two branches meet one branch, it will, if produced, meet the other also.

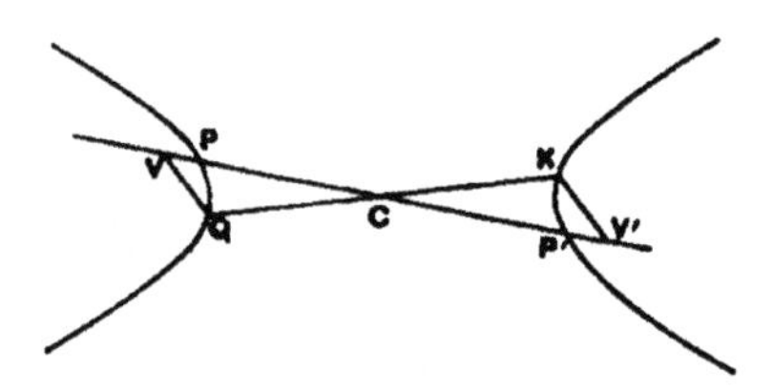

Let PP' be the given diameter and C the centre. Let CQ meet one branch in Q. Draw the ordinate QV to PP', and set off CV' along PP' on the other side of the centre equal to CV. Let $V'K$ be the ordinate to PP' through V'. We shall prove that QCK is a straight line.

Since $CV = CV'$, and $CP = CP'$, it follows that $PV = P'V'$;

$$\therefore PV.P'V = P'V'.PV'.$$

But $QV^2 : KV'^2 = PV.P'V : P'V'.PV'$. [Prop. 8]

$\therefore QV = KV'$; and QV, KV' are parallel, while $CV = CV'$.

Therefore QCK is a straight line.

Hence QC, if produced, will cut the opposite branch.

Proposition 10.

[I. 30.]

In a hyperbola or an ellipse any chord through the centre is bisected at the centre.

Let PP' be the diameter and C the centre; and let QQ' be any chord through the centre. Draw the ordinates QV, $Q'V'$ to the diameter PP'.

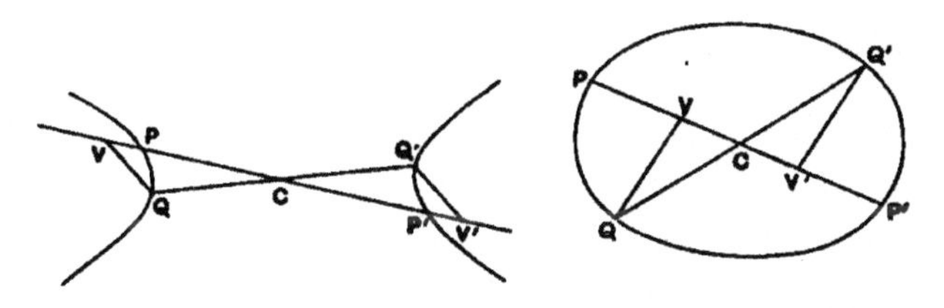

Then

$$PV.P'V : P'V'.PV' = QV^2 : Q'V'^2$$
$$= CV^2 : CV'^2, \text{ by similar triangles.}$$
$$\therefore CV^2 \pm PV.P'V : CV^2 = CV'^2 \pm P'V'.PV' : CV'^2$$

(where the upper sign applies to the ellipse and the lower to the hyperbola).

$$\therefore CP^2 : CV^2 = CP'^2 : CV'^2.$$

But $$CP^2 = CP'^2;$$

$$\therefore CV^2 = CV'^2, \text{ and } CV = CV'.$$

And QV, $Q'V'$ are parallel;

$$\therefore CQ = CQ'.$$

Proposition 11.

[I. 17, 32.]

If a straight line be drawn through the extremity of the diameter of any conic parallel to the ordinates to that diameter, the straight line will touch the conic, and no other straight line can fall between it and the conic.

It is first proved that the straight line drawn in the manner described will fall without the conic.

For, if not, let it fall within it, as PK, where PM is the given diameter. Then KP, being drawn from a point K on the conic parallel to the ordinates to PM, will meet PM and will be bisected by it. But KP produced falls without the conic; therefore it will not be bisected at P.

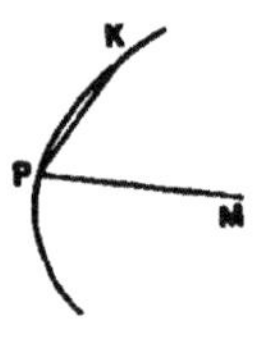

Therefore the straight line PK must fall without the conic and will therefore touch it.

It remains to be proved that no straight line can fall between the straight line drawn as described and the conic.

(1) Let the conic be a *parabola*, and let PF be parallel to the ordinates to the diameter PV. If possible, let PK fall between PF and the parabola, and draw KV parallel to the ordinates, meeting the curve in Q.

Then $$KV^2 : PV^2 > QV^2 : PV^2$$
$$> PL . PV : PV^2$$
$$> PL : PV.$$

Let V' be taken on PV such that

$$KV^2 : PV^2 = PL : PV',$$

and let $V'Q'M$ be drawn parallel to QV, meeting the curve in Q' and PK in M.

Then $KV^2 : PV^2 = PL : PV'$
$$= PL . PV' : PV'^2$$
$$= Q'V'^2 : PV'^2,$$

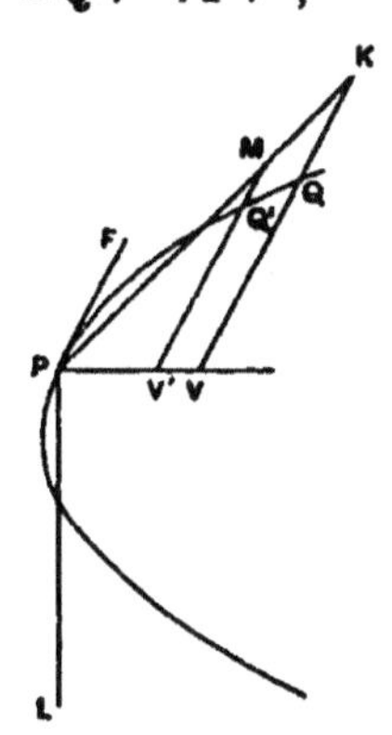

and $\qquad KV^2 : PV^2 = MV'^2 : PV'^2$, by parallels.

Therefore $MV'^2 = Q'V'^2$, and $MV' = Q'V'$.

Thus PK cuts the curve in Q', and therefore does not fall outside it: which is contrary to the hypothesis.

Therefore no straight line can fall between PF and the curve.

(2) Let the curve be a *hyperbola* or an *ellipse* or a *circle*.

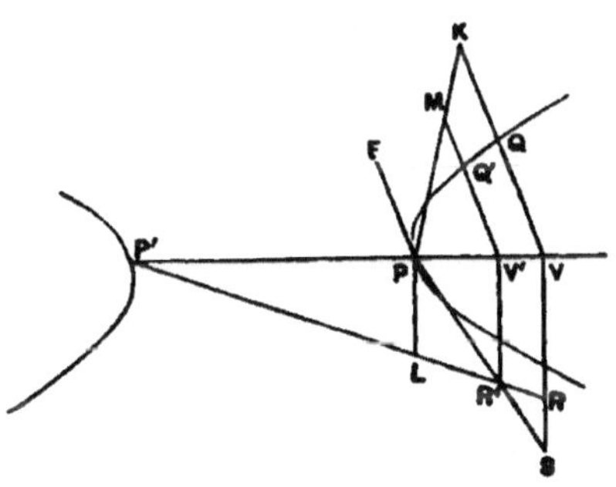

Let PF be parallel to the ordinates to PP', and, if possible, let PK fall between PF and the curve. Draw KV parallel to the ordinates, meeting the curve in Q, and draw VR perpendicular to PV. Join $P'L$ and let it (produced if necessary) meet VR in R.

Then $QV^2 = PV \,.\, VR$, so that $KV^2 > PV \,.\, VR$.

Take a point S on VR produced such that $KV^2 = PV \,.\, VS$. Join PS and let it meet $P'R$ in R'. Draw $R'V'$ parallel to PL meeting PV in V', and through V' draw $V'Q'M$ parallel to QV, meeting the curve in Q' and PK in M.

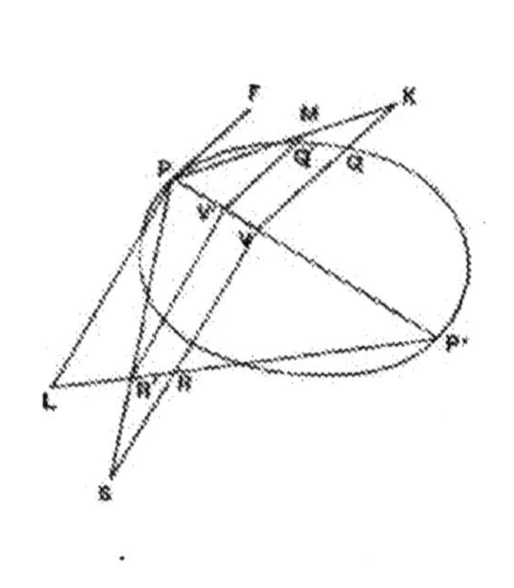

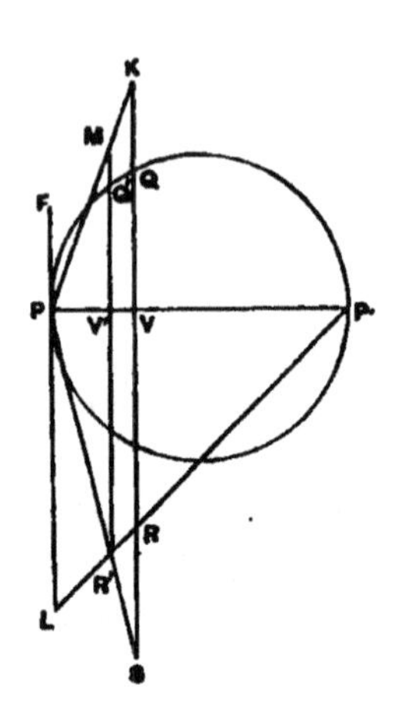

Now $$KV^2 = PV \,.\, VS,$$

$$\therefore\ VS : KV = KV : PV,$$

so that $$VS : PV = KV^2 : PV^2.$$

Hence, by parallels,

$$V'R' : PV' = MV'^2 : PV'^2,$$

or MV' is a mean proportional between PV', $V'R'$,

i.e. $$MV'^2 = PV' \,.\, V'R'$$

$$= Q'V'^2, \text{ by the property of the conic.}$$

$$\therefore\ MV' = Q'V'.$$

Thus PK cuts the curve in Q', and therefore does not fall outside it: which is contrary to the hypothesis.

Hence no straight line can fall between PF and the curve.

Proposition 12.

[I. 33, 35.]

If a point T be taken on the diameter of a parabola outside the curve and such that $TP = PV$, where V is the foot of the ordinate from Q to the diameter PV, the line TQ will touch the parabola.

We have to prove that the straight line TQ or TQ produced does not fall within the curve on either side of Q.

For, if possible, let K, a point on TQ or TQ produced, fall within the curve*, and through K draw $Q'KV'$ parallel to an ordinate and meeting the diameter in V' and the curve in Q'.

Then $Q'V'^2 : QV^2$

$$> KV'^2 : QV^2, \text{ by hypothesis,}$$

$$> TV'^2 : TV^2.$$

$$\therefore PV' : PV > TV'^2 : TV^2.$$

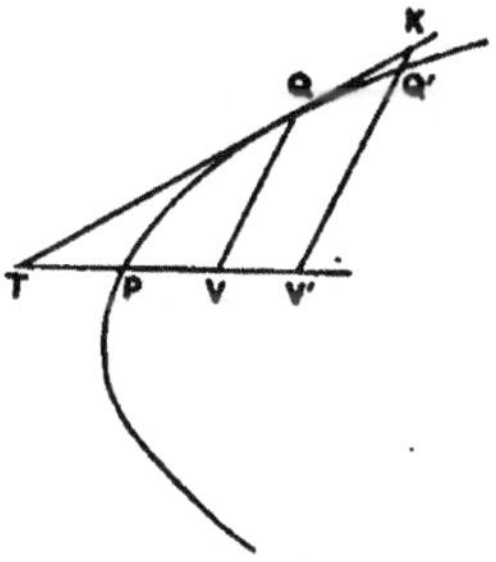

Hence

$$4TP.PV' : 4TP.PV > TV'^2 : TV^2,$$

and, since $TP = PV$,

$$4TP.PV = TV^2,$$

$$\therefore 4TP.PV' > TV'^2.$$

But, since by hypothesis TV' is not bisected in P,

$$4TP.PV' < TV'^2,$$

which is absurd.

Therefore TQ does not at any point fall within the curve, and is therefore a tangent.

* Though the proofs of this proposition and the next follow *in form* the method of *reductio ad absurdum*, it is easily seen that they give in fact the direct demonstration that, if K is any point on the tangent other than Q, the point of contact, K lies outside the curve because, if $KQ'V'$ be parallel to QV, it is proved that $KV' > Q'V'$. The figures in both propositions have accordingly been drawn in accordance with the facts instead of representing the incorrect assumption which leads to the absurdity in each case.

Conversely, *if the tangent at Q meet the diameter produced outside the curve in the point T, $TP = PV$. Also no straight line can fall between TQ and the curve.*

[Apollonius gives a separate proof of this, using the method of *reductio ad absurdum*.]

Proposition 13.

[I. 34, 36.]

In a hyperbola, an ellipse, or a circle, if PP' be the diameter and QV an ordinate to it from a point Q, and if a point T be taken on the diameter but outside the curve such that $TP : TP' = PV : VP'$, then the straight line TQ will touch the curve.

We have to prove that no point on TQ or TQ produced falls within the curve.

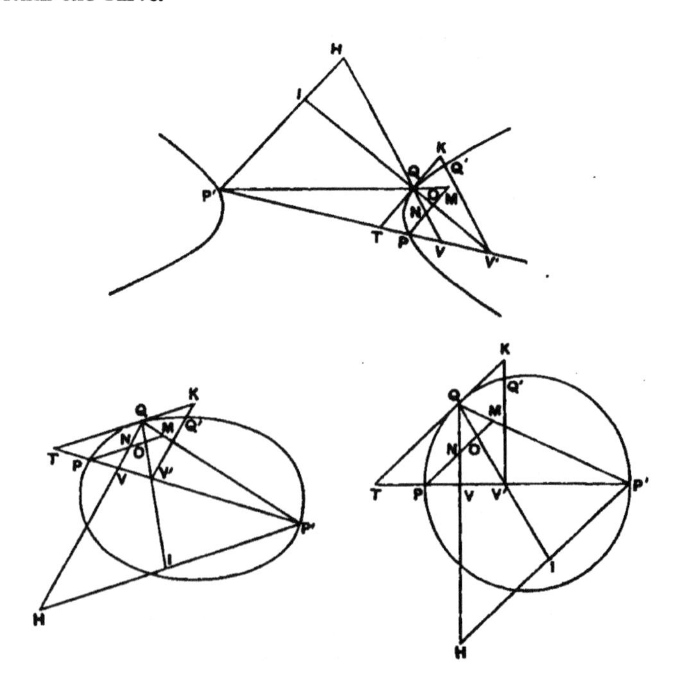

If possible, let a point K on TQ or TQ produced fall within the curve*; draw $Q'KV'$ parallel to an ordinate meeting the curve in Q'. Join $P'Q$, $V'Q$, producing them if necessary, and draw through P', P parallels to TQ meeting $V'Q$, VQ in I, O and H, N respectively. Also let the parallel through P meet $P'Q$ in M.

Now, by hypothesis, $P'V : PV = TP' : TP$;

$$\therefore \text{ by parallels, } P'H : PN = P'Q : QM$$
$$= P'H : NM.$$

Therefore $$PN = NM.$$

Hence $$PN . NM > PO . OM,$$

or $$NM : MO > OP : PN;$$

$$\therefore P'H : P'I > OP : PN,$$

or $$P'H . PN > P'I . OP.$$

It follows that $P'H . PN : TQ^2 > P'I . OP : TQ^2$;

$\therefore$ by similar triangles

$$P'V . PV : TV^2 > P'V' . PV' : TV'^2,$$

or $$P'V . PV : P'V' . PV' > TV^2 : TV'^2;$$

$$\therefore QV^2 : Q'V'^2 > TV^2 : TV'^2$$
$$> QV^2 : KV'^2.$$

$\therefore Q'V' < KV'$, which is contrary to the hypothesis.

Thus TQ does not cut the curve, and therefore it touches it.

Conversely, *if the tangent at a point Q meet the diameter PP' outside the section in the point T, and QV is the ordinate from Q,*

$$TP : TP' = PV : VP'.$$

Also no other straight line can fall between TQ and the curve.

[This again is separately proved by Apollonius by a simple *reductio ad absurdum.*]

* See the note on the previous proposition.

Proposition 14.

[I. 37, 39.]

In a hyperbola, an ellipse, or a circle, if QV be an ordinate to the diameter PP', and the tangent at Q meet PP' in T, then

(1) $CV \cdot CT = CP^2$,

(2) $QV^2 : CV \cdot VT = p : PP'$ [or $CD^2 : CP^2$].

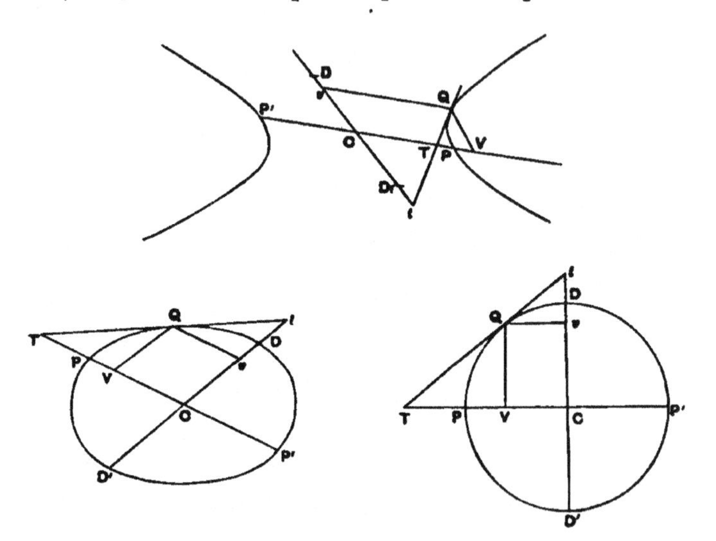

(1) Since QT is the tangent at Q,

$$TP : TP' = PV : P'V, \qquad \text{[Prop. 13]}$$

$$\therefore TP + TP' : TP \sim TP' = PV + P'V : PV \sim P'V;$$

thus, for the *hyperbola*,

$$2CP : 2CT = 2CV : 2CP;$$

and for the *ellipse* or *circle*,

$$2CT : 2CP = 2CP : 2CV;$$

therefore for all three curves

$$CV \cdot CT = CP^2.$$

(2) Since $CV : CP = CP : CT$,

$$CV \sim CP : CV = CP \sim CT : CP,$$

whence $PV : CV = PT : CP$,

or $PV : PT = CV : CP$.

$$\therefore PV : PV + PT = CV : CV + CP,$$

or $PV : VT = CV : P'V$,

and $CV . VT = PV . P'V$.

But $QV^2 : PV . P'V = p : PP'$ (or $CD^2 : CP^2$). [Prop. 8]

$$\therefore QV^2 : CV . VT = p : PP' \text{ (or } CD^2 : CP^2\text{)}.$$

COR. It follows at once that $QV : VT$ is equal to the ratio compounded of the ratios $p : PP'$ (or $CD^2 : CP^2$) and $CV : QV$.

Proposition 15.

[I. 38, 40.]

If Qv be the ordinate to the diameter conjugate to PP', and QT, the tangent at Q, meet that conjugate diameter in t, then

(1) $Cv . Ct = CD^2$,

(2) $Qv^2 : Cv . vt = PP' : p$ [*or* $CP^2 : CD^2$],

(3) $tD : tD' = vD' : vD$ *for the* hyperbola,

and $tD : tD' = vD : vD'$ *for the* ellipse *and* circle.

Using the figures drawn for the preceding proposition, we have (1)

$QV^2 : CV . VT = CD^2 : CP^2$. [Prop. 14]

But $QV : CV = Cv : CV$,

and $QV : VT = Ct : CT$;

$$\therefore QV^2 : CV . VT = Cv . Ct : CV . CT.$$

Hence $Cv . Ct : CV . CT = CD^2 : CP^2$.

And $CV . CT = CP^2$; [Prop. 14]

$$\therefore Cv . Ct = CD^2.$$

(2) As before,

$$QV^2 : CV . VT = CD^2 : CP^2 \text{ (or } p : PP'\text{)}.$$

But $QV : CV = Cv : Qv$,

and $$QV : VT = vt : Qv;$$

$$\therefore QV^2 : CV . VT = Cv . vt : Qv^2.$$

Hence $$Qv^2 : Cv . vt = CP^2 : CD^2$$
$$= PP' : p.$$

(3) Again,

$$Ct . Cv = CD^2 = CD . CD';$$

$$\therefore Ct : CD = CD' : Cv,$$

and $\therefore Ct + CD : Ct \sim CD = CD' + Cv : CD' \sim Cv.$

Thus $tD : tD' = vD' : vD$ for the *hyperbola*,

and $tD' : tD = vD' : vD$ for the *ellipse* and *circle*.

COR. It follows from (2) that $Qv : Cv$ is equal to the ratio compounded of the ratios $PP' : p$ (or $CP^2 : CD^2$) and $vt : Qv$.

PROPOSITIONS LEADING TO THE REFERENCE OF A CONIC TO ANY NEW DIAMETER AND THE TANGENT AT ITS EXTREMITY.

Proposition 16.

[I. 41.]

In a hyperbola, an ellipse, or a circle, if equiangular parallelograms (VK), (PM) *be described on* QV, CP *respectively, and their sides are such that* $\frac{QV}{QK} = \frac{p}{PP'} \cdot \frac{CP}{CM} \left[\text{i.e. } \frac{CD^2}{CP^2} \cdot \frac{CP}{CM}\right]$, *and if* (VN) *be the parallelogram on* CV *similar and similarly situated to* (PM), *then*

$$(VN) \pm (VK) = (PM),$$

the lower sign applying to the hyperbola.

Suppose O to be so taken on KQ produced that

$$QV : QO = p : PP',$$

so that $$QV^2 : QV.QO = QV^2 : PV.PV'.$$

Thus $$QV.QO = PV.P'V \quad\text{..................}(1).$$

Also $QV : QK = (CP : CM).(p : PP') = (CP : CM).(QV : QO)$,

or $$(QV : QO).(QO : QK) = (CP : CM).(QV : QO);$$

$$\therefore QO : QK = CP : CM \quad\text{..................}(2).$$

But $$QO : QK = QV.QO : QV.QK$$

and $$CP : CM = CP^2 : CP.CM;$$

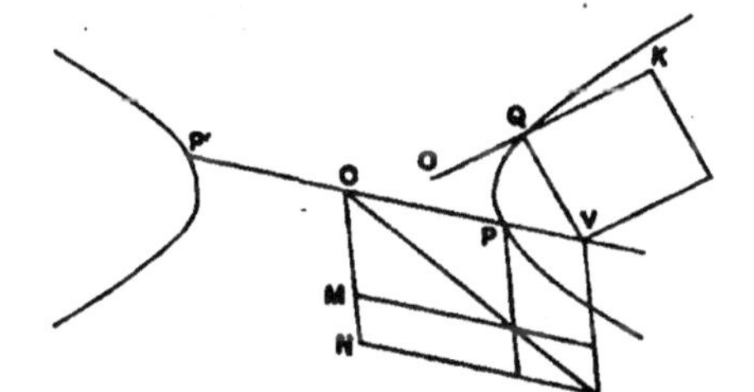

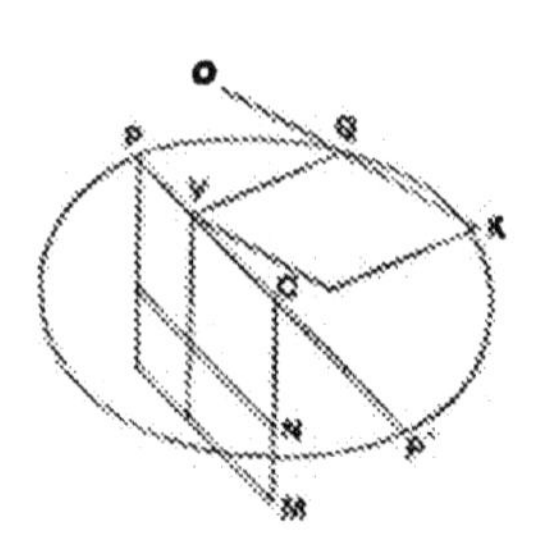

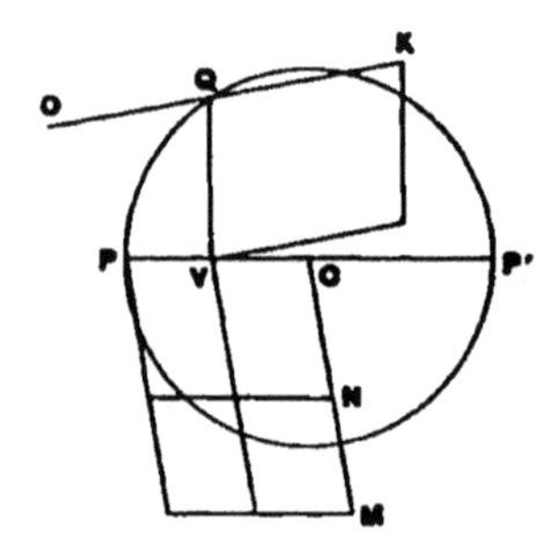

$$\therefore CP^2 : CP.CM = QV.QO : QV.QK$$

$$= PV.P'V : QV.QK, \text{ from (1).}$$

Therefore, since PM, VK are equiangular,

$$CP^2 : PV.P'V = (PM) : (VK) \text{ .... (3).}$$

Hence $CP^2 \mp PV.P'V : CP^2 = (PM) \mp (VK) : (PM)$,
where the upper sign applies to the *ellipse* and *circle* and the lower to the *hyperbola*.

$$\therefore CV^2 : CP^2 = (PM) \mp (VK) : (PM),$$

and hence $\quad (VN) : (PM) = (PM) \mp (VK) : (PM)$,

so that $\quad (VN) = (PM) \mp (VK)$,

or $\quad (VN) \pm (VK) = (PM)$.

[The above proof is reproduced as given by Apollonius in order to show his method of dealing with a somewhat complicated problem by purely geometrical means. The proposition is more shortly proved by a method more akin to algebra as follows.

We have $\quad QV^2 : CV^2 \sim CP^2 = CD^2 : CP^2$,

and $\quad \dfrac{QV}{QK} = \dfrac{CD^2}{CP^2} \cdot \dfrac{CP}{CM}, \quad$ or $\quad QV = QK . \dfrac{CD^2}{CP.CM}$;

$$\therefore QV.QK . \frac{CD^2}{CP.CM} : CV^2 \sim CP^2 = CD^2 : CP^2,$$

or $\quad QV.QK = CP.CM\left(\dfrac{CV^2}{CP^2} \sim 1\right)$.

$$\therefore (VK) = (VN) \sim (PM),$$

or $\quad (VN) \pm (VK) = (PM)$.]

Proposition 17.

[I. 42.]

In a parabola, if QV, RW be ordinates to the diameter through P, and QT, the tangent at Q, and RU parallel to it meet the diameter in T, U respectively; and if through Q a parallel to the diameter be drawn meeting RW produced in F and the tangent at P in E, then

$$\triangle\, RUW = \textit{the parallelogram } (EW).$$

Since QT is a tangent,

$$TV = 2PV; \text{ [Prop. 12]}$$

$$\therefore \triangle\, QTV = (EV) \text{.........} (1).$$

Also $\quad QV^2 : RW^2 = PV : PW;$

$$\therefore \triangle\, QTV : \triangle\, RUW = (EV) : (EW),$$

and $\quad \triangle\, QTV = (EV)$, from (1);

$$\therefore \triangle\, RUW = (EW).$$

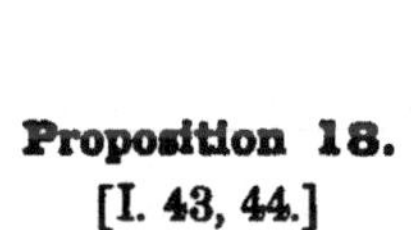

Proposition 18.

[I. 43, 44.]

In a hyperbola, an ellipse, or a circle, if the tangent at Q and the ordinate from Q meet the diameter in T, V, and if RW be the ordinate from any point R and RU be parallel to QT; if also RW and the parallel to it through P meet CQ in F, E respectively, then

$$\triangle\, CFW \sim \triangle\, CPE = \triangle\, RUW.$$

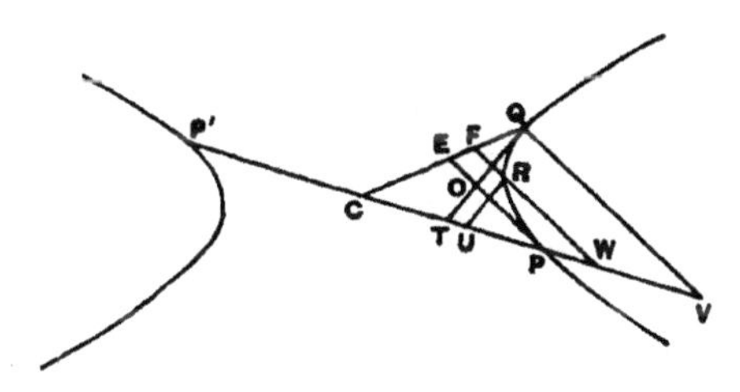

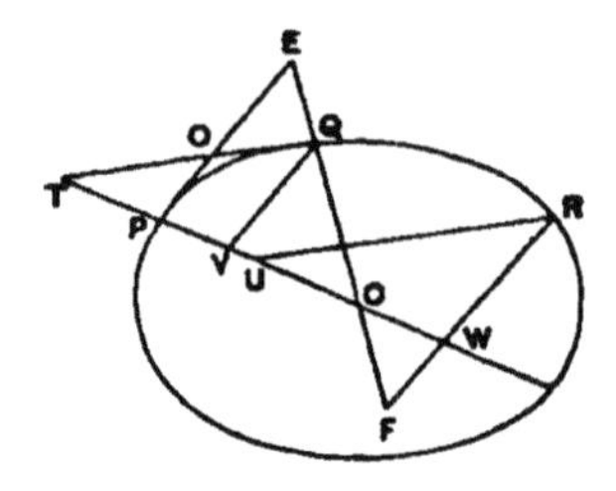

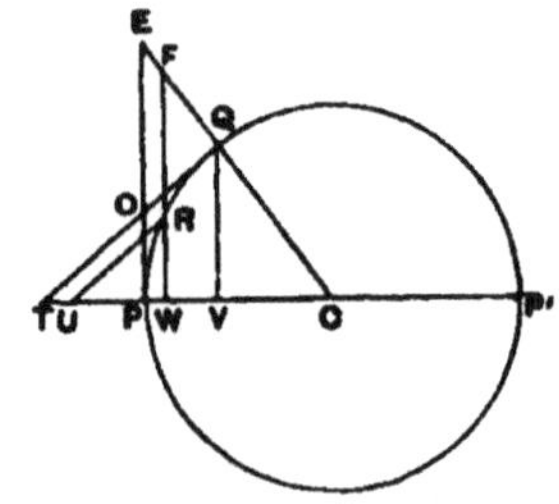

We have $\qquad QV^2 : CV . VT = p : PP'$ [or $CD^2 : CP^2$],

whence $\;QV : VT = (p : PP').(CV : QV)$; [Prop. 14 and Cor.]

therefore, by parallels,

$$RW : WU = (p : PP').(CP : PE).$$

Thus, by Prop. 16, the parallelograms which are the doubles of the triangles RUW, CPE, CWF have the property proved in that proposition. It follows that the same is true of the triangles themselves.

$$\therefore \triangle CFW \sim \triangle CPE = \triangle RUW.$$

[It is interesting to observe the exact significance of this proposition, which is the foundation of Apollonius' method of transformation of coordinates. The proposition amounts to this: If CP, CQ are fixed semidiameters and R a variable point, the area of the quadrilateral $CFRU$ is constant for all positions of R on the conic. Suppose now that CP, CQ are taken as axes of coordinates (CP being the axis of x). If we draw RX parallel to CQ to meet CP and RY parallel to CP to meet CQ, the proposition asserts that (subject to the proper convention as to sign)

$$\triangle RYF + \square CXRY + \triangle RXU = \text{(const.)}.$$

But, since RX, RY, RF, RU are in fixed directions,

$$\triangle RYF \propto RY^2,$$

or $$\triangle RYF = ax^2;$$

$$\square CXRY \propto RX . RY,$$

or $$\square CXRY = \beta xy;$$

$$\triangle RXU \propto RX^2,$$

or $$\triangle RXU = \gamma y^2.$$

Hence, if x, y are the coordinates of R,

$$\alpha x^2 + \beta xy + \gamma y^2 = A,$$

which is the Cartesian equation referred to the centre as origin and any two diameters as axes.]

Proposition 19.

[I. 45.]

If the tangent at Q and the straight line through R parallel to it meet the secondary diameter in t, u respectively, and Qv, Rw be parallel to the diameter PP', meeting the secondary diameter in v, w; if also Rw meet CQ in f, then

$$\triangle\, Cfw = \triangle\, Ruw - \triangle\, CQt.$$

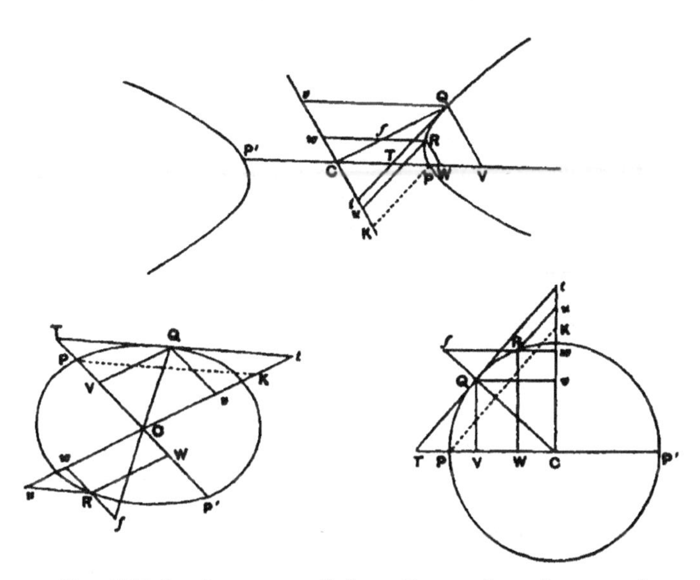

[Let PK be drawn parallel to Qt meeting the secondary diameter in K, so that the triangle CPK is similar to the triangle vQt.]

We have [Prop. 14, Cor.]

$$QV : CV = (p : PP') . (VT : QV)$$
$$= (p : PP') . (Qv : vt),$$

and the triangles QvC, Qvt are the halves of equiangular parallelograms on Cv (or QV) and Qv (or CV) respectively: also CPK is the triangle on CP similar to Qvt.

Therefore [by Prop. 16], $\triangle CQv = \triangle Qvt \sim \triangle CPK$,

and clearly $\triangle CQv = \triangle Qvt - \triangle CQt$;

$$\therefore \triangle CPK = \triangle CQt.$$

Again, the triangle Cfw is similar to the triangle CQv, and the triangle Rwu to the triangle Qvt. Therefore, for the ordinate RW,

$$\triangle Cfw = \triangle Ruw \sim \triangle CPK = \triangle Ruw - \triangle CQt.$$

Proposition 20.

[I. 46.]

In a parabola the straight line drawn through any point parallel to the diameter bisects all chords parallel to the tangent at the point.

Let RR' be any chord parallel to the tangent at Q and let it meet the diameter PV in U. Let QM drawn parallel to PV meet RR' in M, and the straight lines drawn ordinate-wise through R, R', P in F, F', E respectively.

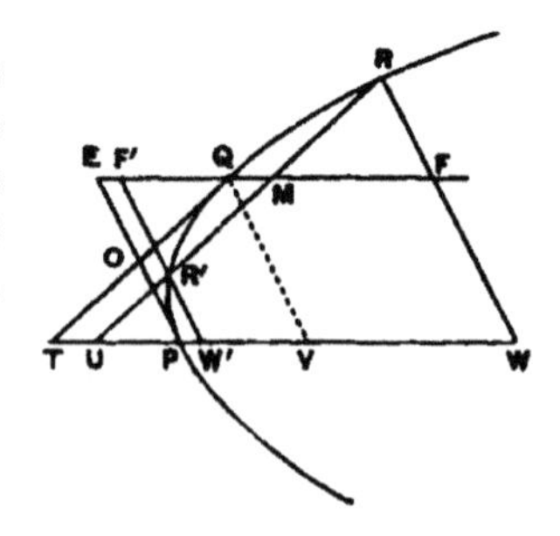

We have then [Prop. 17]

$\triangle RUW = \square EW$,

and $\triangle R'UW' = \square EW'$.

Therefore, by subtraction, the figure $RWW'R' = \square F'W$. Take away the common part $R'W'WFM$, and we have

$$\triangle RMF = \triangle R'MF'.$$

And $R'F'$ is parallel to RF;

$$\therefore RM = MR'.$$

Proposition 21.

[I. 47, 48.]

In a hyperbola, an ellipse, or a circle, the line joining any point to the centre bisects the chords parallel to the tangent at the point.

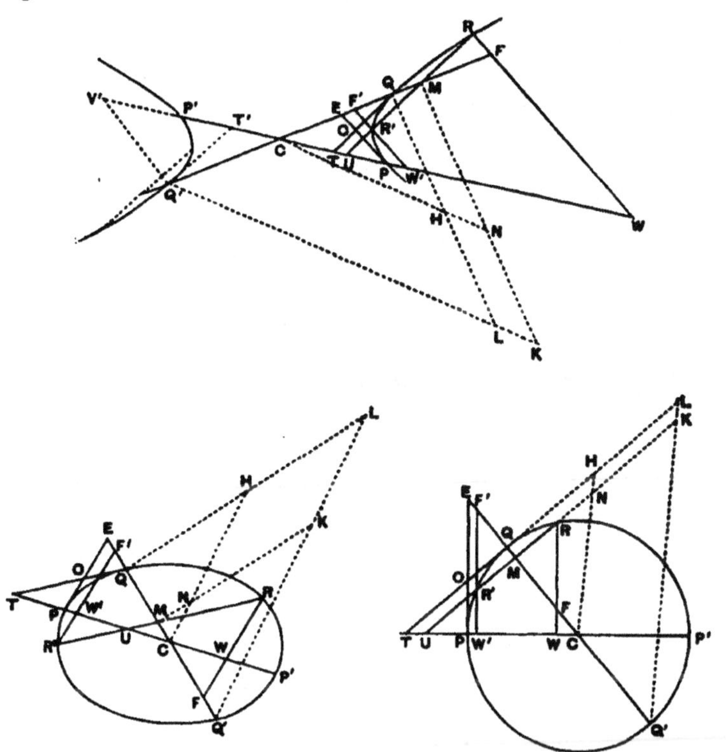

If QT be the given tangent and RR' any parallel chord, let RW, $R'W'$, PE be drawn ordinate-wise to PP', and let CQ meet them in F, F', E respectively. Further let CQ meet RR' in M.

Then we have [by Prop. 18]

$$\triangle CFW \sim \triangle CPE = \triangle RUW,$$

and
$$\triangle CF'W' \sim \triangle CPE = \triangle R'UW'.$$

Thus (1), as the figure is drawn for the *hyperbola*,

$$\Delta RUW = \text{quadrilateral } EPWF,$$

and $$\Delta R'UW' = \text{quadrilateral } EPW'F';$$

$\therefore$, by subtraction, the figure $F'W'WF =$ the figure $R'W'WR$.

Taking away the common part $R'W'WFM$, we obtain

$$\Delta FRM = \Delta F'R'M.$$

And, $\because$ FR, $F'R'$ are parallel,

$$RM = MR'.$$

(2) as the figure is drawn for the *ellipse*,

$$\Delta CPE - \Delta CFW = \Delta RUW,$$

$$\Delta CPE - \Delta CF'W' = \Delta R'UW',$$

$\therefore$, by subtraction,

$$\Delta CF'W' - \Delta CFW = \Delta RUW - \Delta R'UW',$$

or $$\Delta RUW + \Delta CFW = \Delta R'UW' + \Delta CF'W'.$$

Therefore the quadrilaterals $CFRU$, $CF'R'U$ are equal, and, taking away the common part, the triangle CUM, we have

$$\Delta FRM = \Delta F'R'M,$$

and, as before, $$RM = MR'.$$

(3) if RR' is a chord in the opposite branch of a hyperbola, and Q' the point where QC produced meets the said opposite branch, CQ will bisect RR' provided RR' is parallel to the tangent at Q'.

We have therefore to prove that the tangent at Q' is parallel to the tangent at Q, and the proposition follows immediately*.

* Eutocius supplies the proof of the parallelism of the two tangents as follows.

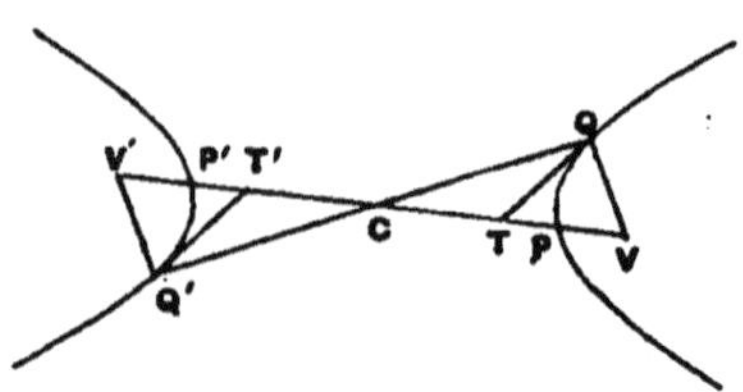

We have $$CV . CT = CP^2 \text{ [Prop. 14]},$$

and $$CV' . CT' = CP'^2;$$

$$\therefore CV . CT = CV' . CT',$$

and $$CV = CV', \because CQ = CQ' \text{ [Prop. 10]};$$

$$\therefore CT = CT'.$$

Hence, from the Δs CQT, $CQ'T'$, it follows that QT, $Q'T'$ are parallel.

Proposition 22.

[I. 49.]

Let the tangent to a parabola at P, the extremity of the original diameter, meet the tangent at any point Q in O, and the parallel through Q to the diameter in E; and let RR' be any chord parallel to the tangent at Q meeting PT in U and EQ produced in M; then, if p' be taken such that

$$OQ : QE = p' : 2QT,$$

it is to be proved that

$$RM^2 = p' . QM.$$

In the figure of Prop. 20 draw the ordinate QV.

Then we have, by hypothesis,

$$OQ : QE = p' : 2TQ.$$

Also $$QE = PV = TP.$$

Therefore the triangles EOQ, POT are equal.

Add to each the figure $QOPWF$;

$\therefore$ the quadrilateral $QTWF = \square(EW) = \triangle\, RUW$. [Prop. 17]

Subtract the quadrilateral $MUWF$;

$$\therefore \square QU = \triangle RMF,$$

and hence $$RM . MF = 2QM . QT \quad\text{..................(1).}$$

But $$RM : MF = OQ : QE = p' : 2QT,$$

or $$RM^2 : RM . MF = p' . QM : 2QM . QT.$$

Therefore, from (1), $$RM^2 = p' . QM.$$

Proposition 23.

[I. 50.]

If in a hyperbola, an ellipse, or a circle, the tangents at P, Q meet in O, and the tangent at P meet the line joining Q to the centre in E; if also a length $QL\,(=p')$ be taken such that

$$OQ : QE = QL : 2TQ$$

and erected perpendicular to QC; if further $Q'L$ be joined (where Q' is on QC produced and $CQ = CQ'$), and MK be drawn parallel to QL to meet $Q'L$ in K (where M is the point of concourse of CQ and RR', a chord parallel to the tangent at Q): then it is to be proved that

$$RM^2 = QM \cdot MK.$$

In the figures of Prop. 21 draw CHN parallel to $Q'L$, meeting QL in H and MK in N, and let RW be an ordinate to PP', meeting CQ in F.

Then, since $CQ = CQ'$, $QH = HL$.

Also $$OQ : QE = QL : 2QT$$
$$= QH : QT;$$
$$\therefore RM : MF = QH : QT \text{ ....(A).}$$

Now

$$\triangle RUW = \triangle CFW \sim \triangle CPE = \triangle CFW \sim \triangle CQT^{*};$$

$\therefore$ in the figures as drawn

(1) for the *hyperbola*,

$$\triangle RUW = QTWF,$$

$\therefore$, subtracting $MUWF$, we have

$$\triangle RMF = QTUM.$$

(2) for the *ellipse* and *circle*,

$$\triangle RUW = \triangle CQT - \triangle CFW;$$

$\therefore \triangle CQT =$ quadrilateral $RUCF$; and, subtracting $\triangle MUC$, we have

$$\triangle RMF = QTUM.$$

$$\therefore RM \cdot MF = QM(QT + MU) \text{ ....(B).}$$

* It will be observed that Apollonius here assumes the equality of the two triangles CPE, CQT, though it is not until Prop. 53 [III. 1] that this equality is actually proved. But Eutocius gives another proof of Prop. 18 which, he says, appears in some copies, and which begins by proving these two triangles to be equal by exactly the same method as is used in our text of the later proof. If then the alternative proof is genuine, we have an explanation of the assumption here. If not, we should be tempted to suppose that Apollonius quoted the property as an obvious *limiting case* of Prop. 18 [I. 43, 44] where R coincides with Q; but this would be contrary to the usual practice of Greek geometers who, no doubt for the purpose of securing greater stringency, preferred to give separate proofs of the limiting cases, though the parallelism of the respective proofs suggests that they were not unaware of the connexion between the general theorem and its limiting cases. Compare Prop. 81 [V. 2], where Apollonius proves separately the case where P coincides with B, though we have for the sake of brevity only mentioned it as a limiting case.

Now $QT : MU = CQ : CM = QH : MN$,

$$\therefore QH + MN : QT + MU = QH : QT$$
$$= RM : MF \text{ [from (A)]};$$
$$\therefore QM(QH + MN) : QM(QT + MU) = RM^2 : RM \, . \, MF;$$

$\therefore$ [by (B)] $RM^2 = QM(QH + MN)$

$$= QM \, . \, MK.$$

The same is true for the opposite branch of the hyperbola. The tangent at Q' is parallel to QT, and $P'E'$ to PE.

[Prop. 21, *Note.*]

$$\therefore O'Q' : Q'E' = OQ : QE = p' : 2QT = p' : 2Q'T',$$

whence the proposition follows.

It results from the propositions just proved that in a *parabola* all straight lines drawn parallel to the original diameter are diameters, and in the *hyperbola* and *ellipse* all straight lines drawn through the centre are diameters; also that the conics can each be referred indifferently to any diameter and the tangent at its extremity as axes.

Proposition 24. (Problem.)

[I. 52, 53.]

Given a straight line in a fixed plane and terminating in a fixed point, and another straight line of a certain length, to find a parabola in the plane such that the first straight line is a diameter, the second straight line is the corresponding parameter, and the ordinates are inclined to the diameter at a given angle.

First, let the given angle be a *right angle*, so that the given straight line is to be the *axis*.

Let AB be the given straight line terminating at A, p_a the given length.

Produce BA to C so that $AC > \frac{p_a}{4}$, and let S be a mean proportional between AC and p_a. (Thus $p_a : AC = S^2 : AC^2$, and $AC > \frac{1}{4}p_a$, whence $AC^2 > \frac{S^2}{4}$, or $2AC > S$, so that it is possible to describe an isosceles triangle having two sides equal to AC and the third equal to S.)

Let AOC be an isosceles triangle in a plane perpendicular to the given plane and such that $AO = AC$, $OC = S$.

Complete the parallelogram $ACOE$, and about AE as diameter, in a plane perpendicular to that of the triangle AOC, describe a circle, and let a cone be drawn with O as

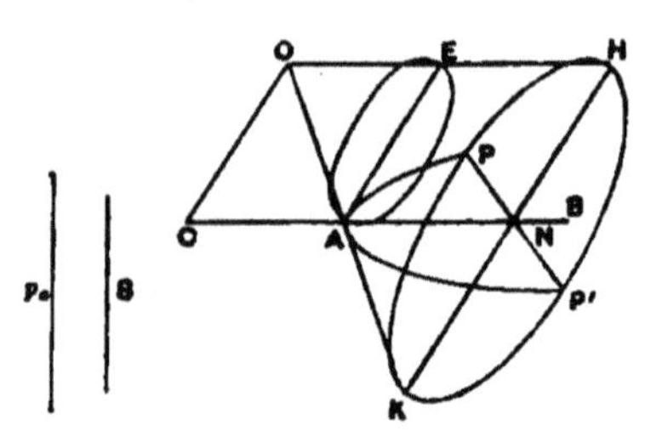

apex and the said circle as base. Then the cone is a right cone because $OE = AC = OA$.

Produce OE, OA to H, K, and draw HK parallel to AE, and let the cone be cut by a plane through HK parallel to the base of the cone. This plane will produce a circular section, and will intersect the original plane in a line PP', cutting AB at right angles in N.

Now $p_a : AE = AE : AO$, since $AE = OC = S$, $AO = AC$;

$$\therefore p_a : AO = AE^2 : AO^2$$
$$= AE^2 : AO.OE.$$

Hence PAP' is a parabola in which p_a is the parameter of the ordinates to AB. [Prop. 1]

Secondly, let the given angle not be right. Let the line which is to be the diameter be PM, let p be the length of the parameter, and let MP be produced to F so that $PF = \frac{1}{4}p$. Make the angle FPT equal to the given angle and draw FT perpendicular to TP. Draw TN parallel to PM, and PN perpendicular to TN; bisect TN in A and draw LAE through A perpendicular to FP meeting PT in O; and let

$$NA.AL = PN^2.$$

Now with axis AN and parameter AL describe a parabola, as in the first case.

This will pass through P since $PN^2 = LA.AN$. Also PT will be a tangent to it since $AT = AN$. And PM is parallel to AN. Therefore PM is a diameter of the parabola bisecting chords parallel to the tangent PT, which are therefore inclined to the diameter at the given angle.

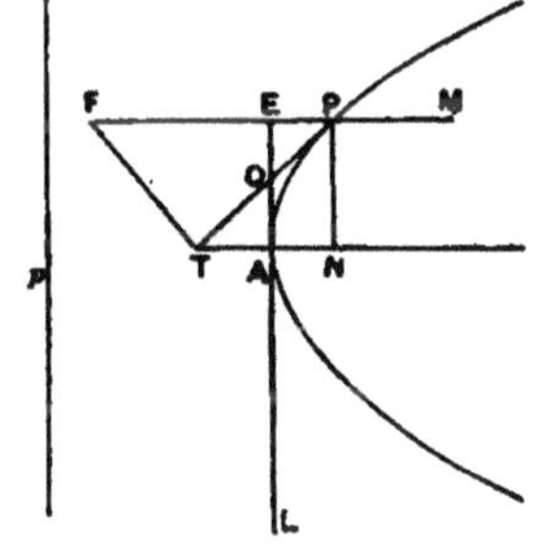

Again the triangles FTP, OEP are similar;

$$\therefore OP : PE = FP : PT,$$
$$= p : 2PT,$$

by hypothesis.

Therefore p is the parameter of the parabola corresponding to the diameter PM. [Prop. 22]

Proposition 25. (Problem.)

[I. 54, 55, 59.]

Given a straight line AA' in a plane, and also another straight line of a certain length; to find a hyperbola in the plane such that the first straight line is a diameter of it and the second equal to the corresponding parameter, while the ordinates to the diameter make with it a given angle.

First, let the given angle be a *right angle.*

Let AA', p_a be the given straight lines, and let a circle be drawn through A, A' in a plane perpendicular to the given plane and such that, if C be the middle point of AA' and DF the diameter perpendicular to AA',

$$DC : CF \not> AA' : p_a.$$

Then, if $DC : CF = AA' : p_a$, we should use the point F for our construction, but, if not, suppose

$$DC : CG = AA' : p_a \text{ (}CG\text{ being less than }CF\text{).}$$

Draw GO parallel to AA', meeting the circle in O. Join AO,

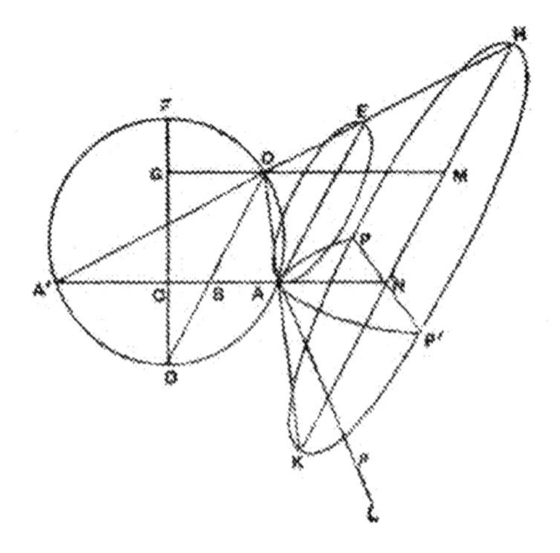

$A'O$, DO. Draw AE parallel to DO meeting $A'O$ produced in E. Let DO meet AA' in B.

Then $\angle OEA = \angle A'OD = \angle AOD = \angle OAE$;

$$\therefore OA = OE.$$

Let a cone be described with O for apex and for base the circle whose diameter is AE and whose plane is perpendicular to that of the circle AOD. The cone will therefore be right, since $OA = OE$.

Produce OE, OA to H, K and draw HK parallel to AE. Draw a plane through HK perpendicular to the plane of the circle AOD. This plane will be parallel to the base of the cone, and the resulting section will be a circle cutting the original plane in PP' at right angles to $A'A$ produced. Let GO meet HK in M.

Then, because NA meets HO produced beyond O, the curve PAP' is a hyperbola.

And
$$\begin{aligned} AA' : p_a &= DC : CG \\ &= DB : BO \\ &= DB . BO : BO^2 \\ &= A'B . BA : BO^2. \end{aligned}$$

But
$$\left.\begin{aligned} A'B : BO &= OM : MH \\ BA : BO &= OM : MK \end{aligned}\right\} \text{by similar triangles.}$$

$$\therefore A'B . BA : BO^2 = OM^2 : HM . MK.$$

Hence
$$AA' : p_a = OM^2 : HM . MK.$$

Therefore p_a is the parameter of the hyperbola PAP' corresponding to the diameter AA'. [Prop. 2]

Secondly, let the given angle not be a right angle. Let PP', p be the given straight lines, CPT the given angle, and C the middle point of PP'. On CP describe a semicircle, and let N be such a point on it that, if NH is drawn parallel to PT to meet CP produced in H,

$$NH^2 : CH . HP = p : PP'^*.$$

* This construction is assumed by Apollonius without any explanation; but we may infer that it was arrived at by a method similar to that adopted for

Join NC meeting PT in T, and take A on CN such that $CA^2 = CT . CN$. Join PN and produce it to K so that

$$PN^2 = AN . NK.$$

Produce AC to A' so that $AC = CA'$, join $A'K$, and draw $EOAM$ through A parallel to PN meeting CP, PT, $A'K$ in E, O, M respectively.

With AA' as axis, and AM as the corresponding parameter, describe a hyperbola as in the first part of the proposition. This will pass through P because $PN^2 = AN . NK$.

a similar case in Prop. 52. In fact the solution given by Eutocius represents sufficiently closely Apollonius' probable procedure.

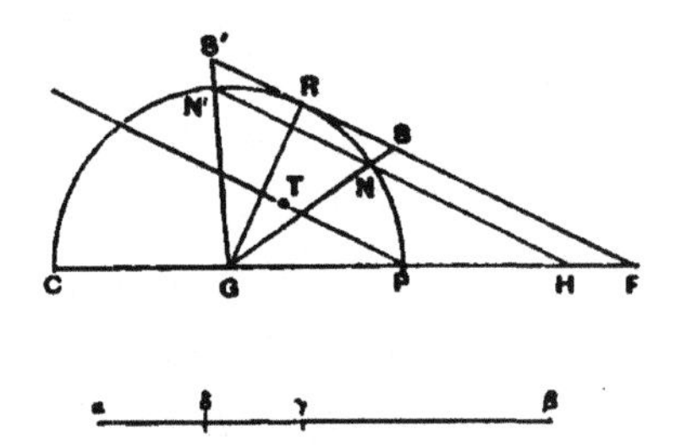

α δ γ β

If HN produced be supposed to meet the curve again in N', then

$$N'H . HN = CH . HP;$$

$$\therefore NH^2 : CH . HP = NH : N'H.$$

Thus we have to draw HNN' at a given inclination to PC and so that

$$N'H : NH = PP' : p.$$

Take any straight line $\alpha\beta$ and divide it at γ so that

$$\alpha\beta : \beta\gamma = PP' : p.$$

Bisect $\alpha\gamma$ in δ. Then draw from G, the centre of the semicircle, GR at right angles to PT which is in the given direction, and let GR meet the circumference in R. Then RF drawn parallel to PT will be the tangent at R. Suppose RF meets CP produced in F. Divide FR at S so that $FS : SR = \beta\gamma : \gamma\delta$, and produce FR to S' so that $RS' = RS$.

Join GS, GS', meeting the semicircle in N, N', and join $N'N$ and produce it to meet CF in H. Then NH is the straight line which it was required to find.

The proof is obvious.

Also PT will be the tangent at P because $CT.CN = CA^2$. Therefore CP will be a diameter of the hyperbola bisecting

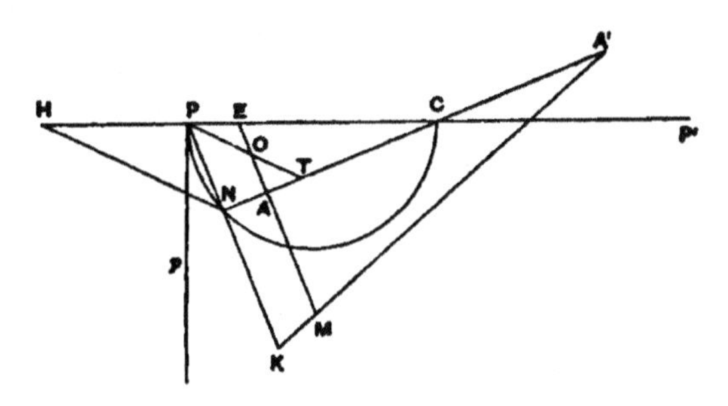

chords parallel to PT and therefore inclined to the diameter at the given angle.

Again we have

$$p : 2CP = NH^2 : CH.HP, \text{ by construction,}$$

and $$2CP : 2PT = CH : NH$$
$$= CH.HP : NH.HP;$$
$$\therefore p : 2PT = NH^2 : NH.HP$$
$$= NH : HP$$
$$= OP : PE, \text{ by similar triangles;}$$

therefore p is the parameter corresponding to the diameter PP'. [Prop. 23]

The opposite branch of the hyperbola with vertex A' can be described in the same way.

Proposition 26. (Problem.)
[I. 60.]

Given two straight lines bisecting one another at any angle, to describe two hyperbolas each with two branches such that the straight lines are conjugate diameters of both hyperbolas.

Let PP', DD' be the two straight lines bisecting each other at C.

From P draw PL perpendicular to PP' and of such a length that $PP'.PL = DD'^2$; then, as in Prop. 25, describe a double hyperbola with diameter PP' and parameter PL and such that the ordinates in it to PP' are parallel to DD'.

Then PP', DD' are conjugate diameters of the hyperbola so constructed.

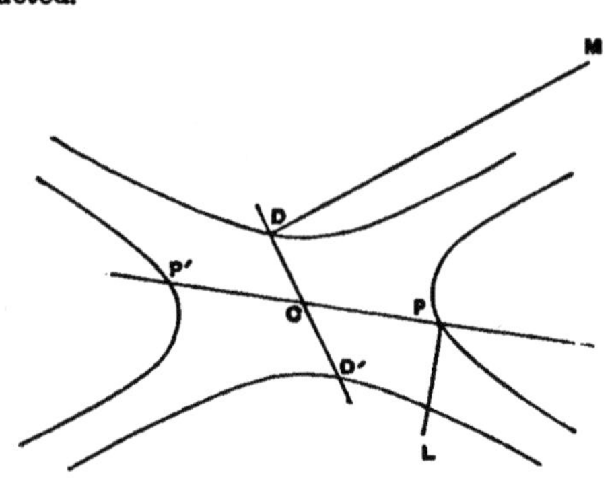

Again, draw DM perpendicular to DD' of such a length that $DM.DD' = PP'^2$; and, with DD' as diameter, and DM as the corresponding parameter, describe a double hyperbola such that the ordinates in it to DD' are parallel to PP'.

Then DD', PP' are conjugate diameters to this hyperbola, and DD' is the transverse, while PP' is the secondary diameter.

The two hyperbolas so constructed are called **conjugate** hyperbolas, and that last drawn is the hyperbola **conjugate** to the first.

Proposition 27. (Problem.)

[I. 56, 57, 58.]

Given a diameter of an ellipse, the corresponding parameter, and the angle of inclination between the diameter and its ordinates: to find the ellipse.

First, let the angle of inclination be a *right angle*, and let the diameter be greater than its parameter.

Let AA' be the diameter and AL, a straight line of length p_a perpendicular to it, the parameter.

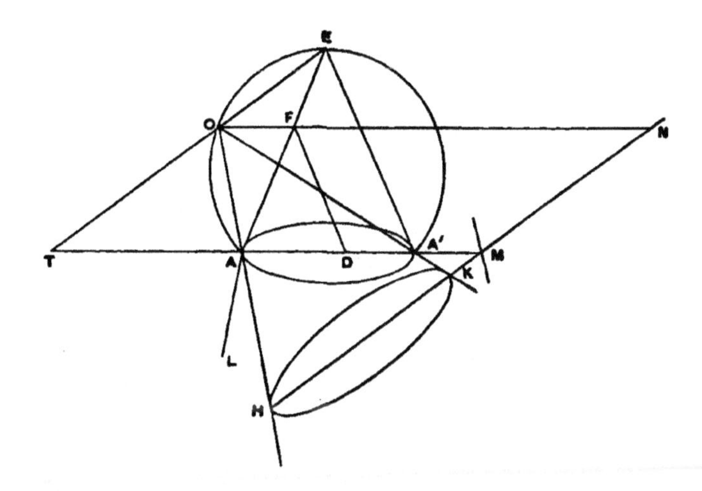

In a plane at right angles to the plane containing the diameter and parameter describe a segment of a circle on AA' as base.

Take AD on AA' equal to AL. Draw AE, $A'E$ to meet at E, the middle point of the segment. Draw DF parallel to $A'E$ meeting AE in F, and OFN parallel to AA' meeting the circumference in O. Join EO and produce it to meet $A'A$ produced in T. Through any point H on OA produced draw $HKMN$ parallel to OE meeting OA', AA', OF in K, M, N respectively.

Now

$$\angle TOA = \angle OEA + \angle OAE = \angle AA'O + \angle OA'E = \angle AA'E$$
$$= \angle EAA' = \angle EOA',$$

and HK is parallel to OE,

whence $$\angle OHK = \angle OKH,$$

and $$OH = OK.$$

With O as vertex, and as base the circle drawn with diameter HK and in a plane perpendicular to that of the triangle OHK, let a cone be described. This cone will be a right cone because $OH = OK$.

Consider the section of this cone by the plane containing AA', AL. This will be an ellipse.

And $$p_a : AA' = AD : AA'$$
$$= AF : AE$$
$$= TO : TE$$
$$= TO^2 : TO . TE$$
$$= TO^2 : TA . TA'.$$

Now $$TO : TA = HN : NO,$$

and $$TO : TA' = NK : NO, \text{ by similar triangles,}$$

$$\therefore TO^2 : TA . TA' = HN . NK : NO^2,$$

so that $$p_a : AA' = HN . NK : NO^2,$$

or p_a is the parameter of the ordinates to AA'. [Prop. 3]

Secondly, if the angle of inclination of the ordinates be still a right angle, but the given diameter less than the parameter, let them be BB', BM respectively.

Let C be the middle point of BB', and through it draw AA', perpendicular to BB' and bisected at C, such that

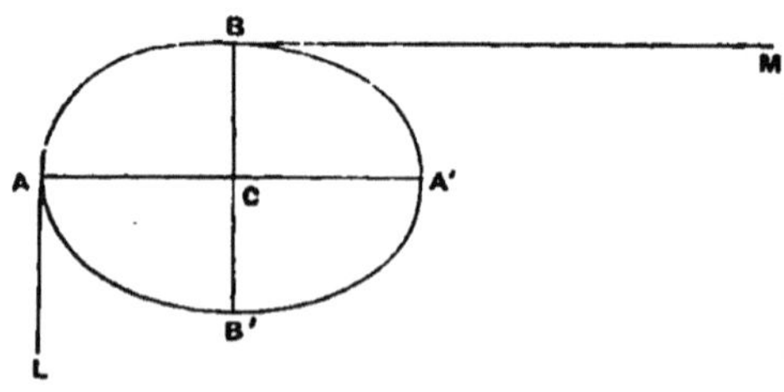

$$AA'^2 = BB' . BM;$$

and draw AL, parallel to BB', such that

$$BM : BB' = AA' : AL;$$

thus $AA' > AL$.

Now with AA' as diameter and AL as the corresponding parameter describe an ellipse in which the ordinates to AA' are perpendicular to it, as above.

This will be the ellipse required, for

(1) it passes through B, B' because

$$\begin{aligned} AL : AA' &= BB' : BM \\ &= BB'^2 : AA'^2 \\ &= BC^2 : AC . CA', \end{aligned}$$

(2)
$$\begin{aligned} BM : BB' &= AC^2 : BC^2 \\ &= AC^2 : BC . CB', \end{aligned}$$

so that BM is the parameter corresponding to BB'.

Thirdly, let the given angle not be a right angle but

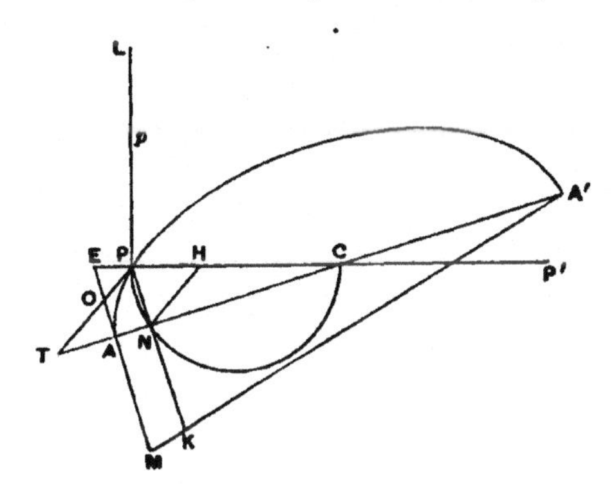

equal to the angle CPT, where C is the middle point of the given diameter PP'; and let PL be the parameter corresponding to PP'.

Take a point N, on the semicircle which has CP for its diameter, such that NH drawn parallel to PT satisfies the relation

$$NH^2 : CH . HP = PL : PP'^*.$$

* This construction like that in Prop. 25 is assumed without explanation. If NH be supposed to meet the other semicircle on CP as diameter in N', the

Join CN and produce it to meet PT in T. Take A, on CT, such that $CT.CN = CA^2$, and produce AC to A' so that $AC = CA'$. Join PN and produce it to K so that $AN.NK = PN^2$. Join $A'K$. Draw EAM through A perpendicular to CA (and therefore parallel to NK) meeting CP produced in E, PT in O, and $A'K$ produced in M.

Then with axis AA' and parameter AM describe an ellipse as in the first part of this proposition. This will be the ellipse required.

For (1) it will pass through P $\because$ $PN^2 = AN.NK$. For a similar reason, it will pass through P' $\because$ $CP' = CP$ and $CA' = CA$.

(2) PT will be the tangent at P $\because$ $CT.CN = CA^2$.

(3) We have $p : 2CP = NH^2 : CH.HP$,

and

$$2CP : 2PT = CH : HN$$
$$= CH.HP : NH.HP;$$

$\therefore$ *ex aequali*

$$p : 2PT = NH^2 : NH.HP$$
$$= NH : HP$$
$$= OP : PE.$$

Therefore p is the parameter corresponding to PP'.

[Prop. 23]

problem here reduces to drawing NHN' in a given direction (parallel to PT) so that

$$N'H : NH = PP' : p,$$

and the construction can be effected by the method shown in the note to Prop. 25 *mutatis mutandis*.

Proposition 28.

[II. 1, 15, 17, 21.]

(1) *If PP' be a diameter of a hyperbola and p the corresponding parameter, and if on the tangent at P there be set off on each side equal lengths PL, PL', such that*

$$PL^2 = PL'^2 = \tfrac{1}{4}p \, . \, PP' \, [= CD^2],$$

then CL, CL' produced will not meet the curve in any finite point and are accordingly defined as **asymptotes**.

(2) *The opposite branches have the same asymptotes.*

(3) *Conjugate hyperbolas have their asymptotes common.*

(1) If possible, let CL meet the hyperbola in Q. Draw the

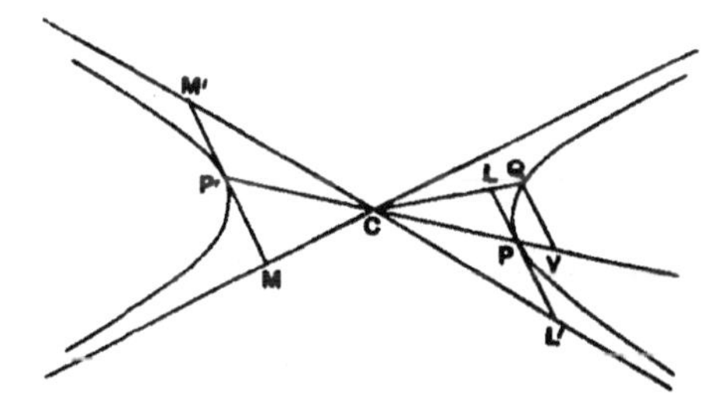

ordinate QV, which will accordingly be parallel to LL'.

Now $$p : PP' = p \, . \, PP' : PP'^2$$
$$= PL^2 : CP^2$$
$$= QV^2 : CV^2.$$

But $$p : PP' = QV^2 : PV . P'V.$$

$$\therefore PV . P'V = CV^2,$$

i.e. $CV^2 - CP^2 = CV^2$, which is absurd.

Therefore CL does not meet the hyperbola in any finite point, and the same is true for CL'.

In other words, CL, CL' are **asymptotes**.

(2) If the tangent at P' (on the opposite branch) be taken, and $P'M$, $P'M'$ measured on it such that $P'M^2 = P'M'^2 = CD^2$, it follows in like manner that CM, CM' are asymptotes.

Now MM', LL' are parallel, $PL = P'M$, and PCP' is a straight line. Therefore LCM is a straight line.

So also is $L'CM'$, and therefore the opposite branches have the same asymptotes.

(3) Let PP', DD' be conjugate diameters of two conjugate

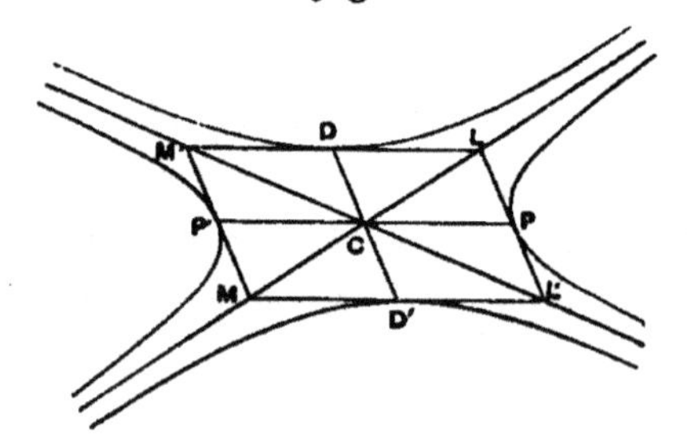

hyperbolas. Draw the tangents at P, P', D, D'. Then [Prop. 11 and Prop. 26] the tangents form a parallelogram, and the diagonals of it, LM, $L'M'$, pass through the centre.

Also $$PL = PL' = P'M = P'M' = CD.$$

Therefore LM, $L'M'$ are the asymptotes of the hyperbola in which PP' is a transverse diameter and DD' its conjugate.

Similarly $DL = DM' = D'L' = D'M = CP$, and LM, $L'M'$ are the asymptotes of the hyperbola in which DD' is a transverse diameter and PP' its conjugate, i.e. the conjugate hyperbola.

Therefore conjugate hyperbolas have their asymptotes common.

Proposition 29.

[II. 2.]

No straight line through C within the angle between the asymptotes can itself be an asymptote.

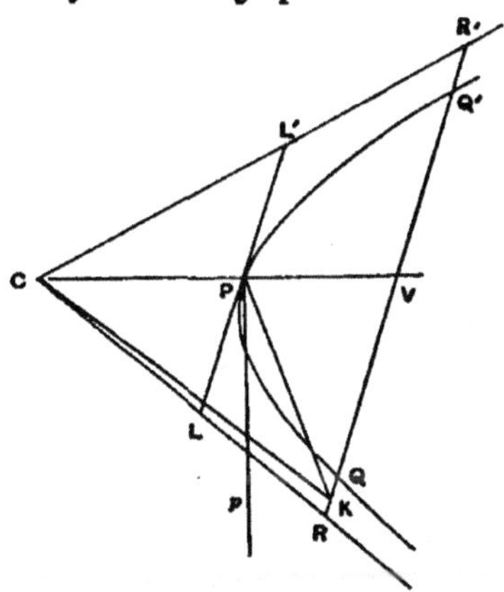

If possible, let CK be an asymptote. Draw from P the straight line PK parallel to CL and meeting CK in K, and through K draw $RKQR'$ parallel to LL', the tangent at P.

Then, since $PL = PL'$, and RR', LL' are parallel, $RV = R'V$, where V is the point of intersection of RR' and CP.

And, since $PKRL$ is a parallelogram, $PK = LR$, $PL = KR$.

Therefore $QR > PL$. Also $R'Q > PL'$;

$$\therefore\ RQ \cdot QR' > PL \cdot PL', \text{ or } PL^2 \quad \ldots\ldots\ldots\ldots(1).$$

Again $\quad RV^2 : CV^2 = PL^2 : CP^2 = p : PP'$, [Prop. 28]

and $\quad p : PP' = QV^2 : PV \cdot P'V$ [Prop. 8]

$$= QV^2 : CV^2 - CP^2;$$

thus $\quad RV^2 : CV^2 = QV^2 : CV^2 - CP^2$

$$= RV^2 - QV^2 : CP^2;$$

$$\therefore\ PL^2 : CP^2 = RV^2 - QV^2 : CP^2,$$

whence $\quad PL^2 = RV^2 - QV^2 = RQ \cdot QR'$,

which is impossible, by (1) above.

Therefore CK cannot be an asymptote.

Proposition 30.

[II. 3.]

If a straight line touch a hyperbola at P, it will meet the asymptotes in two points L, L'; LL' will be bisected at P, and $PL^2 = \frac{1}{4}p \cdot PP'$ $[= CD^2]$.

[This proposition is the converse of Prop. 28 (1) above.]

For, if the tangent at P does not meet the asymptotes in the points L, L' described, take on the tangent lengths PK, PK' each equal to CD.

Then CK, CK' are asymptotes; which is impossible.

Therefore the points K, K' must be identical with the points L, L' on the asymptotes.

Proposition 31. (Problem.)

[II. 4.]

Given the asymptotes and a point P on a hyperbola, to find the curve.

Let CL, CL' be the asymptotes, and P the point. Produce PC to P' so that $CP = CP'$. Draw PK parallel to CL' meeting CL in K, and let CL be made equal to twice CK. Join LP and produce it to L'.

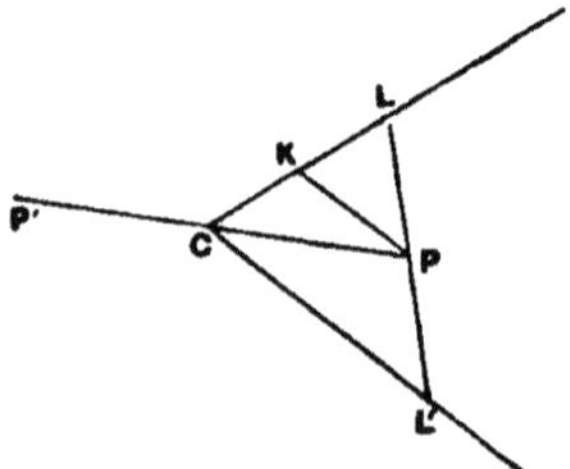

Take a length p such that $LL'^2 = p \cdot PP'$, and with diameter PP' and parameter p describe a hyperbola such that the ordinates to PP' are parallel to LL'. [Prop. 25]

Proposition 32.

[II. 8, 10.]

If Qq be any chord, it will, if produced both ways, meet the asymptotes in two points as R, r, and

(1) *QR, qr will be equal,*

(2) $RQ.Qr = \frac{1}{4}p.PP' [= CD^2]$.

Take V the middle point of Qq, and join CV meeting the curve in P. Then CV is a diameter and the tangent at P is parallel to Qq. [Prop. 11]

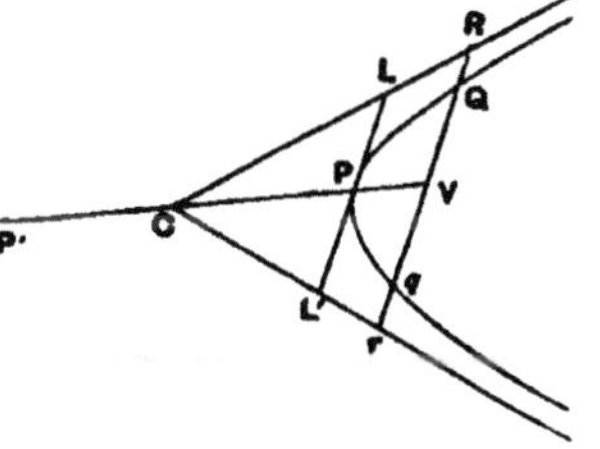

Also the tangent at P meets the asymptotes (in L, L'). Therefore Qq parallel to it also meets the asymptotes.

Then (1), since Qq is parallel to LL', and $LP = PL'$, it follows that $RV = Vr$.

But $\qquad QV = Vq$;

therefore, subtracting, $\qquad QR = qr$.

(2) We have

$$p : PP' = PL^2 : CP^2$$
$$= RV^2 : CV^2,$$

and $\qquad p : PP' = QV^2 : CV^2 - CP^2$; [Prop. 8]

$$\therefore PL^2 : CP^2 = p : PP' = RV^2 - QV^2 : CP^2$$
$$= RQ.Qr : CP^2;$$

thus $\qquad RQ.Qr = PL^2$

$$= \tfrac{1}{4}p.PP' = CD^2.$$

Similarly $\qquad rq.qR = CD^2$.

Proposition 33.

[II. 11, 16.]

If Q, Q' are on opposite branches, and QQ' meet the asymptotes in K, K', and if CP be the semidiameter parallel to QQ', then

$$(1)\quad KQ.QK' = CP^2,$$

$$(2)\quad QK = Q'K'.$$

Draw the tangent at P meeting the asymptotes in L, L', and

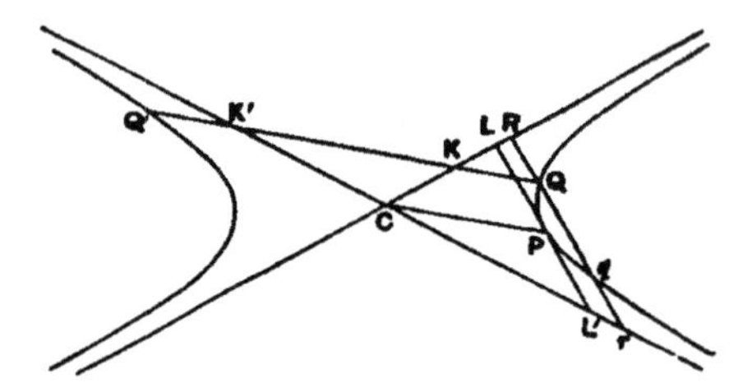

let the chord Qq parallel to LL' meet the asymptotes in R, r. Qq is therefore a double ordinate to CP.

Then we have

$$PL^2 : CP^2 = (PL : CP).(PL' : CP)$$
$$= (RQ : KQ).(Qr : QK')$$
$$= RQ.Qr : KQ.QK'.$$

But $\quad PL^2 = RQ.Qr;$ [Prop. 32]

$$\therefore KQ.QK' = CP^2.$$

Similarly $\quad K'Q'.Q'K = CP^2.$

(2) $\quad KQ.QK' = CP^2 = K'Q'.Q'K;$

$$\therefore KQ.(KQ + KK') = K'Q'(K'Q' + KK'),$$

whence it follows that $\quad KQ = K'Q'.$

Proposition 34.

[II. 12.]

If Q, q be any two points on a hyperbola, and parallel straight lines QH, qh be drawn to meet one asymptote at any angle, and QK, qk (also parallel to one another) meet the other asymptote at any angle, then

$$HQ \cdot QK = hq \cdot qk.$$

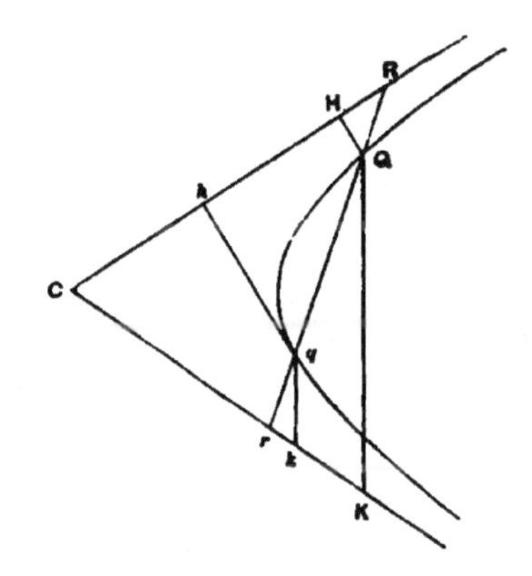

Let Qq meet the asymptotes in R, r.

We have $$RQ \cdot Qr = Rq \cdot qr; \quad \text{[Prop. 32]}$$

$$\therefore RQ : Rq = qr : Qr.$$

But $$RQ : Rq = HQ : hq,$$

and $$qr : Qr = qk : QK;$$

$$\therefore HQ : hq = qk : QK,$$

or $$HQ \cdot QK = hq \cdot qk.$$

Proposition 35.

[II. 13.]

If in the space between the asymptotes and the hyperbola a straight line be drawn parallel to one of the asymptotes, it will meet the hyperbola in one point only.

Let E be a point on one asymptote, and let EF be drawn parallel to the other.

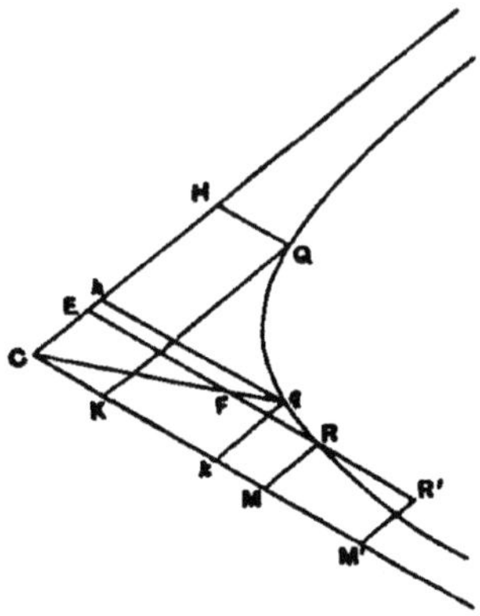

Then EF produced shall meet the curve in one point only.

For, if possible, let it not meet the curve.

Take Q, any point on the curve, and draw QH, QK each parallel to one asymptote and meeting the other; let a point F be taken on EF such that

$$HQ \cdot QK = CE \cdot EF.$$

Join CF and produce it to meet the curve in q; and draw qh, qk respectively parallel to QH, QK.

Then $$hq \cdot qk = HQ \cdot QK, \qquad \text{[Prop. 34]}$$

and $$HQ \cdot QK = CE \cdot EF, \text{ by hypothesis,}$$

$$\therefore hq \cdot qk = CE \cdot EF:$$

which is impossible, $\because hq > EF$, and $qk > CE$.

Therefore EF will meet the hyperbola in one point, as R.

Again, EF will not meet the hyperbola in any other point.

For, if possible, let EF meet it in R' as well as R, and let RM, $R'M'$ be drawn parallel to QK.

Then $$ER \cdot RM = ER' \cdot R'M': \qquad \text{[Prop. 34]}$$

which is impossible, $\because ER' > ER$.

Therefore EF does not meet the hyperbola in a second point R'.

Proposition 36.

[II. 14.]

The asymptotes and the hyperbola, as they pass on to infinity, approach continually nearer, and will come within a distance less than any assignable length.

Let S be the given length.

Draw two parallel chords Qq, $Q'q'$ meeting the asymptotes in R, r and R', r'. Join Cq and produce it to meet $Q'q'$ in F.

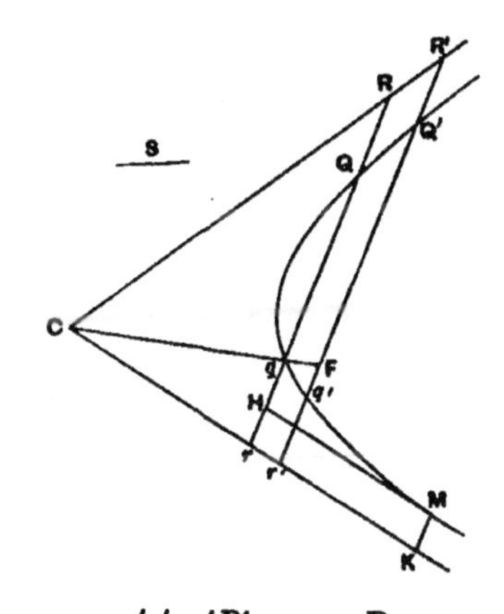

Then $$r'q' \cdot q'R' = rq \cdot qR,$$

and $$q'R' > qR;$$

$$\therefore q'r' < qr,$$

and hence, as successive chords are taken more and more distant from the centre, qr becomes smaller and smaller.

Take now on rq a length rH less than S, and draw HM parallel to the asymptote Cr.

HM will then meet the curve [Prop. 35] in a point M. And, if MK be drawn parallel to Qq to meet Cr in K,

$$MK = rH,$$

whence $$MK < S.$$

Proposition 37.

[II. 19.]

Any tangent to the conjugate hyperbola will meet both branches of the original hyperbola and be bisected at the point of contact.

(1) Let a tangent be drawn to either branch of the conjugate hyperbola at a point D.

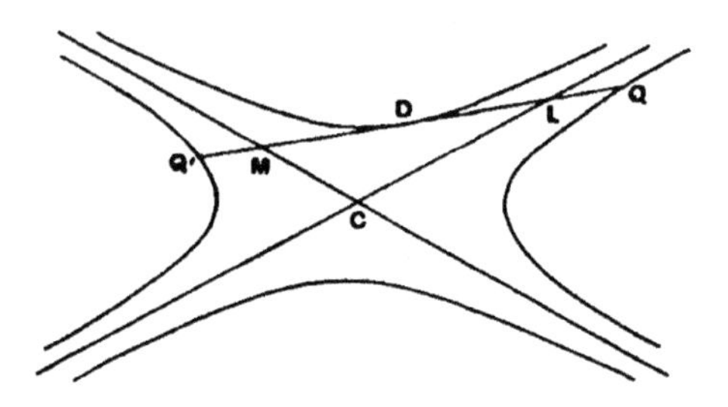

This tangent will then meet the asymptotes [Prop. 30], and will therefore meet both branches of the original hyperbola.

(2) Let the tangent meet the asymptotes in L, M and the original hyperbola in Q, Q'.

Then [Prop. 30] $\quad DL = DM$.

Also [Prop. 33] $\quad LQ = MQ'$;

whence, by addition, $\quad DQ = DQ'$.

Proposition 38.

[II. 23.]

If a chord Qq in one branch of a hyperbola meet the asymptotes in R, r and the conjugate hyperbola in Q', q', then

$$Q'Q \,.\, Qq' = 2CD^2.$$

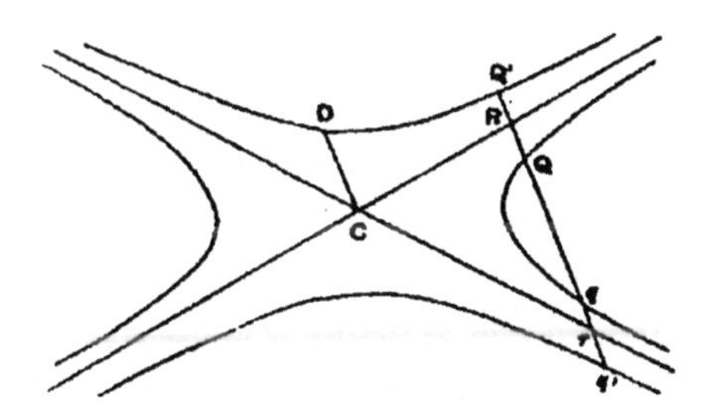

Let CD be the parallel semi-diameter. Then we have [Props. 32, 33]

$$RQ \,.\, Qr = CD^2,$$

$$RQ' \,.\, Q'r = CD^2;$$

$$\begin{aligned}\therefore\ 2CD^2 &= RQ \,.\, Qr + RQ' \,.\, Q'r \\ &= (RQ + RQ')\, Qr + RQ' \,.\, QQ' \\ &= QQ' \,.\, (Qr + RQ') \\ &= QQ'\, (Qr + rq') \\ &= QQ' \,.\, Qq'.\end{aligned}$$

Proposition 39.

[II. 20.]

If Q be any point on a hyperbola, and CE be drawn from the centre parallel to the tangent at Q to meet the conjugate hyperbola in E, then

(1) *the tangent at E will be parallel to CQ, and*

(2) *CQ, CE will be conjugate diameters.*

Let PP', DD' be the conjugate diameters of reference, and let QV be the ordinate from Q to PP', and EW the ordinate

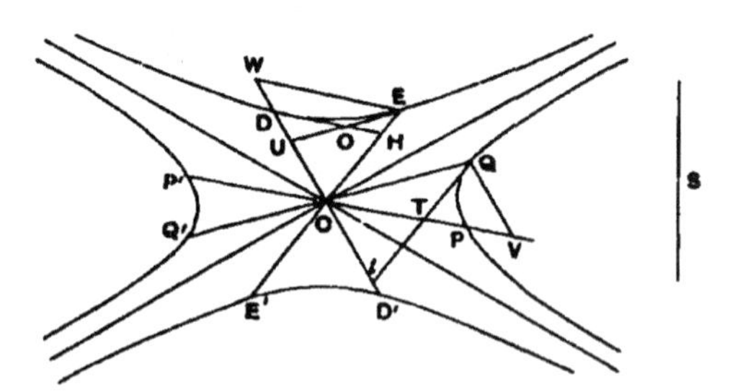

from E to DD'. Let the tangent at Q meet PP', DD' in T, t respectively, let the tangent at E meet DD' in U, and let the tangent at D meet EU, CE in O, H respectively.

Let p, p' be the parameters corresponding to PP', DD' in the two hyperbolas, and we have

(1) $$PP' : p = p' : DD',$$

$$[\because p \,.\, PP' = DD'^2, \quad p' \,.\, DD' = PP'^2]$$

and $$PP' : p = CV . VT : QV^2,$$
$$p' : DD' = EW^2 : CW . WU. \quad \text{[Prop. 14]}$$
$$\therefore CV . VT : QV^2 = EW^2 : CW . WU.$$

But, by similar triangles,
$$VT : QV = EW : CW.$$

Therefore, by division,
$$CV : QV = EW : WU.$$

And in the triangles CVQ, EWU the angles at V, W are equal.

Therefore the triangles are similar, and
$$\angle QCV = \angle UEW.$$

But $\angle VCE = \angle CEW$, since EW, CV are parallel.

Therefore, by subtraction, $\angle QCE = \angle CEU$.

Hence EU is parallel to CQ.

(2) Take a straight line S of such length that
$$HE : EO = EU : S,$$
so that S is equal to half the parameter of the ordinates to the diameter EE' of the conjugate hyperbola. [Prop. 23]

Also $$Ct . QV = CD^2, \text{ (since } QV = Cv),$$
or $$Ct : QV = Ct^2 : CD^2.$$

Now $$Ct : QV = tT : TQ = \Delta tCT : \Delta CQT,$$
and $$Ct^2 : CD^2 = \Delta tCT : \Delta CDH = \Delta tCT : \Delta CEU$$
[as in Prop. 23].

It follows that $$\Delta CQT = \Delta CEU.$$

And $$\angle CQT = \angle CEU.$$
$$\therefore CQ . QT = CE . EU \text{ ....(A).}$$

But $$S : EU = OE : EH$$
$$= CQ : QT.$$
$$\therefore S . CE : CE . EU = CQ^2 : CQ . QT.$$

Hence, by (A), $$S . CE = CQ^2.$$
$$\therefore 2S . EE' = QQ'^2,$$
where $2S$ is the parameter corresponding to EE'.

And similarly it may be proved that EE'^2 is equal to the rectangle contained by QQ' and the corresponding parameter.

Therefore QQ', EE' are conjugate diameters. [Prop. 26]

Proposition 40.

[II. 37.]

If Q, Q' are any points on opposite branches, and v the middle point of the chord QQ', then Cv is the "secondary" diameter corresponding to the transverse diameter drawn parallel to QQ'.

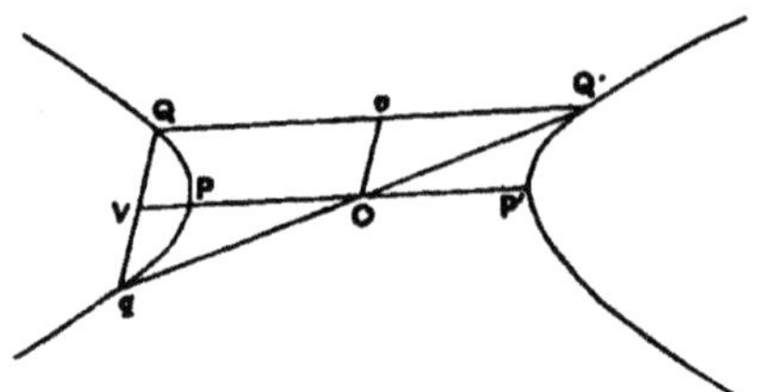

Join $Q'C$ and produce it to meet the hyperbola in q. Join Qq, and draw the diameter PP' parallel to QQ'.

Then we have

$$CQ' = Cq, \quad \text{and} \quad Q'v = Qv.$$

Therefore Qq is parallel to Cv.

Let the diameter PP' produced meet Qq in V.

Now $\qquad QV = Cv = Vq, \quad$ because $CQ' = Cq$.

Therefore the ordinates to PP' are parallel to Qq, and therefore to Cv.

Hence PP', Cv are conjugate diameters. [Prop. 6]

Proposition 41.

[II. 29, 30, 38.]

If two tangents TQ, TQ' be drawn to a conic, and V be the middle point of the chord of contact QQ', then TV is a diameter.

For, if not, let VE be a diameter, meeting TQ' in E. Join EQ meeting the curve in R, and draw the chord RR' parallel to QQ' meeting EV, EQ' respectively in K, H.

Then, since RH is parallel to QQ', and $QV = Q'V$, $RK = KH$.

Also, since RR' is a chord parallel to QQ' bisected by the diameter EV, $RK = KR'$.

Therefore $KR' = KH$: which is impossible.

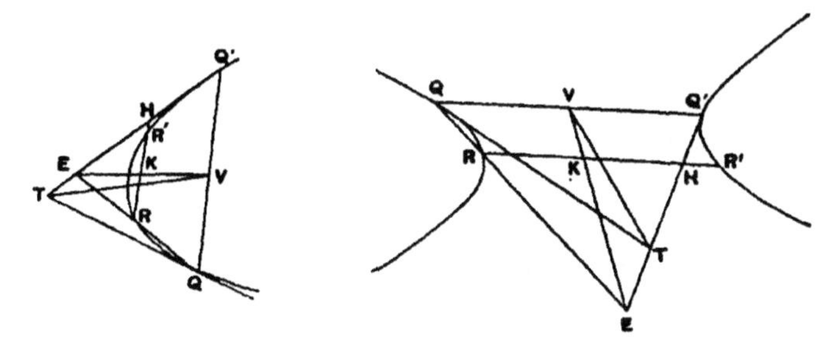

Therefore EV is not a diameter, and it may be proved in like manner that no other straight line through V is a diameter except TV.

Conversely, *the diameter of the conic drawn through T, the point of intersection of the tangents, will bisect the chord of contact QQ'.*

[This is separately proved by Apollonius by means of an easy *reductio ad absurdum.*]

Proposition 42.

[II. 40.]

If tQ, tQ' be tangents to opposite branches of a hyperbola, and a chord RR' be drawn through t parallel to QQ', then the lines joining R, R' to v, the middle point of QQ', will be tangents at R, R'.

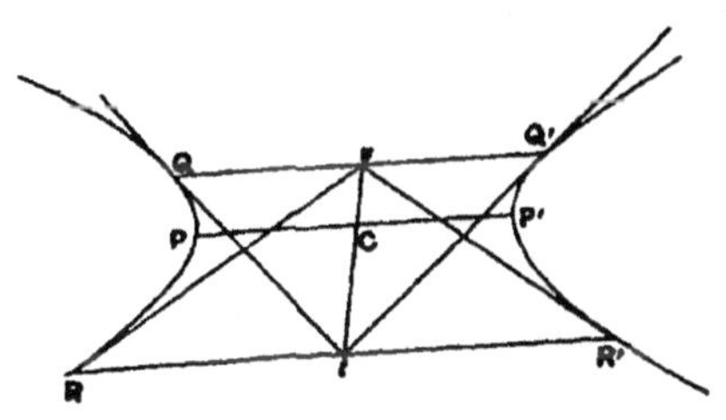

Join vt. vt is then the diameter conjugate to the transverse diameter drawn parallel to QQ', i.e. to PP'.

But, since the tangent Qt meets the secondary diameter in t,

$$Cv \,.\, Ct = \tfrac{1}{4}p \,.\, PP' \,[= CD^2]. \qquad \text{[Prop. 15]}$$

Therefore the relation between v and t is reciprocal, and the tangents at R, R' intersect in v.

Proposition 43.

[II. 26, 41, 42.]

In a conic, or a circle, or in conjugate hyperbolas, if two chords not passing through the centre intersect, they do not bisect each other.

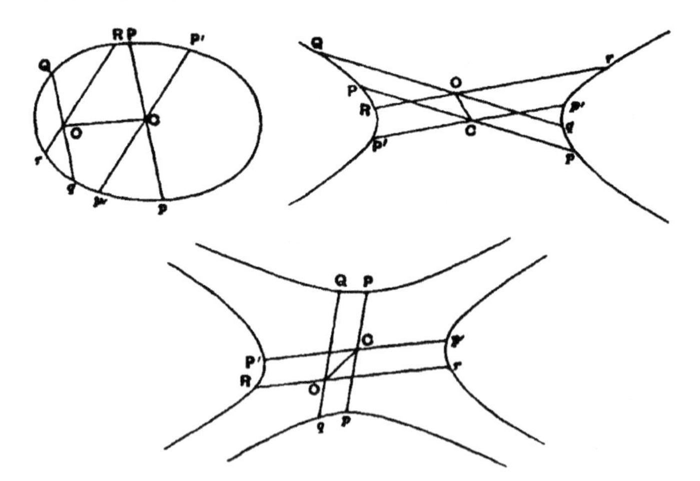

Let Qq, Rr, two chords not passing through the centre, meet in O. Join CO, and draw the diameters Pp, $P'p'$ respectively parallel to Qq, Rr.

Then Qq, Rr shall not bisect one another. For, if possible, let each be bisected in O.

Then, since Qq is bisected in O and Pp is a diameter parallel to it, CO, Pp are conjugate diameters.

Therefore the tangent at P is parallel to CO.

Similarly it can be proved that the tangent at P' is parallel to CO.

Therefore the tangents at P, P' are parallel: which is impossible, since PP' is not a diameter.

Therefore Qq, Rr do not bisect one another.

Proposition 44. (Problem.)
[II. 44, 45.]

To find a diameter of a conic, and the centre of a central conic.

(1) Draw two parallel chords and join their middle points. The joining line will then be a diameter.

(2) Draw any two diameters; and these will meet in, and so determine, the centre.

Proposition 45. (Problem.)
[II. 46, 47.]

To find the axis of a parabola, and the axes of a central conic.

(1) In the case of the *parabola*, let PD be any diameter. Draw any chord QQ' perpendicular to PD, and let N be its middle point. Then AN drawn through N parallel to PD will be the axis.

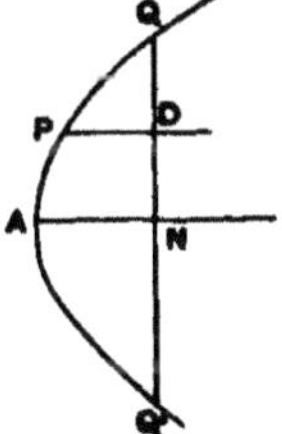

For, being parallel to PD, AN is a diameter, and, inasmuch as it bisects QQ' at right angles, it is the axis.

And there is only *one* axis because there is only one diameter which bisects QQ'.

(2) In the case of a *central conic*, take any point P on the conic, and with centre C and radius CP describe a circle cutting the conic in P, P', Q', Q.

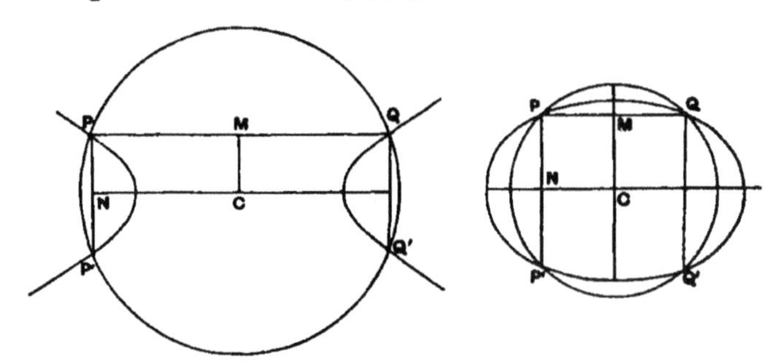

Let PP', PQ be two common chords not passing through the centre, and let N, M be their middle points respectively. Join CN, CM.

Then CN, CM will both be axes because they are both diameters bisecting chords at right angles. They are also conjugate because each bisects chords parallel to the other.

Proposition 46.

[II. 48.]

No central conic has more than two axes.

If possible, let there be another axis CL. Through P' draw $P'L$ perpendicular to CL, and produce $P'L$ to meet the

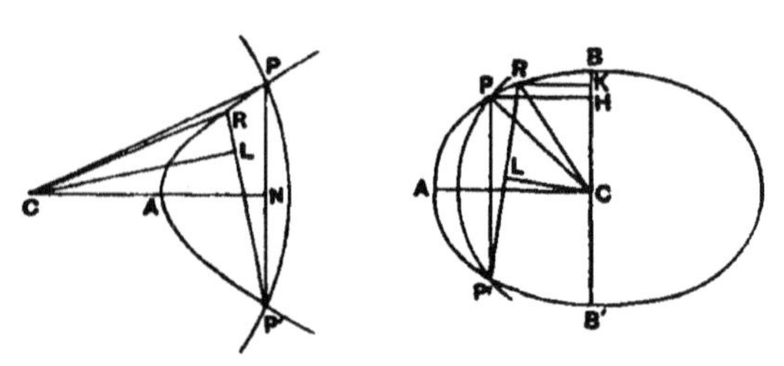

curve again in R. Join CP, CR.

Then, since CL is an axis, $P'L = LR$; therefore also

$$CP = CP' = CR.$$

Now in the case of the *hyperbola* it is clear that the circle PP' cannot meet the same branch of the hyperbola in any other points than P, P'. Therefore the assumption is absurd.

In the *ellipse* draw RK, PH perpendicular to the (minor) axis which is parallel to PP'.

Then, since it was proved that $CP = CR$,

$$CP^2 = CR^2,$$

or $$CH^2 + HP^2 = CK^2 + KR^2.$$

$$\therefore CK^2 - CH^2 = HP^2 - KR^2 \quad\text{..................(1).}$$

Now $BK.KB' + CK^2 = CB^2$,

and $BH.HB' + CH^2 = CB^2$.

$$\therefore CK^2 - CH^2 = BH.HB' - BK.KB'.$$

Hence $HP^2 - KR^2 = BH.HB' - BK.KB'$, from (1).

But, since PH, RK are ordinates to BB',

$$PH^2 : BH.HB' = RK^2 : BK.KB',$$

and the difference between the antecedents has been proved equal to the difference between the consequents.

$$\therefore PH^2 = BH.HB',$$

and $$RK^2 = BK.KB'.$$

$\therefore P$, R are points on a circle with diameter BB': which is absurd.

Hence CL is not an axis.

Proposition 47. (Problem.)

[II. 49.]

To draw a tangent to a parabola through any point on or outside the curve.

(1) Let the point be P on the curve. Draw PN perpendicular to the axis, and produce NA to T so that $AT = AN$. Join PT.

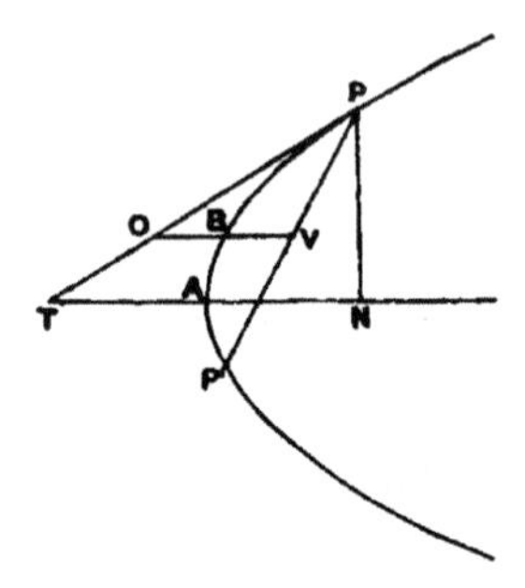

Then, since $AT = AN$, PT is the tangent at P. [Prop. 12]

In the particular case where P coincides with A, the vertex, the perpendicular to the axis through A is the tangent.

(2) Let the given point be any external point O. Draw the diameter OBV meeting the curve at B, and make BV equal to OB. Then draw through V the straight line VP parallel to the tangent at B [drawn as in (1)] meeting the curve in P. Join OP.

OP is the tangent required, because PV, being parallel to the tangent at B, is an ordinate to BV, and $OB = BV$.

[Prop. 12]

[This construction obviously gives the *two* tangents through O.]

Proposition 48. (Problem.)

[II. 49.]

To draw a tangent to a hyperbola through any point on or outside the curve.

There are here four cases.

Case I. Let the point be Q on the curve.

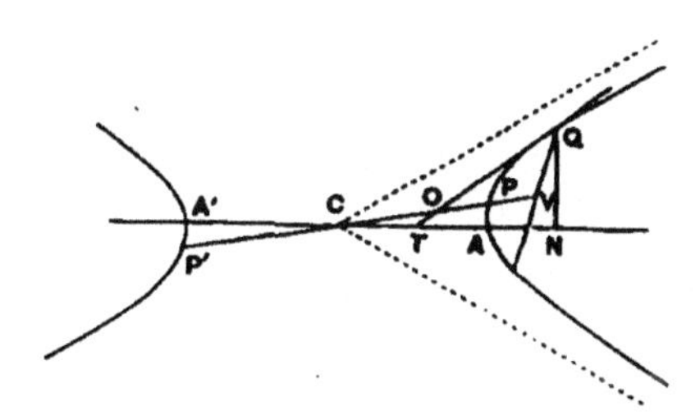

Draw QN perpendicular to the axis AA' produced, and take on AA' a point T such that $A'T : AT = A'N : AN$. Join TQ.

Then TQ is the tangent at Q. [Prop. 13]

In the particular case where Q coincides with A or A' the perpendicular to the axis at that point is the tangent.

Case II. Let the point be any point O within the angle contained by the asymptotes.

Join CO and produce it both ways to meet the hyperbola in P, P'. Take a point V on CP produced such that

$$P'V : PV = OP' : OP,$$

and through V draw VQ parallel to the tangent at P [drawn as in Case I.] meeting the curve in Q. Join OQ.

Then, since QV is parallel to the tangent at P, QV is an ordinate to the diameter $P'P$, and moreover

$$P'V : PV = OP' : OP.$$

Therefore OQ is the tangent at Q. [Prop. 13]

[This construction obviously gives the *two* tangents through O.]

Case III. Let the point O be on one of the asymptotes. Bisect CO at H, and through H draw HP parallel to the other

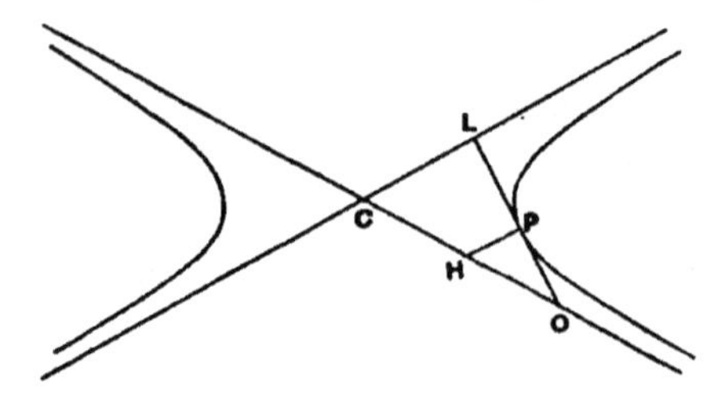

asymptote meeting the curve in P. Join OP and produce it to meet the other asymptote in L.

Then, by parallels,

$$OP : PL = OH : HC,$$

whence $OP = PL$.

Therefore OL touches the hyperbola at P. [Props. 28, 30]

Case IV. Let the point O lie within one of the *exterior* angles made by the asymptotes.

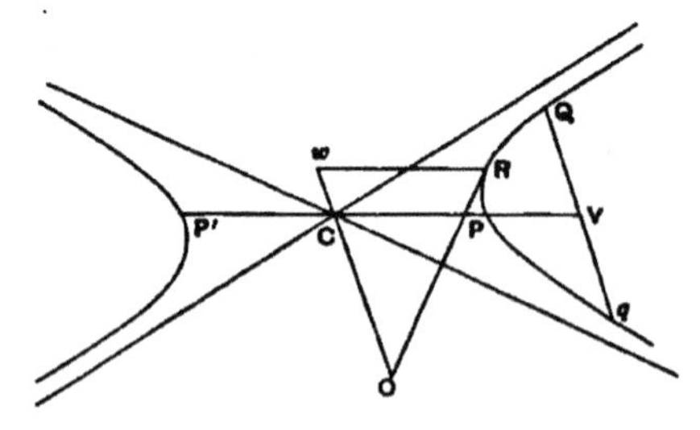

Join CO. Take any chord Qq parallel to CO, and let V be its middle point. Draw through V the diameter PP'. Then PP' is the diameter conjugate to CO. Now take on OC produced a point w such that $CO \cdot Cw = \frac{1}{2}p \cdot PP'$ [$= CD^2$], and draw through w the straight line wR parallel to PP' meeting the curve in R. Join OR. Then, since Rw is parallel to CP and Cw conjugate to it, while $CO \cdot Cw = CD^2$, OR is the tangent at R. [Prop. 15]

Proposition 49. (Problem.)
[II. 49.]

To draw a tangent to an ellipse through any point on or outside the curve.

There are here two cases, (1) where the point is on the curve, and (2) where it is outside the curve; and the con-

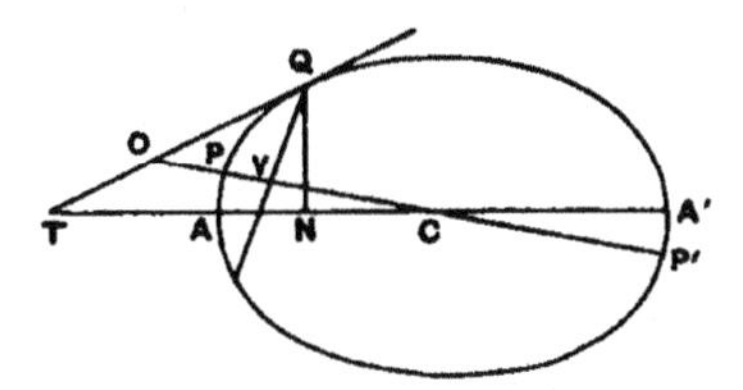

structions correspond, *mutatis mutandis*, with Cases I. and II. of the hyperbola just given, depending as before on Prop. 13.

When the point is external to the ellipse, the construction gives, as before, the *two* tangents through the point.

Proposition 50. (Problem.)
[II. 50.]

To draw a tangent to a given conic making with the axis an angle equal to a given acute angle.

I. Let the conic be a *parabola*, and let DEF be the given acute angle. Draw DF perpendicular to EF, bisect EF at H, and join DH.

Now let AN be the axis of the parabola, and make the angle NAP equal to the angle DHF. Let AP meet the curve in P. Draw PN perpendicular to AN. Produce NA to T so that $AN = AT$, and join PT.

Then PT is a tangent, and we have to prove that

$$\angle PTN = \angle DEF.$$

Since $\angle DHF = \angle PAN,$

$$HF : FD = AN : NP.$$

$$\therefore 2HF : FD = 2AN : NP,$$

or $EF : FD = TN : NP.$

$$\therefore \angle PTN = \angle DEF.$$

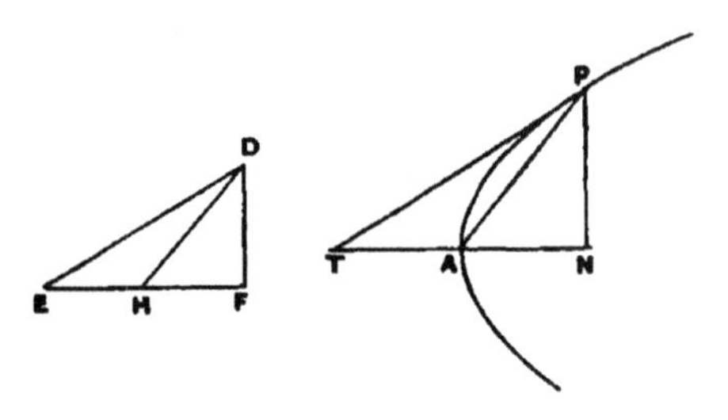

II. Let the conic be a central conic.

Then, for the *hyperbola*, it is a necessary condition of the possibility of the solution that the given angle DEF must be

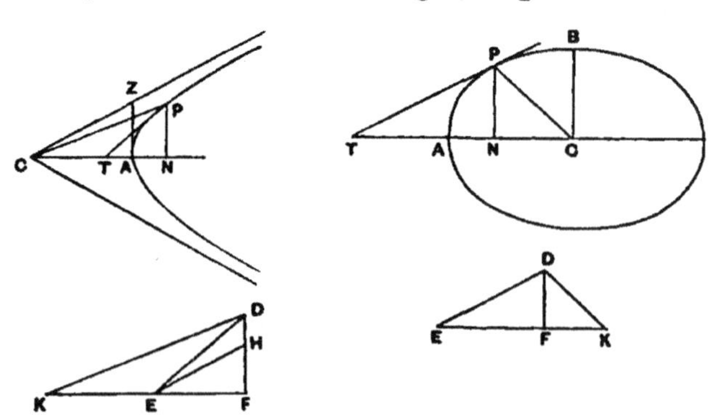

greater than the angle between the axis and an asymptote, or half that between the asymptotes. If DEF be the given angle and DF be at right angles to EF, let H be so taken on DF that $\angle HEF = \angle ACZ$, or half the angle between the asymptotes. Let AZ be the tangent at A meeting an asymptote in Z.

We have then $CA^2 : AZ^2$ (or $CA^2 : CB^2$) $= EF^2 : FH^2$.

$$\therefore CA^2 : CB^2 > EF^2 : FD^2.$$

Take a point K on FE produced such that

$$CA^2 : CB^2 = KF . FE : FD^2.$$

Thus $$KF^2 : FD^2 > CA^2 : AZ^2.$$

Therefore, if DK be joined, the angle DKF is less than the angle ACZ. Hence, if the angle ACP be made equal to the angle DKF, CP must meet the hyperbola in some point P.

In the case of the *ellipse* K has to be taken on EF produced so that $CA^2 : CB^2 = KF . FE : FD^2$, and from this point the constructions are similar for both the central conics, the angle ACP being made equal to the angle DKF in each case.

Draw now PN perpendicular to the axis, and draw the tangent PT. [Props. 48, 49]

Then $$PN^2 : CN . NT = CB^2 : CA^2$$ [Prop. 14]

$$= FD^2 : KF . FE, \text{ from above;}$$

and, by similar triangles,

$$CN^2 : PN^2 = KF^2 : FD^2.$$

$$\therefore CN^2 : CN . NT = KF^2 : KF . FE,$$

or $$CN : NT = KF : FE.$$

And $$PN : CN = DF : KF.$$

$$\therefore PN : NT = DF : FE.$$

Hence $$\angle PTN = \angle DEF.$$

Proposition 51.

[II. 52.]

In an ellipse, if the tangent at any point P meet the major axis in T, the angle CPT is not greater than the angle ABA' (where B is one extremity of the minor axis).

Taking P in the quadrant AB, join PC.

Then PC is either parallel to BA' or not parallel to it.

First, let PC be parallel to BA'. Then, by parallels, CP bisects AB. Therefore the tangent at P is parallel to AB, and $\angle CPT = \angle A'BA$.

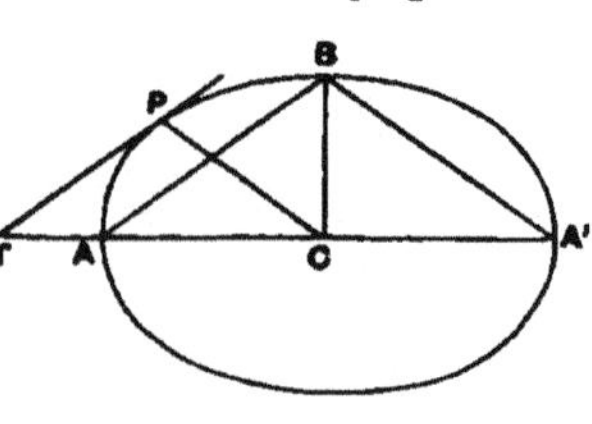

Secondly, suppose that PC is not parallel to BA', and we have in that case, drawing PN perpendicular to the axis,

$$\angle PCN \neq \angle BA'C, \text{ or } \angle BAC.$$

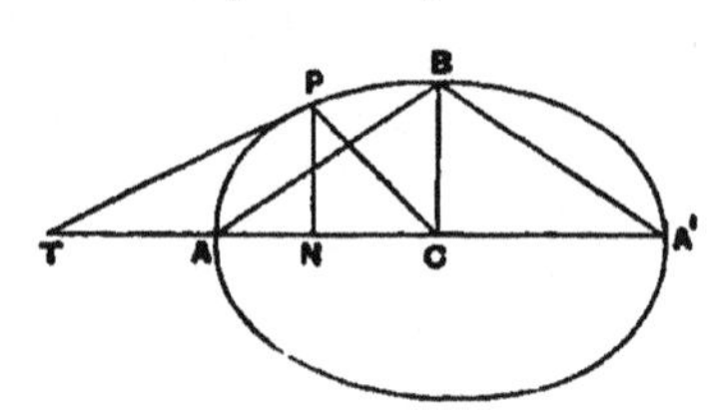

$$\therefore\ PN^2 : CN^2 \neq BC^2 : AC^2,$$

whence $$PN^2 : CN^2 \neq PN^2 : CN . NT. \qquad \text{[Prop. 14]}$$

$$\therefore\ CN \neq NT.$$

Let FDE be a segment in a circle containing an angle FDE equal to the angle ABA', and let DG be the diameter of the circle bisecting FE at right angles in I. Divide FE in M so that

$$EM : MF = CN : NT,$$

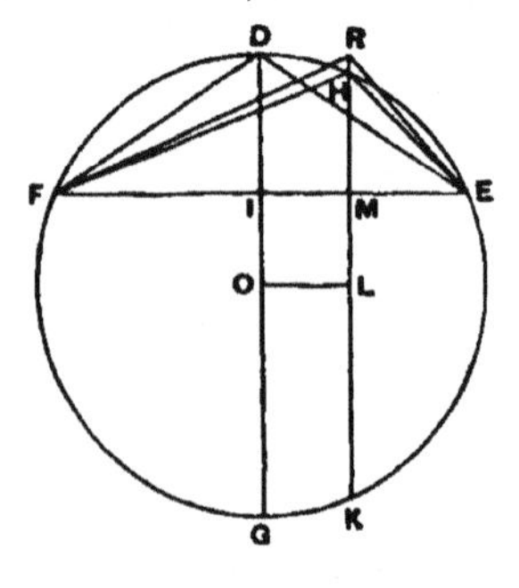

and draw through M the chord HK at right angles to EF. From O, the centre of the circle, draw OL perpendicular to HK, and join EH, HF.

The triangles DFI, BAC are then similar, and

$$FI^2 : ID^2 = CA^2 : CB^2.$$

Now $$OD : OI > LH : LM, \text{ since } OI = LM.$$

$$\therefore\ OD : DI < LH : HM$$

and, doubling the antecedents,

$$DG : DI < HK : HM,$$

whence $$GI : ID < KM : MH.$$

But $$GI : ID = FI^2 : ID^2 = CA^2 : CB^2$$
$$= CN . NT : PN^2.$$

$$\therefore\ CN . NT : PN^2 < KM : MH$$
$$< KM . MH : MH^2$$
$$< EM . MF : MH^2.$$

Let $$CN . NT : PN^2 = EM . MF : MR^2,$$

where R is some point on HK or HK produced.

It follows that $MR > MH$, and R lies on KH produced. Join ER, RF.

Now $$CN . NT : EM . MF = PN^2 : RM^2,$$

and $$CN^2 : EM^2 = CN . NT : EM . MF$$

(since $CN : NT = EM : MF$).

$$\therefore\ CN : EM = PN : RM.$$

Therefore the triangles CPN, ERM are similar.

In like manner the triangles PTN, RFM are similar.

Therefore the triangles CPT, ERF are similar,

and $$\angle CPT = \angle ERF;$$

whence it follows that

$\angle CPT$ is less than $\angle EHF$, or $\angle ABA'$.

Therefore, whether CP is parallel to BA' or not, the $\angle CPT$ is not greater than the $\angle ABA'$.

Proposition 52. (Problem.)
[II. 51, 53.]

To draw a tangent to any given conic making a given angle with the diameter through the point of contact.

I. In the case of the *parabola* the given angle must be an acute angle, and, since any diameter is parallel to the axis, the problem reduces itself to Prop. 50 (1) above.

II. In the case of a central conic, the angle CPT must be acute for the *hyperbola*, and for the *ellipse* it must not be less than a right angle, nor greater than the angle ABA', as proved in Prop. 51.

Suppose θ to be the given angle, and take first the particular case for the *ellipse* in which the angle θ is equal to the angle ABA'. In this case we have simply, as in Prop. 51, to draw CP parallel to BA' (or AB) and to draw through P a parallel to the chord AB (or $A'B$).

Next suppose θ to be any acute angle for the *hyperbola*, and for the *ellipse* any obtuse angle less than ABA'; and suppose the problem solved, the angle CPT being equal to θ.

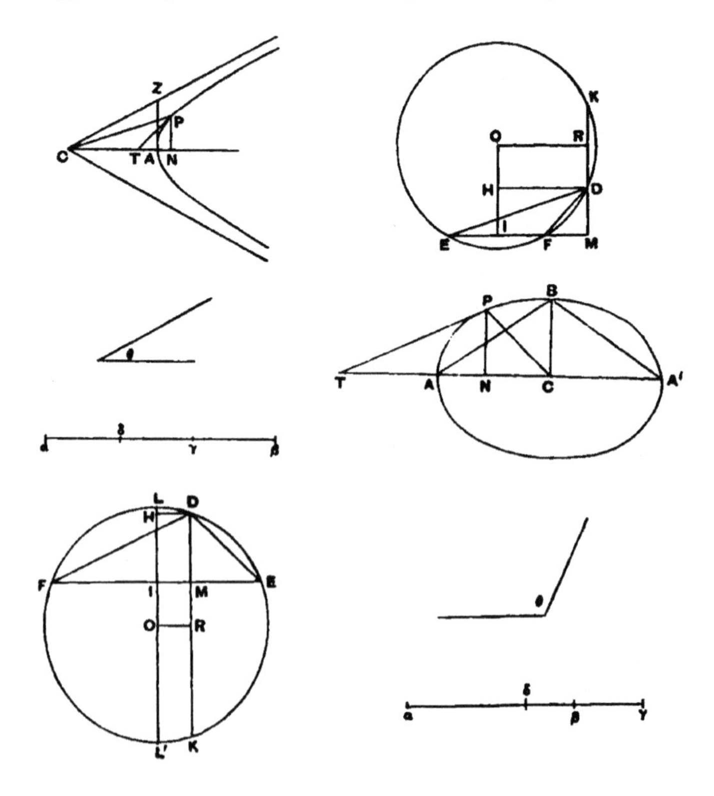

Imagine a segment of a circle taken containing an angle (EDF) equal to the angle θ. Then, if a point D on the circumference of the segment could be found such that, if DM be the perpendicular on the base EF, the ratio $EM.MF : DM^2$ is equal to the ratio $CA^2 : CB^2$, *i.e.* to the ratio $CN.NT : PN^2$, we should have

$$\angle CPT = \angle \theta = \angle EDF,$$

and $$CN.NT : PN^2 = EM.MF : DM^2,$$

and it would follow that triangles PCN, PTN are respectively similar to DEM, DFM*. Thus the angle DEM would be equal to the angle PCN.

The construction would then be as follows:

Draw CP so that the angle PCN is equal to the angle DEM, and draw the tangent at P meeting the axis AA' in T. Also let PN be perpendicular to the axis AA'.

Then $CN.NT : PN^2 = CA^2 : CB^2 = EM.MF : DM^2$,

and the triangles PCN, DEM are similar, whence it follows that the triangles PTN, DFM are similar, and therefore also the triangles CPT, EDF*.

$$\therefore \angle CPT = \angle EDF = \angle \theta.$$

It only remains to be proved for the *hyperbola* that, if the angle PCN be made equal to the angle DEM, CP must necessarily meet the curve, i.e. that the angle DEM is less than half the angle between the asymptotes. If AZ is perpendicular to the axis and meets an asymptote in Z, we have

$$EM.MF : DM^2 = CA^2 : CB^2 = CA^2 : AZ^2,$$

$$\therefore EM^2 : DM^2 > CA^2 : AZ^2,$$

and the angle DEM is less than the angle ZCA.

We have now shown that the construction reduces itself to finding the point D on the segment of the circle, such that

$$EM.MF : DM^2 = CA^2 : CB^2.$$

* These conclusions are taken for granted by Apollonius, but they are easily proved.

This is effected as follows:

Take lengths $\alpha\beta$, $\beta\gamma$ in one straight line such that

$$\alpha\beta : \beta\gamma = CA^2 : CB^2,$$

$\beta\gamma$ being measured towards α for the *hyperbola* and away from α for the *ellipse*; and let $\alpha\gamma$ be bisected in δ.

Draw OI from O, the centre of the circle, perpendicular to EF; and on OI or OI produced take a point H such that

$$OH : HI = \delta\gamma : \gamma\beta,$$

(the points O, H, I occupying positions relative to one another corresponding to the relative positions of δ, γ, β).

Draw HD parallel to EF to meet the segment in D. Let DK be the chord through D at right angles to EF and meeting it in M.

Draw OR bisecting DK at right angles.

Then $$RD : DM = OH : HI = \delta\gamma : \gamma\beta.$$

Therefore, doubling the two antecedents,

$$KD : DM = \alpha\gamma : \gamma\beta;$$

so that $$KM : DM = \alpha\beta : \beta\gamma.$$

Thus

$$KM . MD : DM^2 = EM . MF : DM^2 = \alpha\beta : \beta\gamma = CA^2 : CB^2.$$

Therefore the required point D is found.

In the particular case of the hyperbola where $CA^2 = CB^2$, *i.e.* for the *rectangular* hyperbola, we have $EM . MF = DM^2$, or DM is the tangent to the circle at D.

Note. Apollonius proves incidentally that, in the second figure applying to the case of the ellipse, H falls between I and the middle point (L) of the segment as follows:

$$\angle FLI = \tfrac{1}{2} \angle CPT, \text{ which is less than } \tfrac{1}{2} \angle ABA':$$

$$\therefore \angle FLI \text{ is less than } \angle ABC,$$

whence $$CA^2 : CB^2 > FI^2 : IL^2$$

$$> L'I : IL.$$

It follows that $$\alpha\beta : \beta\gamma > L'I : IL,$$

so that $$\alpha\gamma : \gamma\beta > L'L : IL,$$

and, halving the antecedents,

$$\delta\gamma : \gamma\beta > OL : LI,$$

so that $$\delta\beta : \beta\gamma > OI : IL.$$

Hence, if H be such a point that

$$\delta\beta : \beta\gamma = OI : IH,$$

IH is less than IL.

EXTENSIONS OF PROPOSITIONS 17—19.

Proposition 53.

[III. 1, 4, 13.]

(1) *P, Q being any two points on a conic, if the tangent at P and the diameter through Q meet in E, and the tangent at Q and the diameter through P in T, and if the tangents intersect at O, then* $\triangle OPT = \triangle OQE$.

(2) *If P be any point on a hyperbola and Q any point on the conjugate hyperbola, and if T, E have the same significance as before, then* $\triangle CPE = \triangle CQT$.

(1) Let QV be the ordinate from Q to the diameter through P.

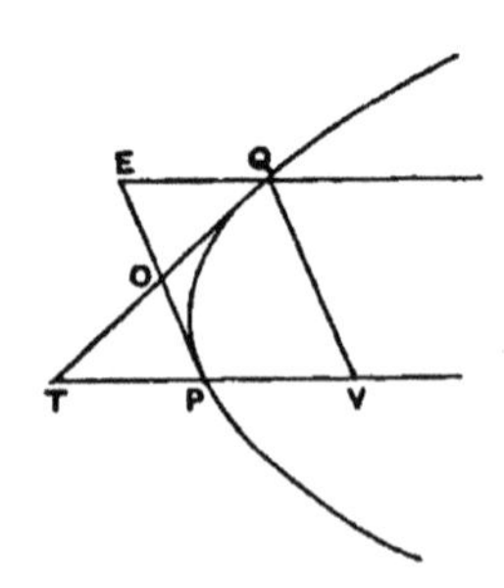

Then for the *parabola* we have

$$TP = PV, \qquad \text{[Prop. 12]}$$

so that $$TV = 2PV,$$

and $$\square EV = \triangle QTV.$$

Subtracting the common area $OPVQ$,

$$\triangle OQE = \triangle OPT.$$

For the *central conic* we have

$$CV . CT = CP^2,$$

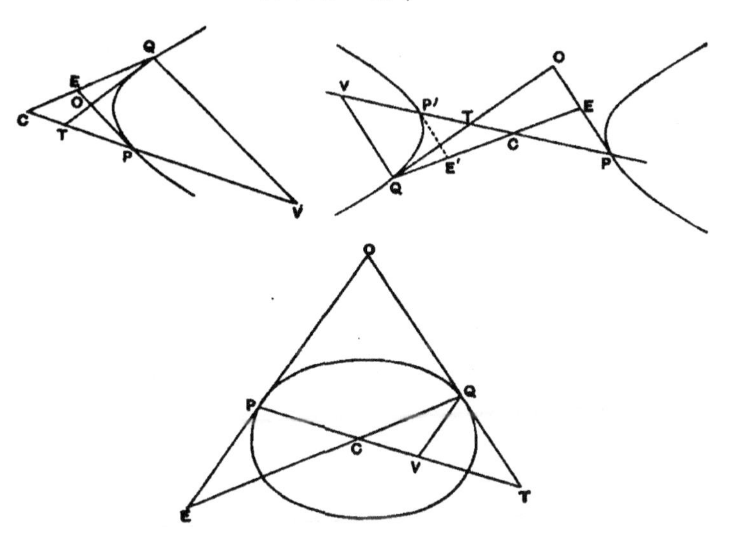

or $$CV : CT = CV^2 : CP^2;$$

$$\therefore \triangle CQV : \triangle CQT = \triangle CQV : \triangle CPE;$$

$$\therefore \triangle CQT = \triangle CPE.$$

Hence the sums or differences of the area $OTCE$ and each triangle are equal, or

$$\triangle OPT = \triangle OQE.$$

(2) In the *conjugate hyperbolas* draw CD parallel to the

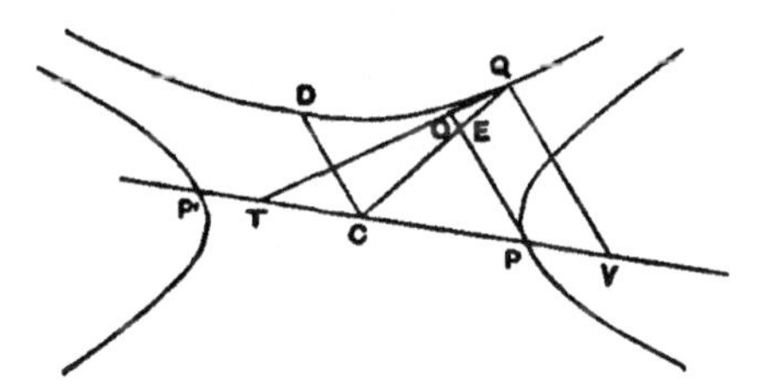

tangent at P to meet the conjugate hyperbola in D, and draw QV also parallel to PE meeting CP in V. Then CP, CD are conjugate diameters of both hyperbolas, and QV is drawn ordinate-wise to CP.

Therefore [Prop. 15]

$$CV \cdot CT = CP^2,$$

or

$$CP : CT = CV : CP$$

$$= CQ : CE;$$

$$\therefore CP \cdot CE = CQ \cdot CT.$$

And the angles PCE, QCT are supplementary;

$$\therefore \triangle CQT = \triangle CPE.$$

Proposition 54.

[III. 2, 6.]

If we keep the notation of the last proposition, and if R be

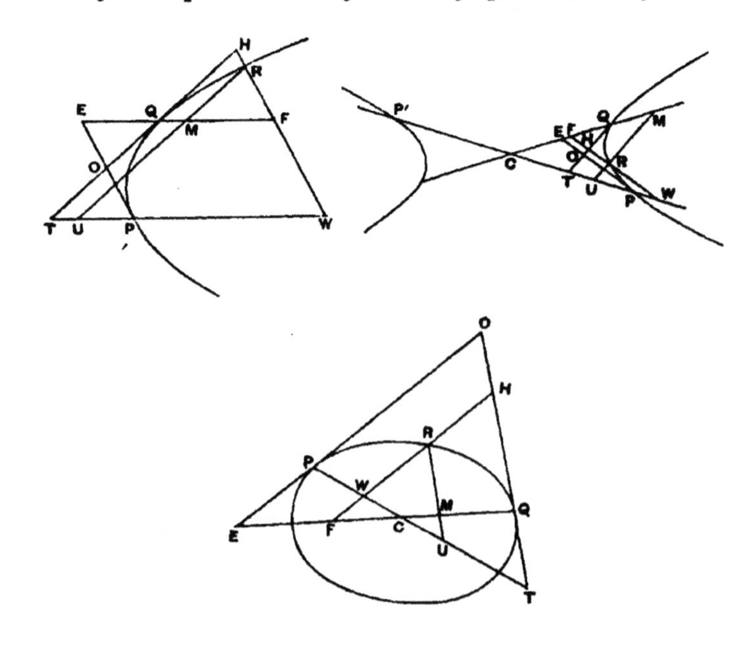

any other point on the conic, let RU be drawn parallel to QT to meet the diameter through P in U, and let a parallel through R to the tangent at P meet QT and the diameters through Q, P in H, F, W respectively. Then

$$\triangle HQF = \textit{quadrilateral } HTUR.$$

Let RU meet the diameter through Q in M. Then, as in Props. 22, 23, we have

$$\triangle RMF = \text{quadrilateral } QTUM;$$

∴, adding (or subtracting) the area HM,

$$\triangle HQF = \text{quadrilateral } HTUR.$$

Proposition 55.

[III. 3, 7, 9, 10.]

If we keep the same notation as in the last proposition and take two points R', R on the curve with points H', F', etc. corresponding to H, F, etc. and if, further, RU, $R'W'$ intersect in I and $R'U'$, RW in J, then the quadrilaterals $F'IRF$, $IUU'R'$ are equal, as also the quadrilaterals $FJR'F'$, $JU'UR$.

[N.B. It will be seen that in some cases (according to the positions of R, R') the quadrilaterals take a form like that in the margin, in which case $F'IRF$ must be taken as meaning the difference between the triangles $F'MI$, RMF.]

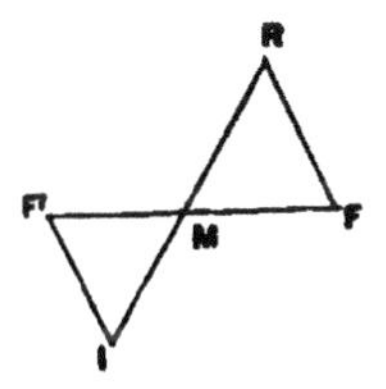

I. We have in figs. 1, 2, 3

$$\triangle HFQ = \text{quadrilateral } HTUR, \qquad \text{[Prop. 54]}$$

$$\triangle H'F'Q = \text{quadrilateral } H'TU'R',$$

$$\therefore\ F'H'HF = H'TU'R' \sim HTUR$$

$$= IUU'R' \mp (IH);$$

whence, adding or subtracting IH,

$$F'IRF = IUU'R' \quad\ldots\ldots\ldots\ldots\ldots\ldots (1),$$

and, adding (IJ) to both,

$$FJR'F' = JU'UR \ldots\ldots\ldots\ldots\ldots\ldots (2).$$

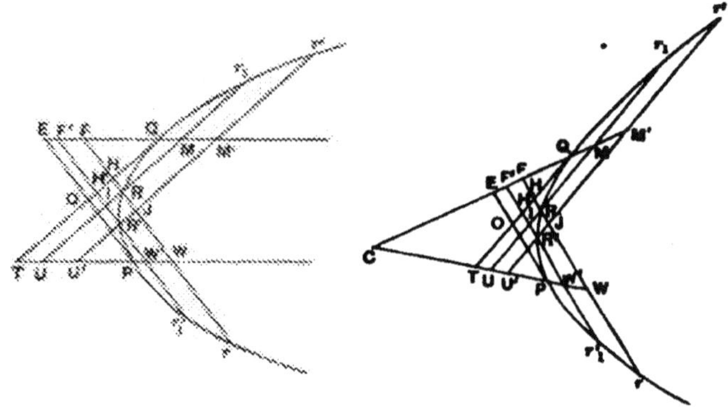

Fig. 1. Fig. 2.

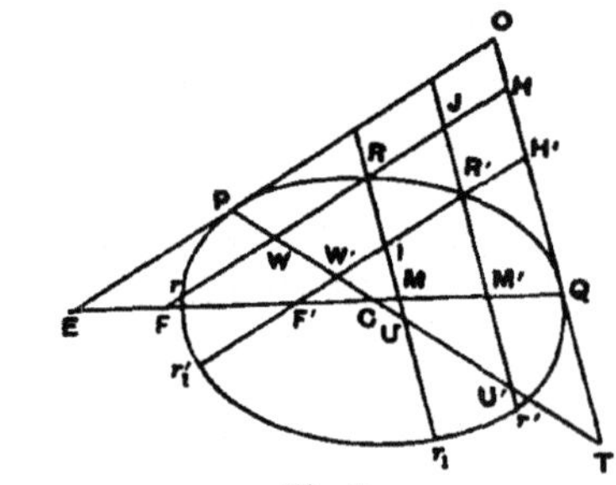

Fig. 3.

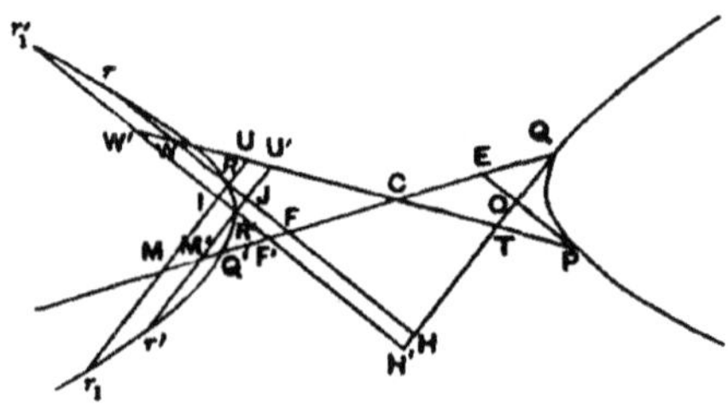

Fig. 4.

II. In figs. 4, 5, 6 we have [Props. 18, 53]

$$\triangle R'U'W' = \triangle CF'W' - \triangle CQT,$$

so that $\triangle CQT =$ quadrilateral $CU'R'F'$,

ding the quadrilateral $CF'H'T$, we have

$$\Delta H'F'Q = \text{quadrilateral } H'TU'R'.$$

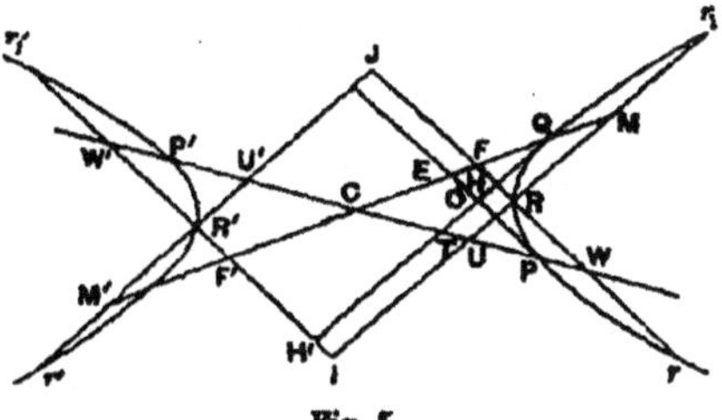

Fig. 5.

Similarly $\Delta HFQ = HTUR$;

and we deduce, as before,

$$F'IRF = IUU'R' \ldots\ldots\ldots\ldots\ldots\ldots\ldots\ldots (1).$$

Thus e.g. in fig. 4,

$$\Delta H'F'Q' - \Delta HFQ = H'TU'R' - HTUR;$$

$$\therefore\ F'H'HF = (R'H) - (RU'),$$

and, subtracting each from (IH),

$$F'IRF = IUU'R'.$$

In fig. 6,

$$F'H'HF = H'TU'R' - \Delta HTW + \Delta RUW,$$

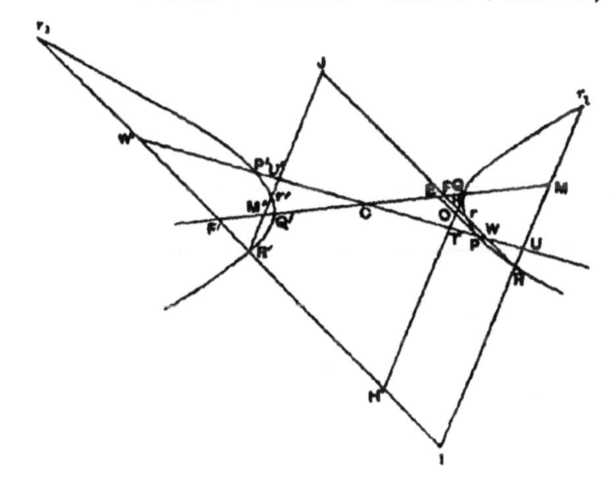

Fig. 6.

and, adding (IH) to each side,

$$F'IRF = H'TU'R' + H'TUI$$
$$= IUU'R' \quad \ldots\ldots\ldots\ldots\ldots\ldots\ldots\ldots\ldots (1).$$

Then, subtracting (IJ) from each side in fig. 4, and subtracting each side from (IJ) in figs. 5, 6, we obtain

$$FJR'F' = JU'UR \quad \ldots\ldots\ldots\ldots\ldots\ldots\ldots\ldots\ldots (2),$$

(the quadrilaterals in fig. 6 being the differences between the triangles FJM', $F'R'M'$ and between the triangles $JU'W$, RUW respectively).

III. The same properties are proved in exactly the same manner in the case where P, Q are on opposite branches, and the quadrilaterals take the same form as in fig. 6 above.

COR. In the particular case of this proposition where R' coincides with P the results reduce to

$$EIRF = \triangle PUI,$$
$$PJRU = PJFE.$$

Proposition 56.

[III. 8.]

If PP', QQ' be two diameters and the tangents at P, P', Q, Q' be drawn, the former two meeting QQ' in E, E' and the latter two meeting PP' in T, T', and if the parallel through P' to the tangent at Q meets the tangent at P in K while the parallel through Q' to the tangent at P meets the tangent at Q in K', then the quadrilaterals (EP'), (TQ') are equal, as also the quadrilaterals $(E'K)$, $(T'K')$.

Since the triangles CQT, CPE are equal [Prop. 53] and have a common vertical angle,

$$CQ \,.\, CT = CP \,.\, CE;$$

$$\therefore CQ : CE = CP : CT,$$

whence $QQ' : EQ = PP' : TP$,

and the same proportion is true for the squares;

$$\therefore \triangle QQ'K' : \triangle QEO = \triangle PP'K : \triangle PTO.$$

And the consequents are equal;

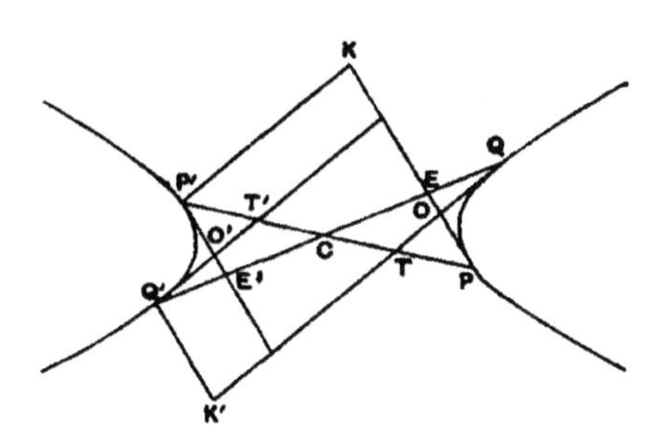

$$\therefore \triangle QQ'K' = \triangle PP'K,$$

and, subtracting the equal triangles CQT, CPE, we obtain

$$(EP') = (TQ') \qquad\qquad (1).$$

Adding the equal triangles $CP'E'$, $CQ'T'$ respectively, we have

$$(E'K) = (T'K') \qquad\qquad (2).$$

Proposition 57.

[III. 5, 11, 12, 14.]

(Application to the case where the ordinates through R, R', the points used in the last two propositions, are drawn to a *secondary* diameter.)

(1) Let Cv be the secondary diameter to which the ordinates are to be drawn. Let the tangent at Q meet it in t, and let the ordinate Rw meet Qt in h and CQ in f. Also let Ru, parallel to Qt, meet Cv in u.

Then [Prop. 19]

$$\triangle Ruw \sim \triangle Cfw = \triangle CQt \qquad\qquad (A)$$

and, subtracting the quadrilateral $CwhQ$,

$$\triangle Ruw - \triangle hQf = \triangle htw;$$

$$\therefore \triangle hQf = \text{quadrilateral } htuR.$$

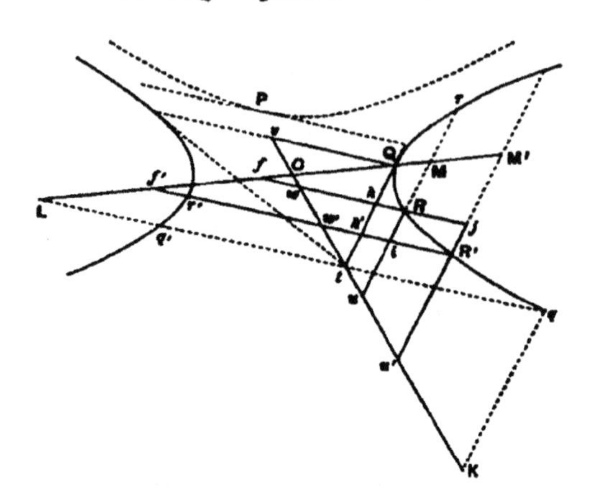

(2) Let $R'w'$ be another ordinate, and h', w' &c. points corresponding to h, w, &c. Also let Ru, $R'w'$ meet in i and Rw, $R'u'$ in j.

Then, from above,

$$\triangle h'Qf' = h'tu'R',$$

and $$\triangle hQf = htuR.$$

Therefore, subtracting,

$$f'h'hf = iuu'R - (hi)$$

and, adding (hi),

$$f'iRf = iuu'R' \text{........................}(1).$$

If we add (ij) to each, we have

$$fjR'f' = ju'uR \text{........................}(2).$$

[This is obviously the case where P is on the conjugate hyperbola, and we deduce from (A) above, by adding the area $CwRM$ to each of the triangles Ruw, Cfw,

$$\triangle CuM \sim \triangle RfM = \triangle CQt,$$

a property of which Apollonius gives a separate proof.]

Proposition 58.

[III. 15.]

In the case where P, Q are on the original hyperbola and R on the conjugate hyperbola, the same properties as those formulated in Propositions 55, 57 still hold, viz.

$$\triangle RMF \sim \triangle CMU = \triangle CQT,$$

and $$F'IRF = IUU'R'.$$

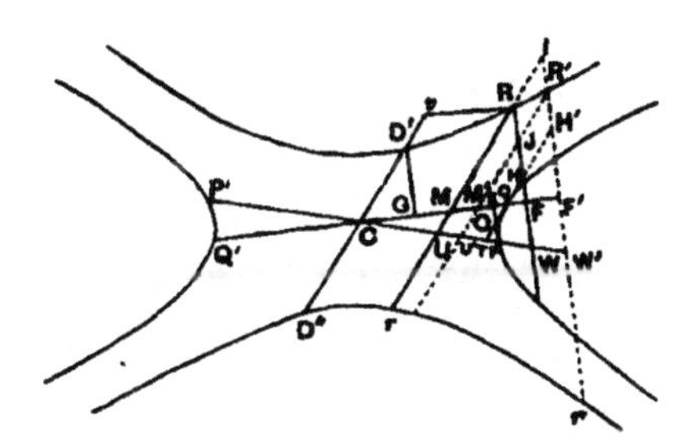

Let $D'D''$ be the diameter of the conjugate hyperbola parallel to RU, and let QT be drawn; and from D' draw $D'G$ parallel to PE to meet CQ in G. Then $D'D''$ is the diameter conjugate to CQ.

Let p' be the parameter in the conjugate hyperbola corresponding to the transverse diameter $D'D''$, and let p be the parameter corresponding to the transverse diameter QQ' in the original hyperbola, so that

$$\frac{p}{2}.CQ = CD'^2, \text{ and } \frac{p'}{2}.CD' = CQ^2.$$

Now we have [Prop. 23]

$$OQ : QE = p : 2QT = \frac{p}{2} : QT;$$

$$\therefore\ D'C : CG = \frac{p}{2} : QT$$

$$= \frac{p}{2} . CQ : CQ . QT$$

$$= CD'^2 : CQ . QT.$$

Hence $$D'C . CG = CQ . QT,$$

or $$\triangle D'CG = \triangle CQT \ldots\ldots\ldots\ldots(1).$$

Again, $$CM : MU = CQ : QT$$

$$= \left(CQ : \frac{p}{2}\right) . (p : 2QT)$$

$$= (p' : D'D'') . (OQ : QE)$$

$$= (p' : D'D'') . (RM : MF) \ldots\ldots(2).$$

Therefore the triangles CMU, RMF, $D'CG$, being respectively half of equiangular parallelograms on CM (or Rv), RM (or Cv), CD', the last two of which are similar while the sides of the first two are connected by the relation (2), have the property of Prop. 16.

$$\therefore\ \triangle RMF \sim \triangle CMU = \triangle D'CG = \triangle CQT \ldots\ldots(3).$$

If R' be another point on the conjugate hyperbola, we have, by subtraction,

$$R'JFF' - RMM'J = MUU'M', \text{ or } RJFF' = RUU'J.$$

And, adding (IJ),

$$F'IRF = IUU'R' \ldots\ldots\ldots\ldots(4)$$

RECTANGLES UNDER SEGMENTS OF INTERSECTING CHORDS.

Proposition 59.

[III. 16, 17, 18, 19, 20, 21, 22, 23.]

Case I. *If OP, OQ be two tangents to any conic and Rr, $R'r'$ two chords parallel to them respectively and intersecting in J, an internal or external point, then*

$$OP^2 : OQ^2 = RJ \, . \, Jr : R'J \, . \, Jr'.$$

(*a*) Let the construction and figures be the same as in Prop. 55.

We have then

$$RJ \, . \, Jr = RW^2 \sim JW^2,$$

and
$$RW^2 : JW^2 = \triangle RUW : \triangle JU'W;$$

$$\therefore \; RW^2 \sim JW^2 : RW^2 = JU'UR : \triangle RUW.$$

But
$$RW^2 : OP^2 = \triangle RUW : \triangle OPT;$$

$$\therefore \; RJ \, . \, Jr : OP^2 = JU'UR : \triangle OPT \; \ldots\ldots\ldots\ldots(1).$$

Again
$$R'J \, . \, Jr' = R'M'^2 \sim JM'^2$$

and
$$R'M'^2 : JM'^2 = \triangle R'F'M' : \triangle JFM',$$

or
$$R'M'^2 \sim JM'^2 : R'M'^2 = FJR'F' : \triangle R'F'M'.$$

But
$$R'M'^2 : OQ^2 = \triangle R'F'M' : \triangle OQE;$$

$$\therefore \; R'J \, . \, Jr' : OQ^2 = FJR'F' : \triangle OQE \; \ldots\ldots\ldots(2).$$

Comparing (1) and (2), we have

$$JU'UR = FJR'F', \text{ by Prop. 55,}$$

and $$\triangle OPT = \triangle OQE', \text{ by Prop. 53.}$$

Thus $$RJ.Jr : OP^2 = R'J.Jr' : OQ^2,$$

or $$OP^2 : OQ^2 = RJ.Jr : R'J.Jr'.$$

(*b*) If we had taken the chords $R'r_1'$, Rr_1 parallel respectively to OP, OQ and intersecting in I, an internal or external point, we should have established in the same manner that

$$OP^2 : OQ^2 = R'I.Ir_1' : RI.Ir_1.$$

Hence the proposition is completely demonstrated.

[COR. If I, or J, which may be any internal or external point be assumed (as a particular case) to be the centre, we have the proposition that the rectangles under the segments of intersecting chords in fixed directions are as the squares of the parallel semi-diameters.]

Case II. *If P be a point on the conjugate hyperbola and the tangent at Q meet CP in t; if further qq' be drawn through t parallel to the tangent at P, and Rr, $R'r'$ be two chords parallel respectively to the tangents at Q, P, and intersecting at i, then*

$$tQ^2 : tq^2 = Ri.ir : R'i.ir'.$$

Using the figure of Prop. 57, we have

$$Ri.ir = Mi^2 \sim MR^2,$$

and $$Mi^2 : MR^2 = \triangle Mf'i : \triangle MfR.$$

Hence $$Ri.ir : MR^2 = f'iRf : \triangle MfR.$$

Therefore, if QC, qq' (both produced) meet in L,

$$Ri.ir : tQ^2 = f'iRf : \triangle QtL \text{ ....(1).}$$

Similarly, $$R'i.ir' : R'w'^2 = iuu'R' : \triangle R'u'w';$$

$$\therefore R'i.ir' : tq^2 = iuu'R' : \triangle tqK \text{ ....(2),}$$

where qK is parallel to Qt and meets Ct produced in K.

But, comparing (1) and (2), we have

$$f'iRf = iuu'R', \qquad \text{[Prop. 57]}$$

and $$\Delta tqK = \Delta CLt + \Delta CQt = \Delta QtL. \qquad \text{[Prop. 19]}$$

$$\therefore\ Ri \,.\, ir : tQ^2 = R'i \,.\, ir' : tq^2,$$

or $$tQ^2 : tq^2 = Ri \,.\, ir : R'i \,.\, ir'.$$

Case III. *If PP' be a diameter and Rr, $R'r'$ be chords parallel respectively to the tangent at P and the diameter PP' and intersecting in I, then*

$$RI \,.\, Ir : R'I \,.\, Ir' = p : PP'.$$

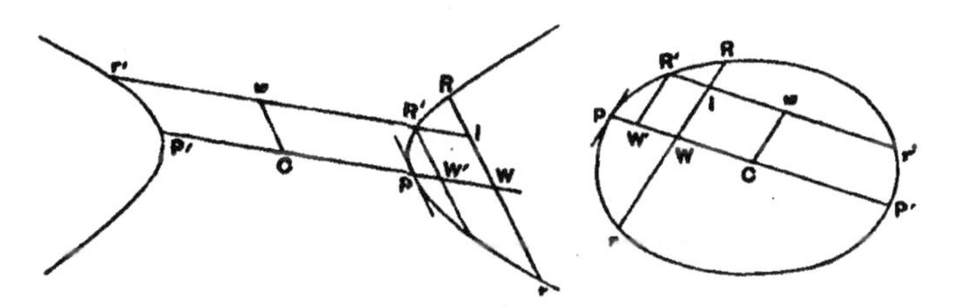

If RW, $R'W'$ are ordinates to PP',

$$p : PP' = RW^2 : CW^2 \sim CP^2 \qquad \text{[Prop. 8]}$$
$$= R'W'^2 : CW'^2 \sim CP^2$$
$$= RW^2 \sim R'W'^2 : CW^2 \sim CW'^2$$
$$= RI \,.\, Ir : R'I \,.\, Ir'.$$

Case IV. *If OP, OQ be tangents to a hyperbola and Rr, $R'r'$ be two chords of the conjugate hyperbola parallel respectively to OQ, OP, and meeting in I, then*

$$OQ^2 : OP^2 = RI \,.\, Ir : R'I \,.\, Ir'.$$

Using the figure of Prop. 58, we have

$$OQ^2 : \Delta OQE = RM^2 : \Delta RMF$$
$$= MI^2 : \Delta MIF'$$
$$= RI \,.\, Ir : \Delta RMF \sim \Delta MIF'$$
$$= RI \,.\, Ir : F'IRF,$$

and, in the same way,

$$OP^2 : \triangle OPT = R'I . Ir' : \triangle R'U'W' \sim \triangle IUW'$$
$$= R'I . Ir' : IUU'R';$$

whence, by Props. 53 and 58, as before,

$$OQ^2 : RI . Ir = OP^2 : R'I . Ir',$$

or $$OQ^2 : OP^2 = RI . Ir : R'I . Ir'.$$

Proposition 60.

[III. 24, 25, 26.]

If Rr, $R'r'$ be chords of conjugate hyperbolas meeting in O and parallel respectively to conjugate diameters PP', DD', then

$$RO . Or + \frac{CP^2}{CD^2} . R'O . Or' = 2CP^2$$

$$\left[\text{or } \frac{RO . Or}{CP^2} \pm \frac{R'O . Or'}{CD^2} = 2\right].$$

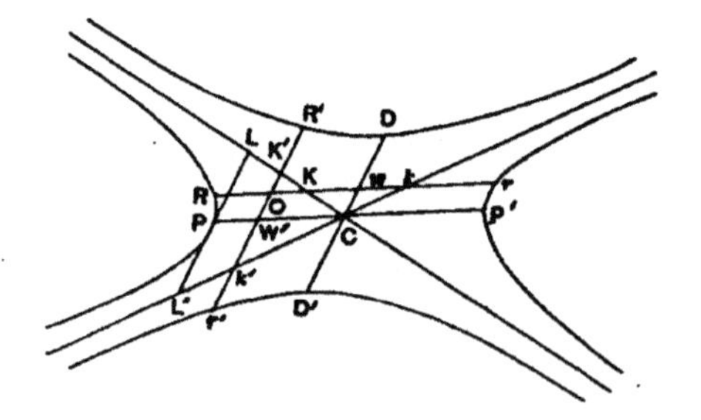

Let Rr, $R'r'$ meet the asymptotes in K, k; K', k', and CD, CP in w, W' respectively. Draw LPL', the tangent at P, meeting the asymptotes in L, L', so that $PL = PL'$.

Then $$LP . PL' = CD^2,$$

and $$LP . PL' : CP^2 = CD^2 : CP^2.$$

Now $$LP : CP = K'O : OK,$$

$$PL' : CP = Ok' : Ok;$$

$$\therefore CD^2 : CP^2 = K'O . Ok' : KO . Ok.$$

[From this point Apollonius distinguishes five cases: (1) where O is in the angle LCL', (2) where O is on one of the asymptotes, (3) where O is in the angle LCk or its opposite, (4) where O is within one of the branches of the original hyperbola, (5) where O lies within one of the branches of the conjugate hyperbola. The proof is similar in all these cases, and it will be sufficient to take case (1), that represented in the accompanying figure.]

We have therefore

$$CD^2 : CP^2 = K'O \,.\, Ok' + CD^2 : KO \,.\, Ok + CP^2$$
$$= K'O \,.\, Ok' + K'R' \,.\, R'k' : KO \,.\, Ok + CP^2$$
$$= K'W'^2 - OW'^2 + R'W'^2 - K'W'^2 : Ow^2 - Kw^2 + CP^2$$
$$= R'W'^2 - OW'^2 : Rw^2 - Kw^2 - Rw^2 + Ow^2 + CP^2$$
$$= R'O \,.\, Or' : RK \,.\, Kr + CP^2 - RO \,.\, Or$$
$$= R'O \,.\, Or' : 2CP^2 - RO \,.\, Or \text{ (since } Kr = Rk\text{)},$$

whence $$RO \,.\, Or + \frac{CP^2}{CD^2} \,.\, R'O \,.\, Or' = 2CP^2,$$

or $$\frac{RO \,.\, Or}{CP^2} + \frac{R'O \,.\, Or'}{CD^2} = 2.$$

[The following proof serves for all the cases: we have

$$R'W'^2 - CD^2 : CW'^2 = CD^2 : CP^2$$

and $$Cw^2 : Rw^2 - CP^2 = CD^2 : CP^2;$$

$$\therefore\ R'W'^2 - Cw^2 - CD^2 : CP^2 - (Rw^2 - CW'^2) = CD^2 : CP^2,$$

so that $$\pm R'O \,.\, Or' - CD^2 : CP^2 \pm RO \,.\, Or = CD^2 : CP^2,$$

whence $$\pm R'O \,.\, Or' : 2CP^2 \pm RO \,.\, Or = CD^2 : CP^2$$

or $$\frac{R'O \,.\, Or'}{CD^2} \pm \frac{RO \,.\, Or}{CP^2} = 2.]$$

Proposition 61.

[III. 27, 28, 29.]

If in an ellipse or in conjugate hyperbolas two chords Rr, $R'r'$ be drawn meeting in O and parallel respectively to two conjugate diameters PP', DD', then

(1) *for the ellipse*

$$RO^2 + Or^2 + \frac{CP^2}{CD^2}(R'O^2 + Or'^2) = 4CP^2,$$

or

$$\frac{RO^2 + Or^2}{CP^2} + \frac{R'O^2 + Or'^2}{CD^2} = 4,$$

and for the hyperbolas

$$RO^2 + Or^2 : R'O^2 + Or'^2 = CP^2 : CD^2.$$

Also, (2) *if $R'r'$ in the hyperbolas meet the asymptotes in K', k', then*

$$K'O^2 + Ok'^2 + 2CD^2 : RO^2 + Or^2 = CD^2 : CP^2.$$

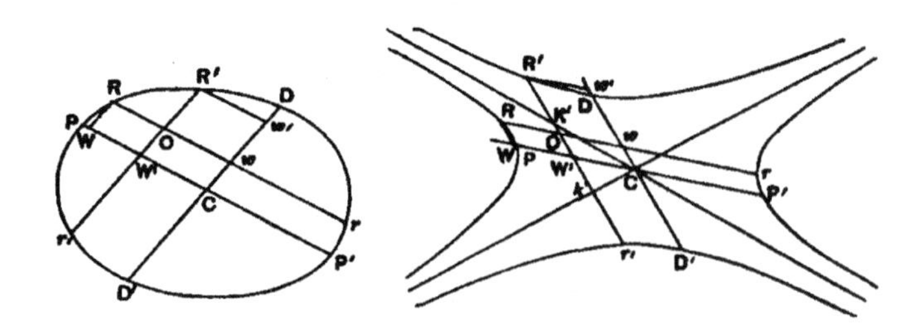

(1) We have for both curves

$$CP^2 : CD^2 = PW \,.\, WP' : RW^2$$

$$= R'w'^2 : Dw' \,.\, w'D'$$

$$= CP^2 + PW \,.\, WP' \pm R'w'^2 : CD^2 + RW^2 \pm Dw' \,.\, w'D',$$

(taking the upper sign for the hyperbolas and the lower for the ellipse);

$$\therefore CP^2 : CD^2 = CP^2 \pm CW'^2 + PW.WP' : CD^2 + Cw^2 \pm Dw'.w'D',$$

whence, for the *hyperbolas*,

$$CP^2 : CD^2 = CW'^2 + CW^2 : Cw^2 + Cw'^2$$
$$= \tfrac{1}{2}(RO^2 + Or^2) : \tfrac{1}{2}(R'O^2 + Or'^2),$$

or $$RO^2 + Or^2 : R'O^2 + Or'^2 = CP^2 : CD^2 \quad \text{.........(A)},$$

while, for the *ellipse*,

$$CP^2 : CD^2 = 2CP^2 - (CW'^2 + CW^2) : Cw'^2 + Cw^2$$
$$= 4CP^2 - (RO^2 + Or^2) : (R'O^2 + Or'^2),$$

whence $$\frac{RO^2 + Or^2}{CP^2} + \frac{R'O^2 + Or'^2}{CD^2} = 4 \quad \text{............(B)}.$$

(2) We have to prove that, in the *hyperbolas*,

$$R'O^2 + Or'^2 = K'O^2 + Ok'^2 + 2CD^2.$$

Now $$R'O^2 - K'O^2 = R'K'^2 + 2R'K'.K'O,$$

and $$Or'^2 - Ok'^2 = r'k'^2 + 2r'k'.k'O$$
$$= R'K'^2 + 2R'K'.k'O.$$

Therefore, by addition,

$$R'O^2 + Or'^2 - K'O^2 - Ok'^2 = 2R'K'(R'K' + K'O + Ok')$$
$$= 2R'K'.R'k'$$
$$= 2CD^2.$$

$$\therefore R'O^2 + Or'^2 = K'O^2 + Ok'^2 + 2CD^2,$$

whence $$K'O^2 + Ok'^2 + 2CD^2 : RO^2 + Or^2 = CD^2 : CP^2,$$

by means of (A) above.

HARMONIC PROPERTIES OF POLES AND POLARS.

Proposition 62.

[III. 30, 31, 32, 33, 34.]

TQ, Tq being tangents to a hyperbola, if V be the middle point of Qq, and if TM be drawn parallel to an asymptote meeting the curve in R and Qq in M, while VN parallel to an asymptote meets the curve in R' and the parallel through T to the chord of contact in N, then

$$TR = RM,$$
$$VR' = R'N\text{*}.$$

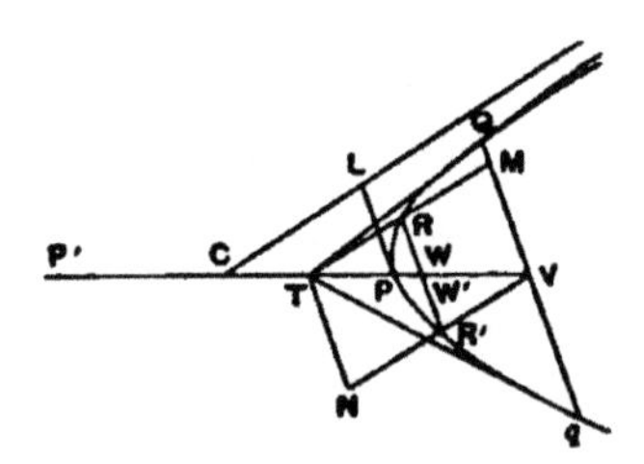

I. Let CV meet the curve in P, and draw the tangent PL, which is therefore parallel to Qq. Also draw the ordinates RW, $R'W'$ to CP.

Then, since the triangles CPL, TWR are similar,

$$RW^2 : TW^2 = PL^2 : CP^2 = CD^2 : CP^2$$
$$= RW^2 : PW . WP';$$
$$\therefore TW^2 = PW . WP'.$$

* It will be observed from this proposition and the next that Apollonius begins with two particular cases of the general property in Prop. 64, namely (*a*) the case where the transversal is parallel to an asymptote, (*b*) the case where the chord of contact is parallel to an asymptote, i.e. where one of the tangents is an asymptote, or a tangent at infinity.

Also $$CV.CT = CP^2;$$

$$\therefore PW.WP' + CP^2 = CV.CT + TW^2,$$

or $$CW^2 = CV.CT + TW^2,$$

whence $$CT(CW + TW) = CV.CT,$$

and $$TW = WV.$$

It follows by parallels that $TR = RM$....................(1).

Again $$CP^2 : PL^2 = W'V^2 : W'R'^2;$$

$$\therefore W'V^2 : W'R'^2 = PW'.W'P' : W'R'^2,$$

so that $$PW'.W'P' = W'V^2.$$

And $$CV.CT = CP^2;$$

$$\therefore CW'^2 = CV.CT + W'V^2,$$

whence, as before, $$TW' = W'V,$$

and $$NR' = R'V$$ ....(2).

II. Next let Q, q be on opposite branches, and let $P'P$ be the diameter parallel to Qq. Draw the tangent PL, and the ordinates from R, R', as before.

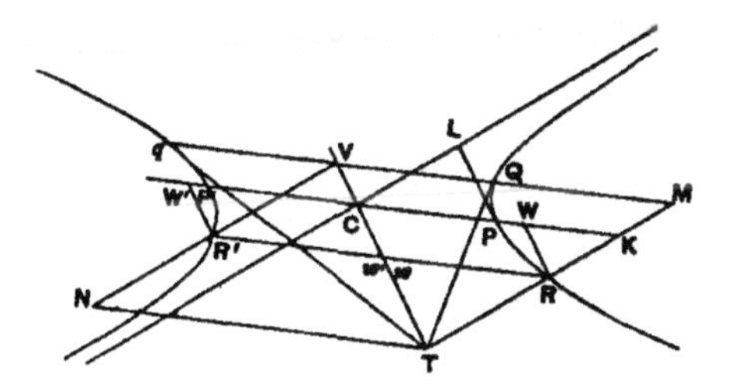

Let TM, CP intersect in K.

Then, since the triangles CPL, KWR are similar,

$$CP^2 : PL^2 = KW^2 : WR^2,$$

and $$CP^2 : CD^2 = PW.WP' : WR^2;$$

$$\therefore KW^2 = PW.WP'.$$

Hence, adding CP^2,

$$CW^2 [= Rw^2] = KW^2 + CP^2.$$

But $$Rw^2 : KW^2 + CP^2 = Tw^2 : RW^2 + PL^2,$$

by similar triangles.

Therefore $$Tw^2 = RW^2 + CD^2$$

$$= Cw^2 + CV.CT,$$

whence $Tw - Cw = CV$, or $Tw = wV$;

$$\therefore TR = RM \text{ .... (1).}$$

Again $CP^2 : PL^2 = PW' . W'P' : R'W'^2$

$$= PW' . W'P' + CP^2 : R'W'^2 + CD^2$$

$$= CW'^2 : Cw'^2 + CV . CT.$$

Also $CP^2 : PL^2 = R'w'^2 : w'V^2$;

$$\therefore w'V^2 = Cw'^2 + CV . CT,$$

whence, as before, $Tw' = w'V$,

and, by parallels, $NR' = R'V$ .... (2).

III. The particular case in which one of the tangents is a tangent at infinity, or an asymptote, is separately proved as follows.

Let LPL' be the tangent at P. Draw PD, LM parallel to CL', and let LM meet the curve in R and the straight line PF drawn through P parallel to CL in M. Also draw RE parallel to CL.

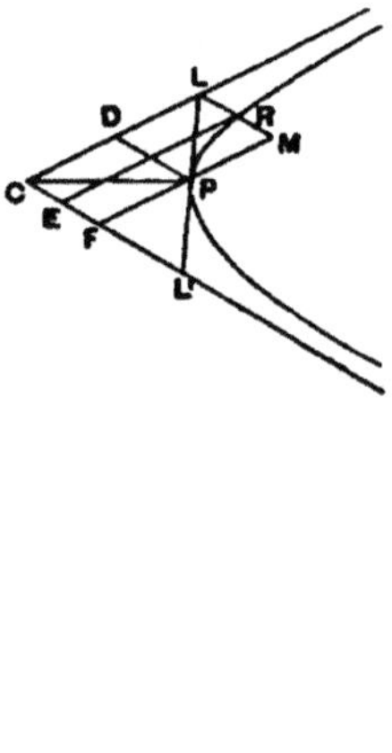

Now $LP = PL'$;

$\therefore PD = CF = FL'$, $FP = CD = DL$.

And $FP . PD = ER . RL$. [Prop. 34]

But $ER = LC = 2CD = 2FP$;

$$\therefore PD = 2LR,$$

or

$$LR = RM.$$

Proposition 63.

[III. 35, 36.]

If PL, the tangent to a hyperbola at P, meet the asymptote in L, and if PO be parallel to that asymptote, and any straight line $LQOQ'$ be drawn meeting the hyperbola in Q, Q' and PO in O, then

$$LQ' : LQ = Q'O : OQ.$$

We have, drawing parallels through L, Q, P, Q' to both asymptotes as in the figures,

$LQ = Q'L'$; whence, by similar triangles, $DL = IQ' = CF$

$$\therefore CD = FL,$$

and
$$CD : DL = FL : LD$$
$$= Q'L : LQ$$
$$= MD : DQ.$$

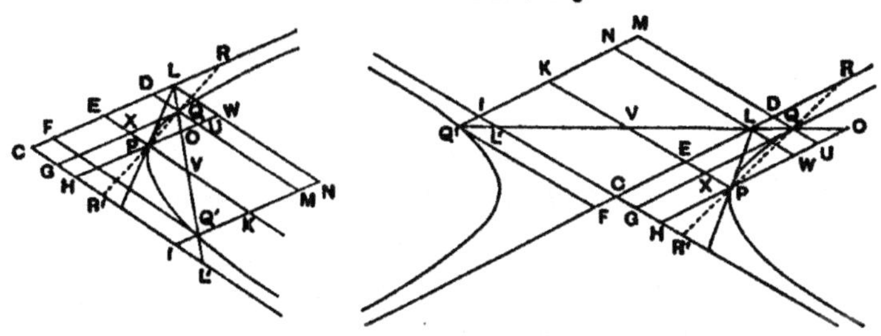

Hence
$$(HD) : (DW) = (MC) : (CQ)$$
$$= (MC) : (EW),$$

since
$$(CQ) = (CP) = (EW). \quad \text{[Prop. 34]}$$

Therefore
$$(MC) : (EW) = (MC) \pm (HD) : (EW) \pm (DW)$$
$$= (MH) : (EU) \ldots\ldots\ldots\ldots\ldots\ldots\ldots\ldots (1).$$

Now
$$(DG) = (HE). \quad \text{[Prop. 34]}$$

Therefore, subtracting CX from both,
$$(DX) = (XH),$$

and, adding (XU) to each, $(EU) = (HQ)$.

Hence, from (1), since $(EW) = (CQ)$,
$$(MC) : (CQ) = (MH) : (HQ),$$

or
$$LQ' : LQ = Q'O : OQ.$$

[Apollonius gives separate proofs of the above for the two cases in which Q, Q' are (1) on the same branch, and (2) on opposite branches, but the second proof is omitted for the sake of brevity.

Eutocius gives two simpler proofs, of which the following is one.

Join PQ and produce it both ways to meet the asymptotes in R, R'. Draw PV parallel to CR' meeting QQ' in V.

Then $$LV = VL'.$$

But $$QL = Q'L'; \quad \therefore \; QV = VQ'.$$

Now
$$QV : VL' = QP : PR'$$
$$= PQ : QR$$
$$= OQ : QL.$$
$$\therefore \; 2QV : 2VL' = OQ : QL,$$
or
$$QQ' : OQ = LL' : QL;$$
$$\therefore \; Q'O : OQ = LQ' : LQ.]$$

Proposition 64.

[III. 37, 38, 39, 40.]

(1) *If TQ, Tq be tangents to a conic and any straight line be drawn through T meeting the conic and the chord of contact, the straight line is divided harmonically;*

(2) *If any straight line be drawn through V, the middle point of Qq, to meet the conic and the parallel through T to Qq* [or the polar of the point V], *this straight line is also divided harmonically;*

i.e. in the figures drawn below

(1) $RT : TR' = RI : IR'$,

(2) $RO : OR' = RV : VR'$.

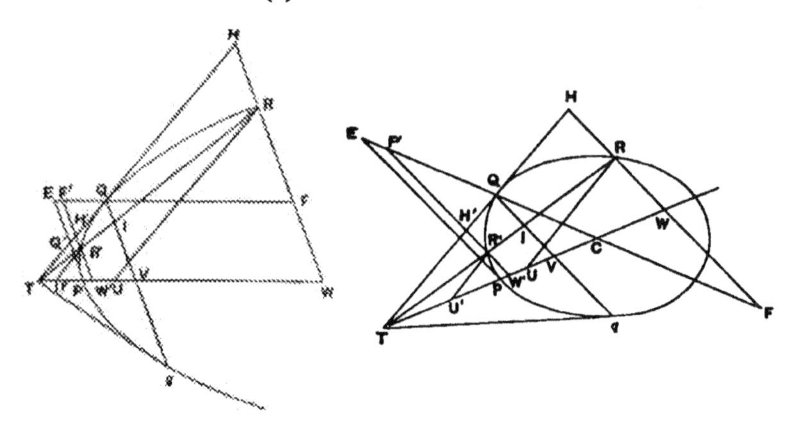

Let TP be the diameter bisecting Qq in V. Draw as usual $HRFW$, $H'R'F'W'$, EP ordinate-wise to the diameter TP; and draw RU, $R'U'$ parallel to QT meeting TP in U, U'.

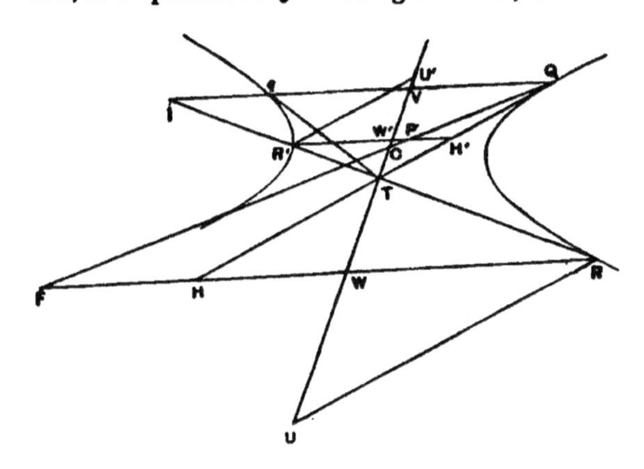

(1) We have then

$$R'I^2 : IR^2 = H'Q^2 : HQ^2$$
$$= \triangle H'F'Q : \triangle HFQ$$
$$= H'TU'R' : HTUR. \quad \text{[Props. 54, 55]}$$

Also $$R'T^2 : TR^2 = R'U'^2 : RU^2$$
$$= \triangle R'U'W' : \triangle RUW;$$

and at the same time

$$R'T^2 : TR^2 = TW'^2 : TW^2$$
$$= \triangle TH'W' : \triangle THW;$$

$$\therefore\ R'T^2 : TR^2 = \triangle R'U'W' \sim \triangle TH'W' : \triangle RUW \sim \triangle THW$$
$$= H'TU'R' : HTUR$$
$$= R'I^2 : IR^2, \text{ from above.}$$
$$\therefore\ RT : TR' = RI : IR'.$$

(2) We have in this case (it is unnecessary to give more than two figures)

$$RV^2 : VR'^2 = RU^2 : R'U'^2$$
$$= \triangle RUW : \triangle R'U'W'.$$

Also $$RV^2 : VR'^2 = HQ^2 : QH'^2$$
$$= \Delta HFQ : \Delta H'F'Q = HTUR : H'TU'R'.$$
$$\therefore\ RV^2 : VR'^2 = HTUR \pm \Delta RUW : H'TU'R' \pm \Delta R'U'W'$$
$$= \Delta THW : \Delta TH'W'$$
$$= TW^2 : TW'^2$$
$$= RO^2 : OR'^2;$$
that is, $$RO : OR' = RV : VR'.$$

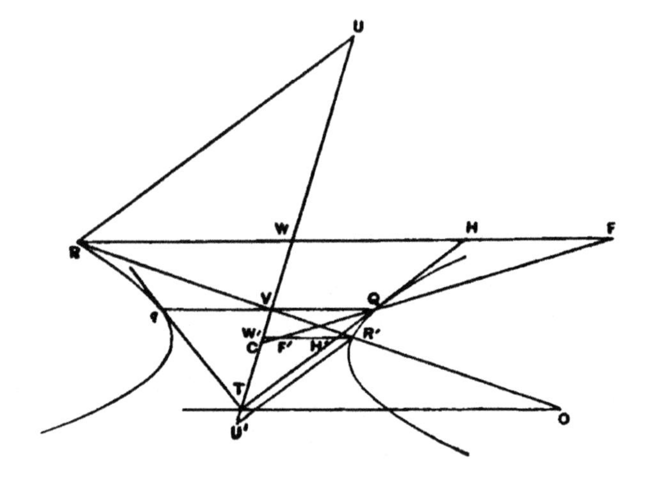

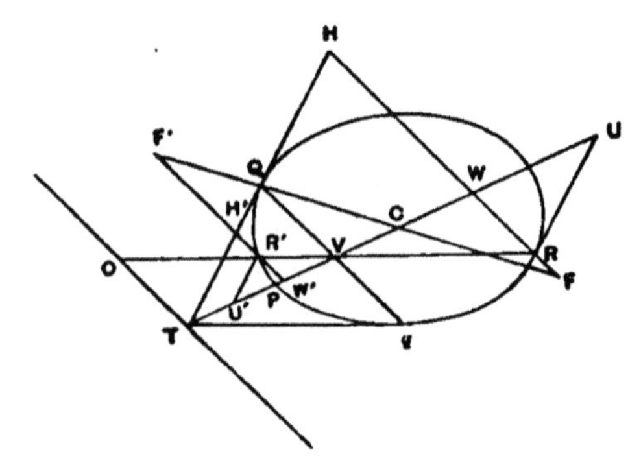

INTERCEPTS MADE ON TWO TANGENTS BY A THIRD.

Proposition 65.

[III. 41.]

If the tangents to a parabola at three points P, Q, R form a triangle pqr, all three tangents are divided in the same proportion, or

$$Pr : rq = rQ : Qp = qp : pR.$$

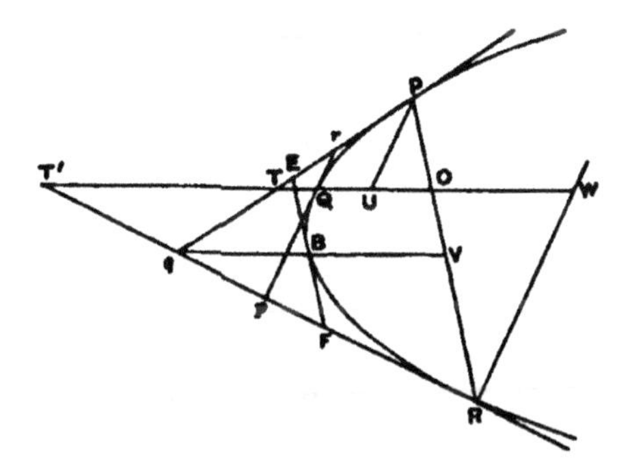

Let V be the middle point of PR, and join qV, which is therefore a diameter. Draw $T'TQW$ parallel to it through Q, meeting Pq in T and qR in T'. Then QW is also a diameter. Draw the ordinates to it from P, R, viz. PU, RW, which are therefore parallel to pQr.

Now, if qV passes through Q, the proposition is obvious, and the ratios will all be ratios of equality.

If not, we have, by the properties of tangents, drawing EBF the tangent at the point B where qV meets the curve,

$$TQ = QU, \quad T'Q = QW, \quad qB = BV,$$

whence, by parallels,

$$Pr = rT, \quad T'p = pR, \quad qF = FR.$$

Then (1) $\quad rP : PT = EP : Pq = 1 : 2,$

and, alternately, $\quad rP : PE = TP : Pq$

$$= OP : PV,$$

whence, doubling the consequents,

$$rP : Pq = OP : PR,$$

and $\quad Pr : rq = PO : OR \quad \text{..............} (1).$

(2) $\quad rQ : Qp = PU : RW,$

since $PU = 2rQ$, and $RW = 2pQ$;

$$\therefore rQ : Qp = PO : OR \quad \text{..............} (2).$$

(3) $\quad FR : Rq = pR : RT',$

and, alternately, $\quad FR : Rp = qR : RT'$

$$= VR : RO.$$

Therefore, doubling the antecedents,

$$qR : Rp = PR : RO,$$

whence $\quad qp : pR = PO : OR \quad \text{..............} (3).$

It follows from (1), (2) and (3) that

$$Pr : rq = rQ : Qp = qp : pR.$$

Proposition 66.

[III. 42.]

If the tangents at the extremities of a diameter PP' of a central conic be drawn, and any other tangent meet them in r, r' respectively, then

$$Pr \cdot P'r' = CD^2.$$

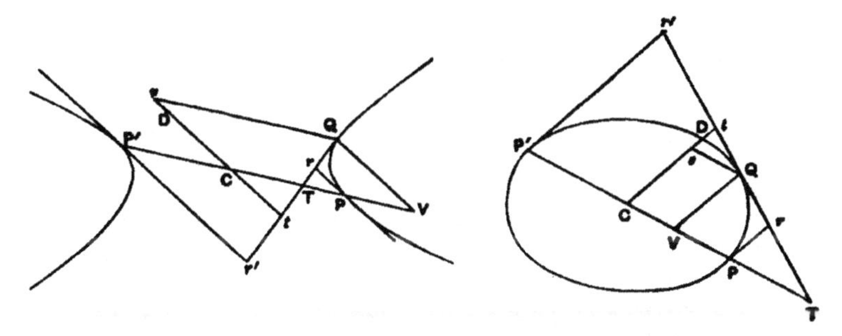

Draw the ordinates QV, Qv to the conjugate diameters PP' and DD'; and let the tangent at Q meet the diameters in T, t respectively.

If now, in the case of an *ellipse* or *circle*, CD pass through Q, the proposition is evident, since in that case rP, CD, $r'P'$ will all be equal.

If not, we have for all three curves

$$CT \cdot CV = CP^2,$$

so that $$CT : CP = CP : CV$$
$$= CT \sim CP : CP \sim CV$$
$$= PT : PV;$$
$$\therefore \; CT : CP' = PT : PV,$$

whence $$CT : P'T = PT : VT.$$

Hence, by parallels, $Ct : P'r' = Pr : QV$
$$= Pr : Cv;$$
$$\therefore \; Pr \cdot P'r' = Cv \cdot Ct = CD^2.$$

Proposition 67.

[III. 43.]

If a tangent to a hyperbola, LPL', meet the asymptotes in L, L', the triangle LCL' has a constant area, or the rectangle $LC \cdot CL'$ is constant.

Draw PD, PF parallel to the asymptotes (as in the third figure of Prop. 62).

Now $$LP = PL';$$

$$\therefore\ CL = 2CD = 2PF,$$

$$CL' = 2CF = 2PD.$$

$$\therefore\ LC \cdot CL' = 4DP \cdot PF,$$

which is constant for all positions of P. [Prop. 34]

Proposition 68.

[III. 44.]

If the tangents at P, Q to a hyperbola meet the asymptotes respectively in L, L'; M, M', then LM', $L'M$ are each parallel to PQ, the chord of contact.

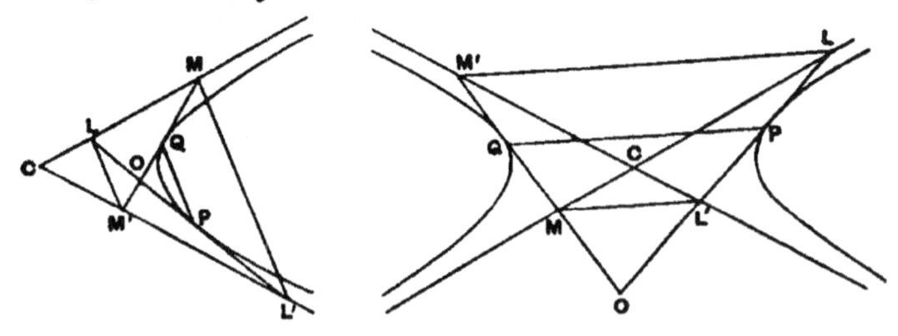

Let the tangents meet at O.

We have then [Prop. 67]

$$LC \cdot CL' = MC \cdot CM',$$

so that $$LC : CM' = MC : CL';$$

$\therefore\ LM'$, $L'M$ are parallel.

It follows that $OL : LL' = OM' : M'M$,

or, halving the consequents,

$$OL : LP = OM' : M'Q;$$

$\therefore\ LM'$, PQ are parallel.

FOCAL PROPERTIES OF CENTRAL CONICS.

The foci are not spoken of by Apollonius under any equivalent of that name, but they are determined as the two points on the axis of a central conic (lying in the case of the ellipse between the vertices, and in the case of the hyperbola within each branch, or on the axis produced) such that the rectangles $AS \cdot SA'$, $AS' \cdot S'A'$ are each equal to "one-fourth part of the figure of the conic," *i.e.* $\frac{1}{4}p_a \cdot AA'$ or CB^2. The shortened expression by which S, S' are denoted is *τὰ ἐκ τῆς παραβολῆς γινόμενα σημεῖα*, "the points arising out of the application." The meaning of this will appear from the full description of the method by which they are arrived at, which is as follows: *ἐὰν τῷ τετάρτῳ μέρει τοῦ εἴδους ἴσον παρὰ τὸν ἄξονα παραβληθῇ ἐφ' ἑκάτερα ἐπὶ μὲν τῆς ὑπερβολῆς καὶ τῶν ἀντικειμένων ὑπερβάλλον εἴδει τετραγώνῳ, ἐπὶ δὲ τῆς ἐλλείψεως ἐλλεῖπον*, "if there be applied along the axis in each direction [a rectangle] equal to one-fourth part of the figure, in the case of the hyperbola and opposite branches exceeding, and in the case of

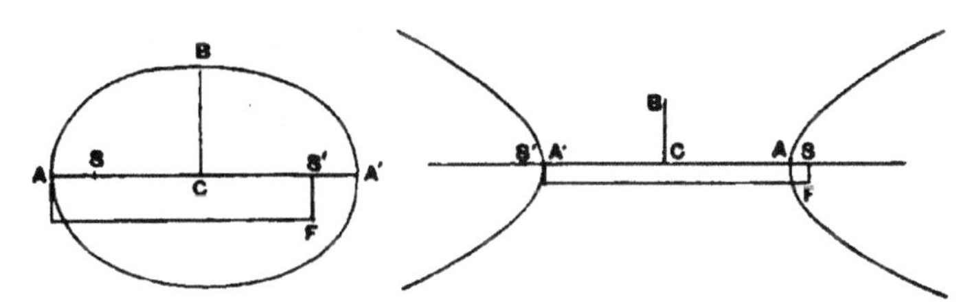

the ellipse falling short, by a square figure." This determines two points, which are accordingly *τὰ ἐκ τῆς παραβολῆς γενηθέντα* σημεῖα. That is, we are to suppose a rectangle applied to the axis as base which is equal to CB^2 but which exceeds or falls short of the rectangle of equal altitude described on the *whole* axis by a square. Thus in the figures drawn the rectangles AF, $A'F$ are respectively to be equal to CB^2, the base AS' falling short of AA' in the ellipse, and the base $A'S$ exceeding $A'A$ in the hyperbola, while $S'F$ or SF is equal to $S'A'$ or SA respectively.

The focus of a *parabola* is not used or mentioned by Apollonius.

Proposition 69.

[III. 45, 46.]

If Ar, $A'r'$, the tangents at the extremities of the axis of a central conic, meet the tangent at any point P in r, r' respectively, then

(1) *rr' subtends a right angle at each focus, S, S';*

(2) *the angles $rr'S$, $A'r'S'$ are equal, as also are the angles $r'rS'$, ArS.*

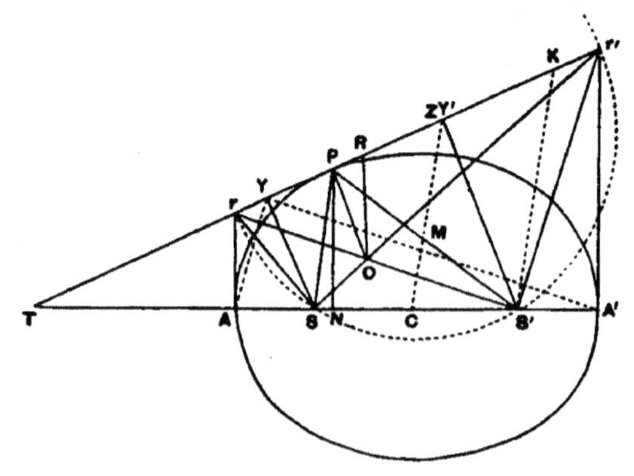

(1) Since [Prop. 66]

$$rA \,.\, A'r' = CB^2$$

$$= AS \,.\, SA', \text{ by definition,}$$

$$rA : AS = SA' : A'r'.$$

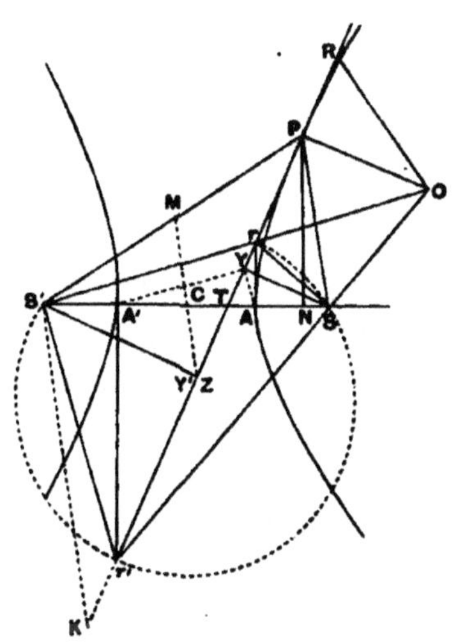

Hence the triangles rAS, $SA'r'$ are similar, and

$$\angle ArS = \angle A'Sr';$$

$\therefore$ the angles rSA, $A'Sr'$ are together equal to a right angle, so that the angle rSr' is a right angle.

And similarly the angle $rS'r'$ is a right angle.

(2) Since rSr', $rS'r'$ are right angles, the circle on rr' as diameter passes through S, S';

$\therefore$ $\angle rr'S = \angle rS'S$, in the same segment,

$= \angle S'r'A'$, by similar triangles.

In like manner $\angle r'rS' = \angle ArS$.

Proposition 70.

[III. 47.]

If, in the same figures, O be the intersection of rS', $r'S$, then OP will be perpendicular to the tangent at P.

Suppose that OR is the perpendicular from O to the tangent at P. We shall show that R must coincide with P.

For $\angle Or'R = \angle S'r'A'$, and the angles at R, A' are right; $\therefore$ the triangles $Or'R$, $S'r'A'$ are similar.

Therefore $A'r' : r'R = S'r' : r'O$

$= Sr : rO$, by similar triangles,

$= Ar : rR$,

because the triangles ArS, RrO are similar;

$\therefore\ r'R : Rr = A'r' : Ar$

$= A'T : TA$..................(1).

Again, if PN be drawn perpendicular to the axis, we have

[Prop. 13] $A'T : TA = A'N : NA$

$= r'P : Pr$, by parallels.

Hence, from (1), $r'R : Rr = r'P : Pr$,

and therefore R coincides with P.

It follows that OP is perpendicular to the tangent at P.

Proposition 71.

[III. 48.]

The focal distances of P make equal angles with the tangent at that point.

In the above figures, since the angles rSO, OPr are right [Props. 69, 70] the points O, P, r, S are concyclic;

$\therefore \angle SPr = \angle SOr$, in the same segment.

In like manner $\angle S'Pr' = \angle S'Or'$,

and the angles SOr, $S'Or'$ are equal, being the same or opposite angles.

Therefore $\angle SPr = \angle S'Pr'$.

Proposition 72.

[III. 49, 50.]

(1) *If, from either focus, as S, SY be drawn perpendicular to the tangent at any point P, the angle AYA' will be a right angle, or the locus of Y is a circle on the axis AA' as diameter.*

(2) *The line drawn through C parallel to either of the focal distances of P to meet the tangent will be equal in length to CA, or CA'.*

Draw SY perpendicular to the tangent, and join AY, YA'. Let the rest of the construction be as in the foregoing propositions.

We have then

(1) the angles rAS, rYS are right;

$\therefore$ A, r, Y, S are concyclic, and

$$\angle AYS = \angle ArS$$

$$= \angle r'SA', \text{ since } \angle rSr' \text{ is right}$$

$$= \angle r'YA', \text{ in the same segment,}$$

S, Y, r', A' being concyclic;

$\therefore$, adding the angle SYA', or subtracting each angle from it,

$$\angle AYA' = \angle SYr' = \text{a right angle.}$$

Therefore Y lies on the circle having AA' for diameter.

Similarly for Y'.

(2) Draw CZ parallel to SP meeting the tangent in Z, and draw $S'K$ also parallel to SP, meeting the tangent in K.

Now $$AS.SA' = AS'.S'A',$$

whence $AS = S'A'$, and therefore $CS = CS'$.

Therefore, by parallels, $PZ = ZK$.

Again $\angle S'KP = \angle SPY$, since SP, $S'K$ are parallel,

$= \angle S'PK$; [Prop. 71]

$\therefore$ $S'P = S'K$.

And $PZ = ZK$;

$\therefore$ $S'Z$ is at right angles to the tangent, or Z coincides with Y'.

But Y' is on the circle having AA' for diameter;

$$\therefore CY' = CA, \text{ or } CA'.$$

And similarly for CY.

Proposition 73.

[III. 51, 52.]

In an ellipse the sum, and in a hyperbola the difference, of the focal distances of any point is equal to the axis AA'.

We have, as in the last proposition, if SP, CY', $S'K$ are parallel, $S'K = S'P$. Let $S'P$, CY' meet in M.

Then, since $SC = CS'$,

$$SP = 2CM,$$

$$S'P = S'K = 2MY';$$

$$\therefore SP \pm S'P = 2(CM \pm MY')$$

$$= 2CY'$$

$$= AA'. \qquad \text{[Prop. 72]}$$

Proposition 74.

[III. 53.]

If PP' be a diameter of a central conic, and Q any other point on it, and if PQ, $P'Q$ respectively meet the tangents at P', P in R', R, then

$$PR.P'R' = DD'^2.$$

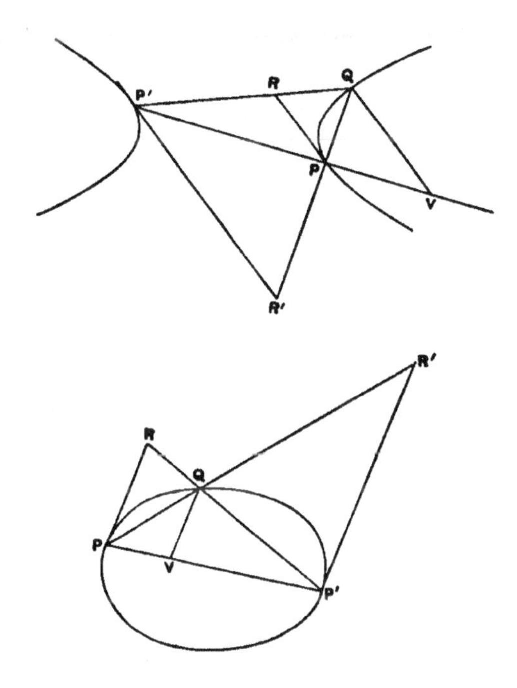

Draw the ordinate QV to PP'.

Now $p : PP' = QV^2 : PV . P'V$ [Prop. 8]

$= (QV : PV) . (QV : P'V)$

$= (P'R' : PP') . (PR : PP')$, by similar triangles;

Hence $p : PP' = PR . P'R' : PP'^2$.

Therefore $PR . P'R' = p . PP'$

$= DD'^2$.

Proposition 75.

[III. 54, 56.]

TQ, TQ' being two tangents to a conic, and R any other point on it, if Qr, Q'r' be drawn parallel respectively to TQ', TQ, and if Qr, Q'R meet in r and Q'r', QR in r', then

$$Qr . Q'r' : QQ'^2 = (PV^2 : PT^2) \times (TQ . TQ' : QV^2),$$

where P is the point of contact of a tangent parallel to QQ'.

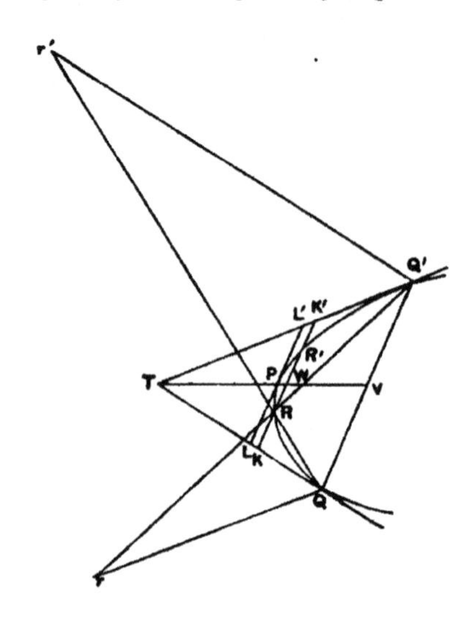

Draw through R the ordinate RW (parallel to QQ') meeting the curve again in R' and meeting TQ, TQ' in K, K' respectively; also let the tangent at P meet TQ, TQ' in L, L'. Then, since PV bisects QQ', it bisects LL', KK', RR' also.

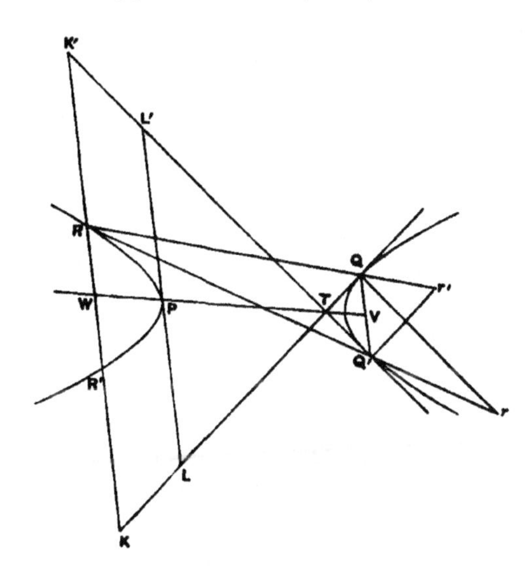

Now $\quad QL^2 : LP . PL' = QL^2 : LP^2$

$$= QK^2 : RK . KR' \qquad \text{[Prop. 59]}$$

$$= QK^2 : RK . RK'.$$

But $\quad QL . Q'L' : QL^2 = QK . Q'K' : QK^2.$

Therefore, *ex aequali*,

$$QL . Q'L' : LP . PL' = QK . Q'K' : RK . RK'$$

$$= (Q'K' : K'R) . (QK : KR)$$

$$= (Qr : QQ') . (Q'r' : QQ')$$

$$= Qr . Q'r' : QQ'^2;$$

$$\therefore \; Qr . Q'r' : QQ'^2 = QL . Q'L' : LP . PL'$$

$$= (QL . Q'L' : LT . TL') . (LT . TL' : LP . PL')$$

$$= (PV^2 : PT^2) . (TQ . TQ' : QV^2).$$

Proposition 76.

[III. 55.]

If the tangents are tangents to opposite branches and meet in t, and if tq is half the chord through t parallel to QQ', while R, r, r' have the same meaning as before, then

$$Qr \,.\, Q'r' : QQ'^2 = tQ \,.\, tQ' : tq^2.$$

Let RR' be the chord parallel to QQ' drawn through R, and let it meet tQ, tQ' in L, L'. Then QQ', RR', LL' are all bisected by tv.

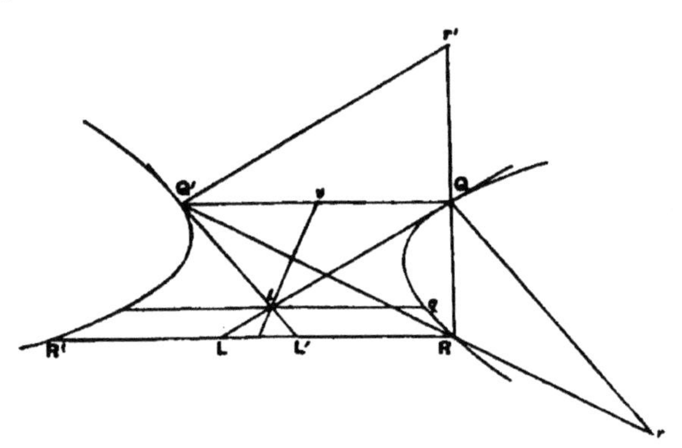

Now $$tq^2 : tQ^2 = R'L \,.\, LR : LQ^2$$ [Prop. 59]

$$= L'R \,.\, RL : LQ^2.$$

But $$tQ^2 : tQ \,.\, tQ' = LQ^2 : LQ \,.\, L'Q'.$$

Therefore, *ex aequali*,

$$tq^2 : tQ \,.\, tQ' = L'R \,.\, RL : LQ \,.\, L'Q'$$

$$= (L'R : L'Q') \,.\, (RL : LQ)$$

$$= (QQ' : Qr) \,.\, (QQ' : Q'r') = QQ'^2 : Qr \,.\, Q'r'.$$

Thus $$Qr \,.\, Q'r' : QQ'^2 = tQ \,.\, tQ' : tq^2.$$

[It is easy to see that the last two propositions give the property of the *three-line locus*. For, since the two tangents and the chord of contact are fixed while the position of R alone varies, the result may be expressed thus,

$$Qr \,.\, Q'r' = (\text{const.}).$$

Now suppose Q_1, Q_2, T_1 in the accompanying figure substituted for Q, Q', T respectively in the first figure of Prop. 75, and we have

$$Q_1r \,.\, Q_2r' = \text{(const.)}$$

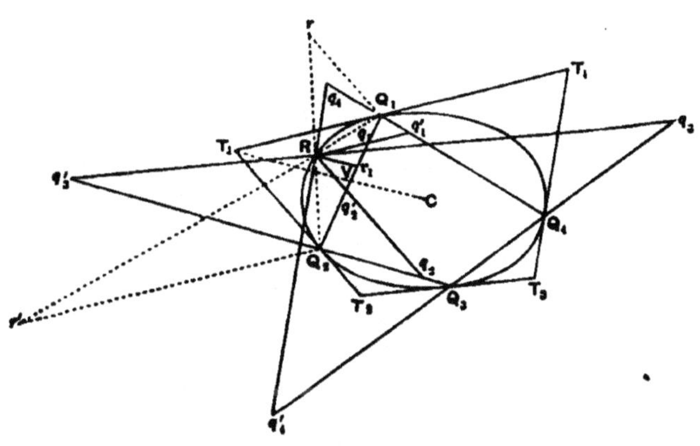

Draw Rq_1, Rq_2' parallel respectively to T_1Q_1, T_1Q_2 and meeting Q_1Q_2 in q_1, q_2'. Also let Rv_1 be drawn parallel to the diameter CT_1 and meeting Q_1Q_2 in v_1.

Then, by similar triangles,

$$Q_1r : Rq_2' = Q_1Q_2 : Q_2q_2',$$

$$Q_2r' : Rq_1 = Q_1Q_2 : Q_1q_1.$$

Hence $\quad Q_1r \,.\, Q_2r' : Rq_1 \,.\, Rq_2' = Q_1Q_2^2 : Q_1q_1 \,.\, Q_2q_2'$.

But $Rq_1 \,.\, Rq_2' : Rv_1^2 = T_1Q_1 \,.\, T_1Q_2 : T_1V^2$, by similar triangles

$$\therefore\ Rq_1 \,.\, Rq_2' : Rv_1^2 = \text{(const.)}.$$

Also $Q_1Q_2^2$ is constant, and $Q_1r \,.\, Q_2r'$ is constant, as proved.

It follows that

$$Rv_1^2 : Q_1q_1 \,.\, Q_2q_2' = \text{(const.)}.$$

But Rv_1 is the distance of R from Q_1Q_2, the chord of contact measured in a fixed direction (parallel to CT_1); and Q_1q_1, Q_2q_2' are equal to the distances of R from the tangents T_1Q_1, T_1Q_2 respectively, measured in a fixed direction (parallel to the chord of contact). If the distances are measured in any other fixed directions, they will be similarly related, and the constant value of the ratio will alone be changed.

Hence R is such a point that, if three straight lines be drawn from it to meet three fixed straight lines at given angles, the rectangle contained by two of the straight lines so drawn bears a constant ratio to the square on the third. In other words, a conic is a "three-line locus" where the three lines are any two tangents and the chord of contact.

The *four-line locus* can be easily deduced from the three-line locus, as presented by Apollonius, in the following manner.

If $Q_1Q_2Q_3Q_4$ be an inscribed quadrilateral, and the tangents at Q_1, Q_2 meet at T_1, the tangents at Q_2, Q_3 at T_2 and so on, suppose Rq_2, Rq_3' drawn parallel to the tangents at Q_2, Q_3 respectively and meeting Q_2Q_3 in q_2, q_3' (in the same way as Rq_1, Rq_2' were drawn parallel to the tangents at Q_1, Q_2 to meet Q_1Q_2), and let similar pairs of lines Rq_3, Rq_4' and Rq_4, Rq_1' be drawn to meet Q_3Q_4 and Q_4Q_1 respectively.

Also suppose Rv_2 drawn parallel to the diameter CT_2, meeting Q_2Q_3 in v_2, and so on.

Then we have

$$\left.\begin{aligned} Q_1q_1 \cdot Q_2q_2' &= k_1 \cdot Rv_1^2 \\ Q_2q_2 \cdot Q_3q_3' &= k_2 \cdot Rv_2^2 \\ Q_3q_3 \cdot Q_4q_4' &= k_3 \cdot Rv_3^2 \\ Q_4q_4 \cdot Q_1q_1' &= k_4 \cdot Rv_4^2 \end{aligned}\right\} \text{ where } k_1, k_2, k_3, k_4 \text{ are constants.}$$

Hence we derive

$$\frac{Rv_1^2 \cdot Rv_3^2}{Rv_2^2 \cdot Rv_4^2} = k \cdot \frac{Q_1q_1}{Q_1q_1'} \cdot \frac{Q_3q_3}{Q_3q_3'} \cdot \frac{Q_2q_2'}{Q_2q_2} \cdot \frac{Q_4q_4'}{Q_4q_4}$$

where k is some constant.

But the triangles $Q_1q_1q_1'$, $Q_2q_2q_2'$ etc. are given in species, as all their sides are in fixed directions. Hence all the ratios $\frac{Q_1q_1}{Q_1q_1'}$, etc. are constant;

$$\therefore \frac{Rv_1 \cdot Rv_3}{Rv_2 \cdot Rv_4} = \text{(const.)}.$$

But Rv_1, Rv_2, Rv_3, Rv_4 are straight lines drawn in fixed directions (parallel to CT_1, etc.) to meet the sides of the inscribed quadrilateral $Q_1Q_2Q_3Q_4$.

Hence *the conic has the property of the four-line locus with respect to the sides of any inscribed quadrilateral.*]

The beginning of Book IV. of Apollonius' work contains a series of propositions, 23 in number, in which he proves the converse of Propositions 62, 63, and 64 above for a great variety of different cases. The method of proof adopted is the *reductio ad absurdum*, and it has therefore been thought unnecessary to reproduce the propositions.

It may, however, be observed that one of them [IV. 9] gives a method of drawing two tangents to a conic from an external point.

Draw any two straight lines through T each cutting the conic in two points as Q, Q' and R, R'. Divide QQ' in O and RR' in O' so that

$$TQ : TQ' = QO : OQ',$$
$$TR : TR' = RO' : O'R'.$$

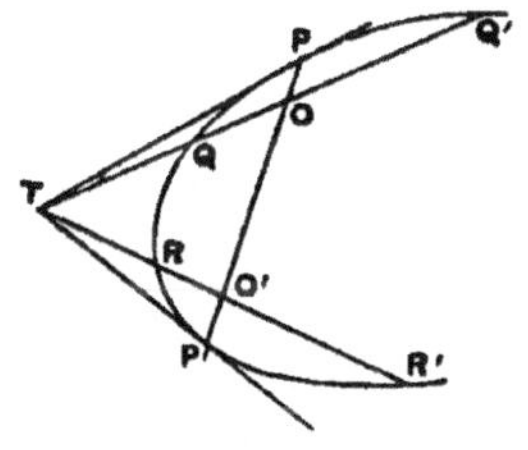

Join OO', and produce it both ways to meet the conic in P, P'. Then P, P' are the points of contact of the two tangents from T.

Proposition 77.

[IV. 24.]

No two conics can intersect in such a way that part of one of them is common to both, while the rest is not.

If possible, let a portion $q'Q'PQ$ of a conic be common to two, and let them diverge at Q. Take Q' any other point on the common portion and join QQ'. Bisect QQ' in V and draw the diameter PV. Draw $rqvq'$ parallel to QQ'.

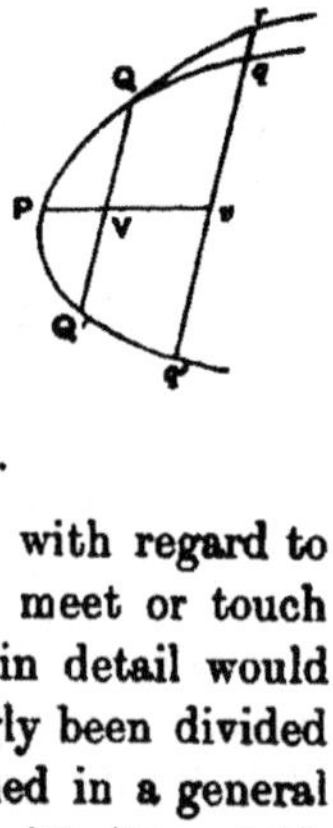

Then the line through P parallel to QQ' will touch both curves and we shall have in one of them $qv = vq'$, and in the other $rv = vq'$;

$$\therefore \; rv = qv, \text{ which is impossible.}$$

There follow a large number of propositions with regard to the number of points in which two conics can meet or touch each other, but to give all these propositions in detail would require too much space. They have accordingly been divided into five groups, three of which can be combined in a general enunciation and are accordingly given as Props. 78, 79 and 80, while indications are given of the proofs by which each particular case under all the five groups is established. The terms "conic" and "hyperbola" in the various enunciations do not (except when otherwise stated) include the double-branch hyperbola but only the single branch. The term "conic" must be understood as including a circle.

Group I. Propositions depending on the more elementary considerations affecting conics.

1. Two conics having their concavities in opposite directions will not meet in more than two points. [IV. 35.]

If possible, let ABC, $ADBEC$ be two such conics meeting in three points, and draw the chords of contact AB, BC. Then AB, BC contain an angle towards the same parts as the concavity of ABC. And for the same reason they contain an angle towards the same parts as the concavity of $ADBEC$.

Therefore the concavity of the two curves is in the same direction: which is contrary to the hypothesis.

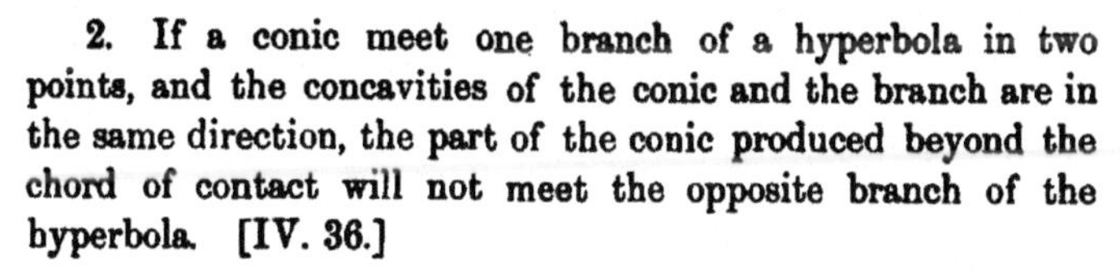

2. If a conic meet one branch of a hyperbola in two points, and the concavities of the conic and the branch are in the same direction, the part of the conic produced beyond the chord of contact will not meet the opposite branch of the hyperbola. [IV. 36.]

The chord joining the two points of intersection will cut both the lines forming one of the angles made by the asymptotes of the double hyperbola. It will not therefore fall within the opposite angle between the asymptotes and so cannot meet the opposite branch. Therefore neither can the part of the conic more remote than the said chord.

3. If a conic meet one branch of a hyperbola, it will not meet the other branch in more points than two. [IV. 37.]

The conic, being a one-branch curve, must have its concavity in the opposite direction to that of the branch which it meets in two points, for otherwise it could not meet the opposite branch in a third point [by the last proposition]. The proposition therefore follows from (1) above. The same is true if the conic touches the first branch.

4. A conic touching one branch of a hyperbola with its concave side will not meet the opposite branch. [IV. 39.]

Both the conic and the branch which it touches must be on the same side of the common tangent and therefore will be

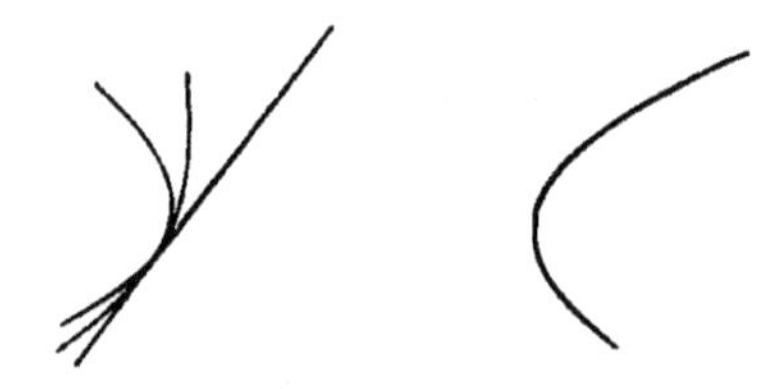

separated by the tangent from the opposite branch. Whence the proposition follows.

5. If one branch of a hyperbola meet one branch of another hyperbola with concavity in the opposite direction in two points, the opposite branch of the first hyperbola will not meet the opposite branch of the second. [IV. 41.]

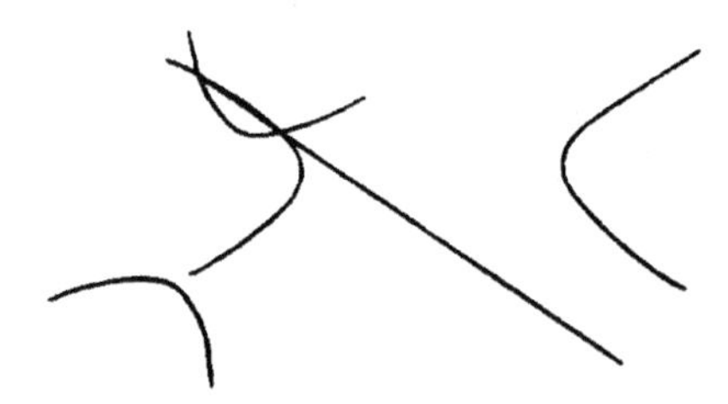

The chord joining the two points of concourse will fall across one asymptotal angle in each hyperbola. It will not therefore fall across the opposite asymptotal angle and therefore will not meet either of the opposite branches. Therefore neither will the opposite branches themselves meet, being separated by the chord referred to.

6. If one branch of a hyperbola meet both branches of another hyperbola, the opposite branch of the former will not meet either branch of the second in two points. [IV. 42.]

For, if possible, let the second branch of the former meet one branch of the latter in D, E. Then, joining DE, we use the same argument as in the last proposition. For *DE* crosses one asymptotal angle of each hyperbola, and it will therefore not meet either of the branches opposite to the branches *DE*. Hence those branches are separated by *DE* and therefore cannot meet one another: which contradicts the hypothesis.

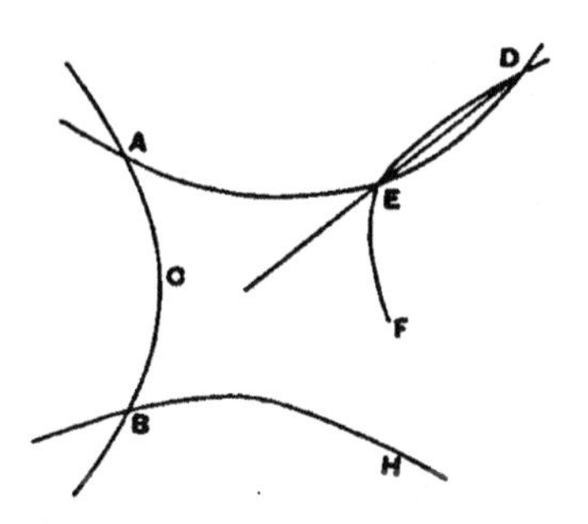

Similarly, if the two branches *DE* touch, the result will be the same, an impossibility.

7. If one branch of a hyperbola meet one branch of another hyperbola with concavity in the same direction, and if it also meet the other branch of the second hyperbola in one point, then the opposite branch of the first hyperbola will not meet either branch of the second. [IV. 45.]

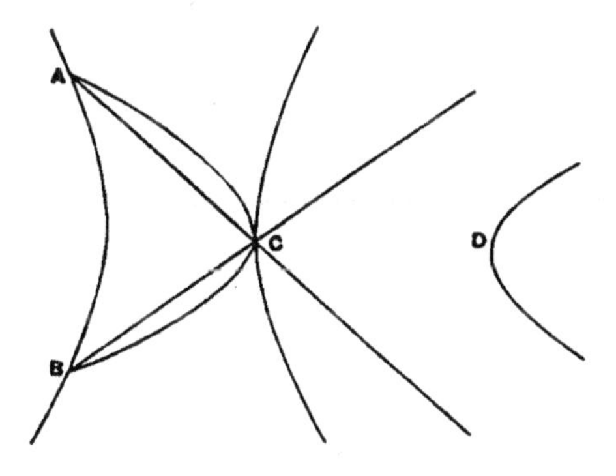

A, *B* being the points of meeting with the first branch and

C that with the opposite branch, by the same principle as before, neither *AC* nor *BC* will meet the branch opposite to *ACB*. Also they will not meet the branch *C* opposite to *AB* in any other point than *C*, for, if either met it in two points, it would not meet the branch *AB*, which, however, it does, by hypothesis.

Hence *D* will be within the angle formed by *AC*, *BC* produced and will not meet *C* or *AB*.

8. If a hyperbola touch one of the branches of a second hyperbola with its concavity in the opposite direction, the opposite branch of the first will not meet the opposite branch of the second. [IV. 54.]

The figure is like that in (6) above except that in this case *D* and *E* are two consecutive points; and it is seen in a similar manner that the second branches of the two hyperbolas are separated by the common tangent to the first branches, and therefore the second branches cannot meet.

Group II. containing propositions capable of being expressed in one general enunciation as follows:

Proposition 78.

No two conics (including under the term a hyperbola with two branches) can intersect in more than four points.

1. Suppose the double-branch hyperbola to be alone excluded. [IV. 25.]

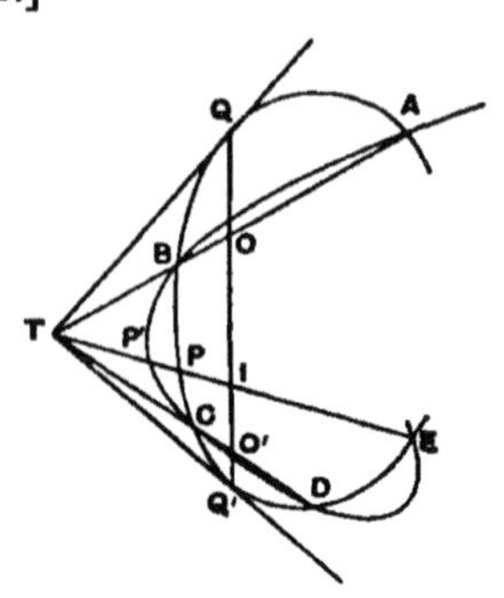

If possible, let there be five points of intersection A, B, C, D, E, being *successive* intersections, so that there are no others between. Join AB, DC and produce them. Then

(*a*) if they meet, let them meet at T. Let O, O' be taken on AB, DC such that TA, TD are harmonically divided. If OO' be joined and produced it will meet each conic, and the lines joining the intersections to T will be tangents to the conics. Then TE cuts the two conics in different points P, P', since it does not pass through any common point except E.

Therefore
$$\left.\begin{aligned} ET : TP &= EI : IP \\ ET : TP' &= EI : IP' \end{aligned}\right\},$$
where OO', TE intersect at I.

But these ratios cannot hold simultaneously; therefore the conics do not intersect in a fifth point E.

(*b*) If AB, DC are parallel, the conics will be either ellipses or circles. Bisect AB, DC at M, M'; MM' is then

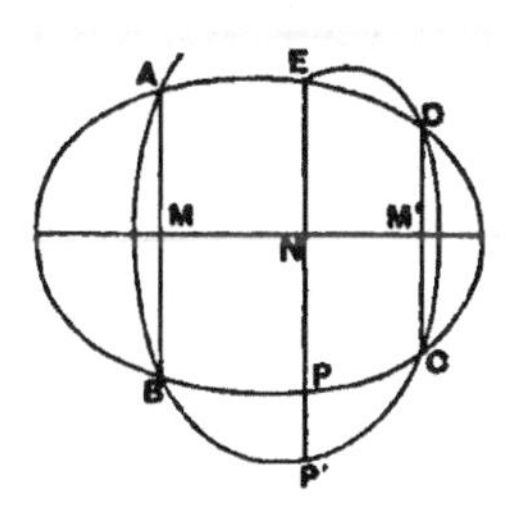

a diameter. Draw $ENPP'$ through E parallel to AB or DC, meeting MM' in N and the conics in P, P'. Then, since MM' is a diameter of both,
$$NP = NE = NP',$$
which is impossible.

Thus the conics do not intersect in more than four points.

2. A conic section not having two branches will not meet a double-branch hyperbola in more than four points. [IV. 38.]

This is clear from the fact that [Group I. 3] the conic meeting one branch will not meet the opposite branch in more points than two.

3. If one branch of a hyperbola cut each branch of a second hyperbola in two points, the opposite branch of the first hyperbola will not meet either branch of the second. [IV. 43.]

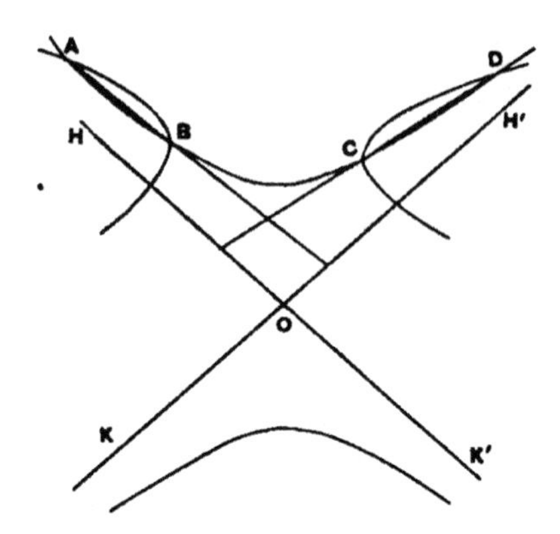

The text of the proof in Apollonius is corrupt, but Eutocius gives a proof similar to that in Group I. 5 above. Let HOH' be the asymptotal angle containing the one branch of the first hyperbola, and KOK' that containing the other branch. Now AB, meeting one branch of the second hyperbola, will not meet the other, and therefore AB separates the latter from the asymptote OK'. Similarly DC separates the former branch from OK. Therefore the proposition follows.

4. If one branch of a hyperbola cut one branch of a second in four points, the opposite branch of the first will not meet the opposite branch of the second. [IV. 44.]

The proof is like that of 1 (a) above. If E is the supposed fifth point and T is determined as before, ET meets the intersecting branches in separate points, whence the harmonic property produces an absurdity.

5. If one branch of a hyperbola meet one branch of a second in three points, the other branch of the first will not meet the other branch of the second in more than one point. [IV. 46.]

Let the first two branches intersect in A, B, C, and (if possible) the other two in D, E. Then

(a) if AB, DE be parallel, the line joining their middle points will be a diameter of both conics, and the parallel chord through C in both conics will be bisected by the diameter; which is impossible.

(b) If AB, DE be not parallel, let them meet in O.

Bisect AB, DE in M, M', and draw the diameters MP, MP' and $M'Q$, $M'Q'$ in the respective hyperbolas. Then the tangents at P', P will be parallel to AO, and the tangents at Q', Q parallel to BO.

Let the tangents at P, Q and P', Q' meet in T, T'.

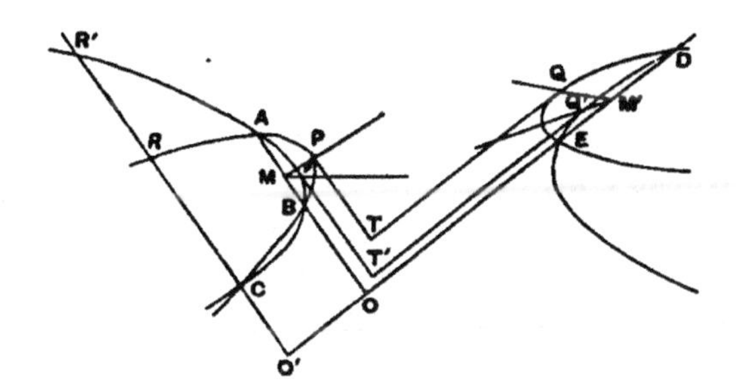

Let CRR' be parallel to AO and meet the hyperbolas in R, R', and DO in O'.

Then $$TP^2 : TQ^2 = AO \,.\, OB : DO \,.\, OE$$
$$= T'P'^2 : T'Q'^2. \qquad \text{[Prop. 59]}$$

It follows that

$$RO' \,.\, O'C : DO' \,.\, O'E = R'O' \,.\, O'C : DO' \,.\, O'E,$$

whence $$RO' \,.\, O'C = R'O' \,.\, O'C;$$

which is impossible.

Therefore, etc.

6. The two branches of a hyperbola do not meet the two branches of another hyperbola in more points than four. [IV. 55.]

Let A, A' be the two branches of the first hyperbola and B, B' the two branches of the second.

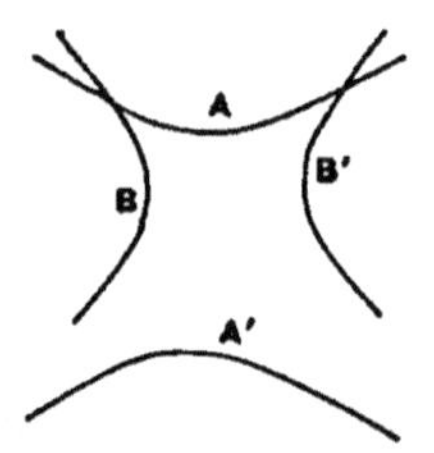

Then (*a*) if A meet B, B' each in two points, the proposition follows from (3) above;

(*b*) if A meet B in two points and B' in one point, A' cannot meet B' at all [Group I. 5], and it can only meet B in one point, for if A' met B in two points A could not have met B' (which it does);

(*c*) if A meet B in two points and A' meet B, A' will not meet B' [Group I. 5], and A' cannot meet B in more points than two [Group I. 3];

(*d*) if A meet B in one point and B' in one point, A' will not meet either B or B' in two points [Group I. 6];

(*e*) if the branches A, B have their concavities in the same direction, and A cut B in four points, A' will not cut B' [case (4) above] nor B [case (2) above];

(*f*) if A meet B in three points, A' will not meet B' in more than one point [case (5) above].

And similarly for all possible cases.

Group III. being particular cases of

Proposition 79.

Two conics (including double hyperbolas) which touch at one point cannot intersect in more than two other points.

1. The proposition is true of all conics excluding hyperbolas with two branches. [IV. 26.]

The proof follows the method of Prop. 78 (1) above.

2. If one branch of a hyperbola touch one branch of another in one point and meet the other branch of the second hyperbola in two points, the opposite branch of the first will not meet either branch of the second. [IV. 47.]

The text of Apollonius' proof is corrupt, but the proof of Prop. 78 (3) can be applied.

3. If one branch of a hyperbola touch one branch of a second in one point and cut the same branch in two other points, the opposite branch of the first does not meet either branch of the second. [IV. 48.]

Proved by the harmonic property like Prop. 78 (4).

4. If one branch of a hyperbola touch one branch of a second hyperbola in one point and meet it in one other point, the opposite branch of the first will not meet the opposite branch of the second in more than one point. [IV. 49.]

The proof follows the method of Prop. 78 (5).

5. If one branch of a hyperbola touch one branch of another hyperbola (having its concavity in the same direction), the opposite branch of the first will not meet the opposite branch of the second in more than two points. [IV. 50.]

The proof follows the method of Prop. 78 (5), like the last case (4).

6. If a hyperbola with two branches touch another hyperbola with two branches in one point, the hyperbolas will not meet in more than two other points. [IV. 56.]

The proofs of the separate cases follow the methods employed in Group I. 3, 5, and 8.

Group IV. merging in

Proposition 80.

No two conics touching each other at two points can intersect at any other point.

1. The proposition is true of all conics excluding hyperbolas with two branches. [IV. 27, 28, 29.]

Suppose the conics touch at A, B. Then, if possible, let them also cut at C.

(*a*) If the tangents are not parallel and C does not lie between A and B, the proposition is proved from the harmonic property;

(*b*) if the tangents are parallel, the absurdity is proved by the bisection of the chord of each conic through C by the chord of contact which is a diameter;

(*c*) if the tangents are not parallel, and C is between A and B, draw TV from the point of intersection of the tangents to the middle point of AB. Then TV cannot pass through C, for then the parallel through C to AB would touch both conics, which is absurd. And the bisection of the chords parallel to AB through C in each conic results in an absurdity.

2. If a single-branch conic touch each branch of a hyperbola, it will not intersect either branch in any other point. [IV. 40.]

This follows by the method employed in Group I. 4.

3. If one branch of a hyperbola touch each branch of a second hyperbola, the opposite branch of the first will not meet either branch of the second. [IV. 51.]

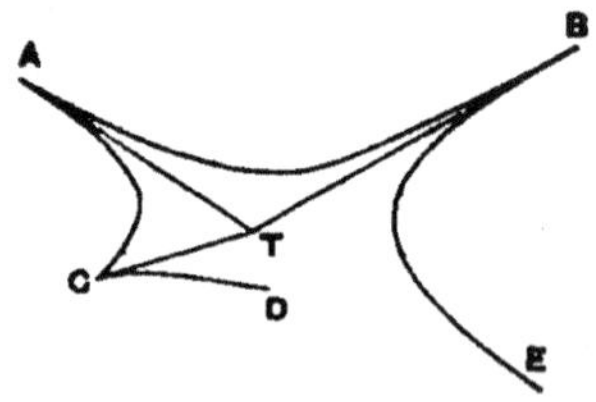

Let the branch AB touch the branches AC, BE in A, B. Draw the tangents at A, B meeting in T. If possible, let CD, the opposite branch to AB, meet AC in C. Join CT.

Then T is within the asymptotes to AB, and therefore CT falls within the angle ATB. But BT, touching BE, cannot meet the opposite branch AC. Therefore BT falls on the side of CT remote from the branch AC, or CT passes through the angle adjacent to ATB; which is impossible, since it falls within the angle ATB.

4. If one branch of one hyperbola touch one branch of another in one point, and if also the other branches touch in one point, the concavities of each pair being in the same direction, there are no other points of intersection. [IV. 52.]

This is proved at once by means of the bisection of chords parallel to the chord of contact.

5. If one branch of a hyperbola touch one branch of another in two points, the opposite branches do not intersect. [IV. 53.]

This is proved by the harmonic property.

6. If a hyperbola with two branches touch another hyperbola with two branches in two points, the hyperbolas will not meet in any other point. [IV. 57.]

The proofs of the separate cases follow those of (3), (4), (5) above and Group I. 8.

Group V. Propositions respecting double contact between conics.

1. A parabola cannot touch another parabola in more points than one. [IV. 30.]

This follows at once from the property that $TP = PV$.

2. A parabola, if it fall outside a hyperbola, cannot have double contact with the hyperbola. [IV. 31.]

For the hyperbola

$$\begin{aligned} CV : CP &= CP : CT \\ &= CV - CP : CP - CT \\ &= PV : PT. \end{aligned}$$

Therefore $$PV > PT.$$

And for the parabola $P'V = P'T$: therefore the hyperbola falls outside the parabola, which is impossible.

3. A parabola cannot have internal double contact with an ellipse or circle. [IV. 32.]

The proof is similar to the preceding.

4. A hyperbola cannot have double contact with another hyperbola having the same centre. [IV. 33.]

Proved by means of $CV.CT = CP^2$.

5. If an ellipse have double contact with an ellipse or a circle having the same centre, the chord of contact will pass through the centre. [IV. 34.]

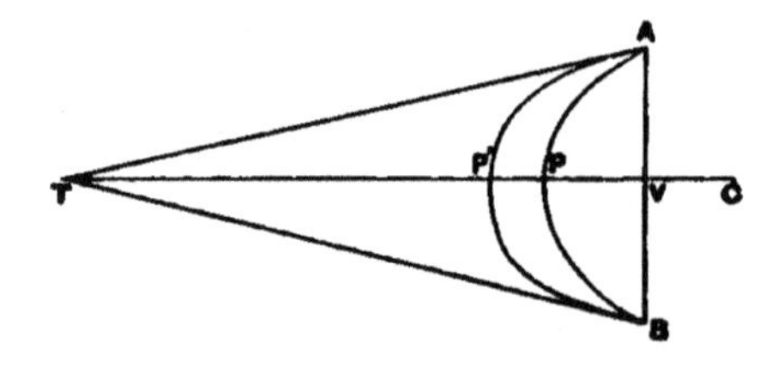

Let (if possible) the tangents at A, B meet in T, and let V be the middle point of AB. Then TV is a diameter. If possible, let C be the centre.

Then $CP'^2 = CV.CT = CP^2$, which is absurd. Therefore the tangents at A, B do not meet, i.e. they are parallel. Therefore AB is a diameter and accordingly passes through the centre.

NORMALS AS MAXIMA AND MINIMA.

Proposition 81. (Preliminary.)

[V. 1, 2, 3.]

If in an ellipse or a hyperbola AM be drawn perpendicular to the axis AA' and equal to one-half its parameter, and if CM meet the ordinate PN of any point P on the curve in H, then

$$PN^2 = 2\,(\text{quadrilateral } MANH).$$

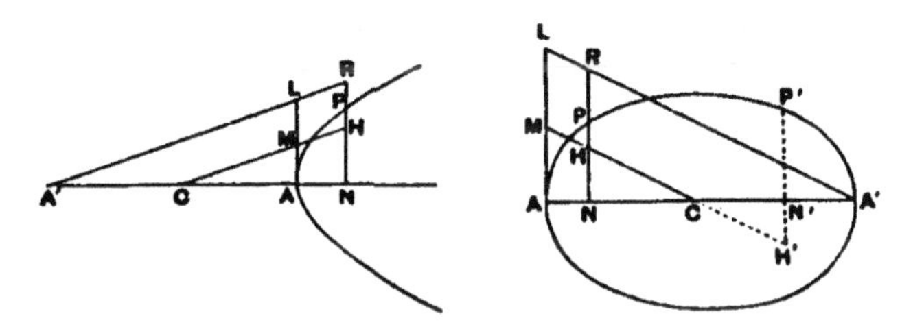

Let AL be twice AM, i.e. let AL be the latus rectum or parameter. Join $A'L$ meeting PN in R. Then $A'L$ is parallel to CM. Therefore $HR = LM = AM$.

Now $$PN^2 = AN.NR; \qquad \text{[Props. 2, 3]}$$

$$\therefore PN^2 = AN(AM + HN)$$

$$= 2\,(\text{quadrilateral } MANH).$$

In the particular case where P is between C and A' in the ellipse, the quadrilateral becomes the difference between two triangles, and

$$P'N'^2 = 2(\triangle CAM - \triangle CN'H').$$

Also, if P be the end of the minor axis of the ellipse, the quadrilateral becomes the triangle CAM, and

$$BC^2 = 2\triangle CAM.$$

[The two last cases are proved by Apollonius in separate propositions. Cf. the note on Prop. 23 above, p. 40.]

Proposition 82.

[V. 4.]

In a parabola, if E be a point on the axis such that AE is equal to half the latus rectum, then the **minimum** *straight line from E to the curve is AE; and, if P be any other point on the curve, PE increases as P moves further from A on either side. Also for any point*

$$PE^2 = AE^2 + AN^2.$$

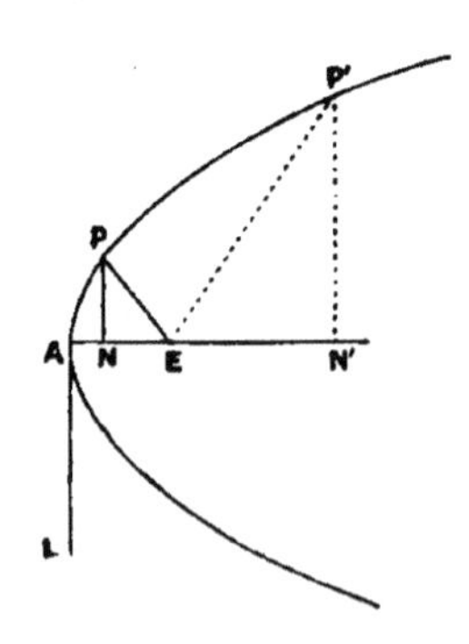

Let AL be the parameter or latus rectum.

Then $$PN^2 = AL . AN$$
$$= 2AE . AN.$$

Adding EN^2, we have

$$PE^2 = 2AE . AN + EN^2$$
$$= 2AE . AN + (AE \sim AN)^2$$
$$= AE^2 + AN^2.$$

Thus $PE^2 > AE^2$ and increases with AN, i.e. as P moves further and further from A.

Also the *minimum* value of PE is AE, or AE is the shortest straight line from E to the curve.

[In this proposition, as in the succeeding propositions, Apollonius takes three cases, (1) where N is between A and E, (2) where N coincides with E and PE is therefore perpendicular to the axis, (3) where AN is greater than AE, and he proves the result separately for each. The three cases will for the sake of brevity be compressed, where possible, into one.]

Proposition 83.

[V. 5, 6.]

If E be a point on the axis of a hyperbola or an ellipse such that AE is equal to half the latus rectum, then AE is the **least** *of all the straight lines which can be drawn from E to the curve; and, if P be any other point on it, PE increases as P moves further from A on either side, and*

$$PE^2 = AE^2 + AN^2 . \frac{AA' \pm p_a}{AA'} \; [= AE^2 + e^2 . AN^2]$$

(where the upper sign refers to the hyperbola).*

Also in the ellipse EA' is the **maximum** *straight line from E to the curve.*

Let AL be drawn perpendicular to the axis and equal to the parameter; and let AL be bisected at M, so that $AM = AE$.

Let P be any point on the curve, and let PN (produced if necessary) meet CM in H and EM in K. Join EP, and draw MI perpendicular to HK. Then, by similar triangles,

$$MI = IK, \text{ and } EN = NK.$$

* The area represented by the second term on the right-hand side of the equation is of course described, in Apollonius' phrase, as the rectangle on the base AN similar to that contained by the axis (as base) and the sum (or difference) of the axis and its parameter. A similar remark applies to the similar expression on the next page.

Now $PN^2 = 2$ (quadrilateral $MANH$), [Prop. 81]

and $EN^2 = 2\triangle ENK$;

$$\therefore PE^2 = 2(\triangle EAM + \triangle MHK)$$
$$= AE^2 + MI.HK$$
$$= AE^2 + MI.(IK \pm IH)$$
$$= AE^2 + MI.(MI \pm IH) \text{.................. (1).}$$

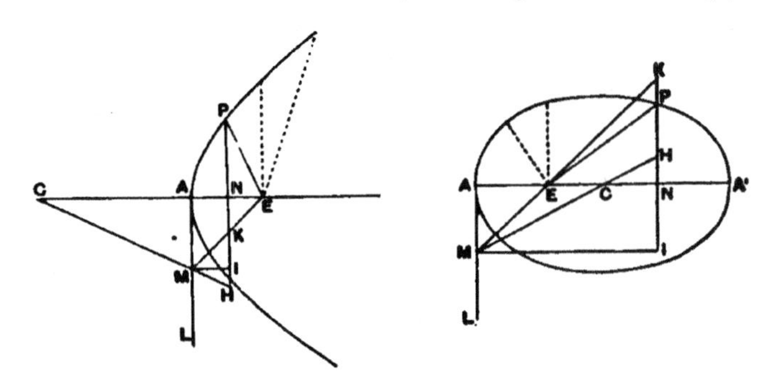

Now $MI : IH = CA : AM = AA' : p_a$.

Therefore $MI.(MI \pm IH) : AA'.(AA' \pm p_a) = MI^2 : AA'^2$,

or $$MI.(MI \pm IH) = \frac{MI^2}{AA'^2}.AA'.(AA' \pm p_a)$$
$$= MI^2.\frac{AA' \pm p_a}{AA'}$$
$$= AN^2.\frac{AA' \pm p_a}{AA'},$$

whence, by means of (1),

$$PE^2 = AE^2 + AN^2.\frac{AA' \pm p_a}{AA'}.$$

It follows that AE is the *minimum* value of PE, and that PE increases with AN, i.e. as the point P moves further from A.

Also in the ellipse the *maximum* value of PE^2 is

$$AE^2 + AA'(AA' - p_a) = AE^2 + AA'^2 - 2AE.AA'$$
$$= EA'^2.$$

Proposition 84.

[V. 7.]

If any point O be taken on the axis of any conic such that $AO < \frac{1}{2}p_a$, then OA is the **minimum** *straight line from O to the curve, and OP (if P is any other point on it) increases as P moves further and further from A.*

Let AE be set off along the axis equal to half the parameter, and join PE, PO, PA.

Then [Props. 82, 83] $PE > AE$,

so that $$\angle PAE > \angle APE;$$

and *a fortiori*

$$\angle PAO > \angle APO,$$

so that $$PO > AO.$$

And, if P' be another point more remote from A,

$$P'E > PE.$$

$$\therefore \angle EPP' > \angle EP'P;$$

and *a fortiori*

$$\angle OPP' > \angle OP'P.$$

$$\therefore OP' > OP,$$

and so on.

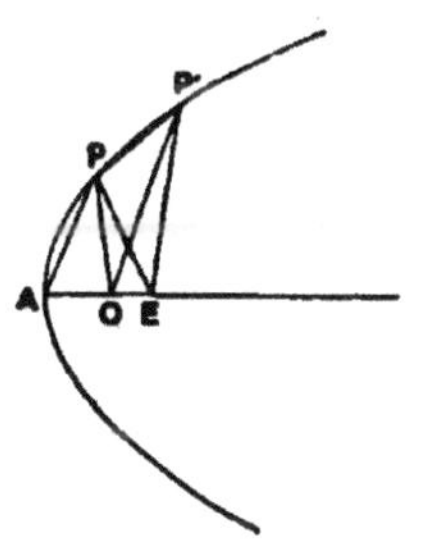

Proposition 85.

[V. 8.]

In a parabola, if G be a point on the axis such that $AG > \frac{1}{2}p_a$, and if N be taken between A and G such that

$$NG = \frac{p_a}{2},$$

then, if NP is drawn perpendicular to the axis meeting the curve in P, PG is the **minimum** *straight line from G to the curve [or the normal at P].*

If P' be any other point on the curve, $P'G$ increases as P' moves further from P in either direction.

Also $$P'G^2 = PG^2 + NN'^2.$$

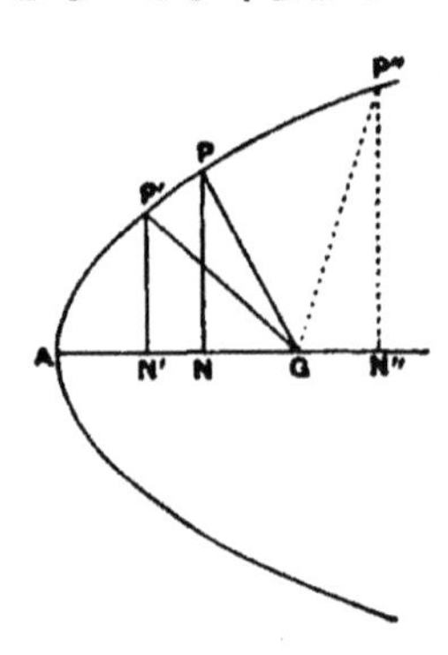

We have $$P'N'^2 = p_a . AN'$$
$$= 2NG . AN'.$$

Also $$N'G^2 = NN'^2 + NG^2 \pm 2NG . NN'$$
(according to the position of N').

Therefore, adding,

$$P'G^2 = 2NG . AN + NN'^2 + NG^2$$
$$= PN^2 + NG^2 + NN'^2$$
$$= PG^2 + NN'^2.$$

Thus it is clear that PG is the *minimum* straight line from G to the curve [or the normal at P].

And $P'G$ increases with NN', i.e. as P' moves further from P in either direction.

Proposition 86.

[V. 9, 10, 11.]

In a hyperbola or an ellipse, if G be any point on AA' (within the curve) such that $AG > \frac{p_a}{2}$, and if GN be measured towards the nearer vertex A so that

$$NG : CN = p_a : AA' \; [= CB^2 : CA^2],$$

then, if the ordinate through N meet the curve in P, PG is the **minimum** *straight line from G to the curve [or PG is the normal at P]; and, if P' be any other point on the curve, $P'G$ increases as P' moves further from P on either side.*

Also $$P'G^2 - PG^2 = NN'^2 \cdot \frac{AA' \pm p_a}{AA'}$$

$$[= e^2 \cdot NN'^2],$$

where $P'N'$ is the ordinate from P'.

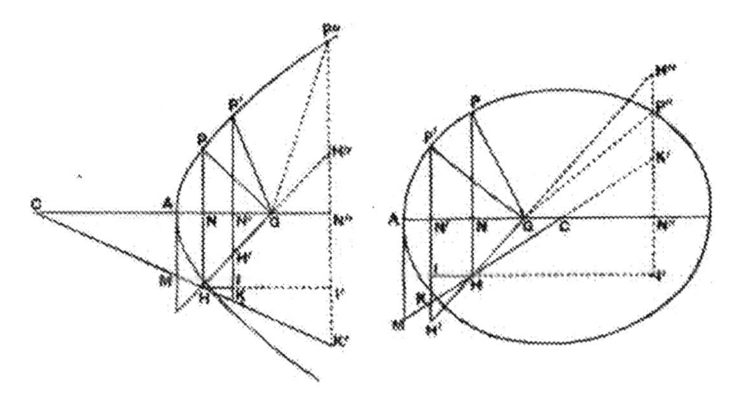

Draw AM perpendicular to the axis and equal to half the parameter. Join CM meeting PN in H and $P'N'$ in K. Join GH meeting $P'N'$ in H'.

Then since, by hypothesis,

$$NG : CN = p_a : AA',$$

and, by similar triangles,

$$NH : CN = AM : AC$$

$$= p_a : AA',$$

it follows that $$NH = NG,$$

whence also $$N'H' = N'G.$$

Now $PN^2 = 2$ (quadrilateral $MANH$), [Prop. 81]

$NG^2 = 2 \triangle HNG$.

Therefore, by addition, $PG^2 = 2$ (quadrilateral $AMHG$).

Also $P'G^2 = P'N'^2 + N'G^2 = 2\,(\text{quadr. } AMKN') + 2\triangle H'N'G$

$$= 2\,(\text{quadr. } AMHG) + 2\triangle HH'K.$$

$$\therefore\ P'G^2 - PG^2 = 2\triangle HH'K$$
$$= HI\,.\,(H'I \pm IK)$$
$$= HI\,.\,(HI \pm IK)$$
$$= HI^2\,.\,\frac{CA \pm AM}{CA} = NN'^2\,.\,\frac{AA' \pm p_a}{AA'}.$$

Thus it follows that PG is the *minimum* straight line from G to the curve, and $P'G$ increases with NN' as P' moves further from P in either direction.

In the ellipse GA' will be the *maximum* straight line from G to the curve, as is easily proved in a similar manner.

COR. In the particular case where G coincides with C, the centre, the two minimum straight lines are proved in a similar manner to be CB, CB', and the two maxima CA, CA', and CP increases continually as P moves from B to A.

Proposition 87.

[V. 12.]

If G be a point on the axis of a conic and GP be the minimum straight line from G to the curve [or the normal at P], and if O be any point on PG, then OP is the minimum straight line from O to the curve, and OP' continually increases as P' moves from P to A [or to A'].

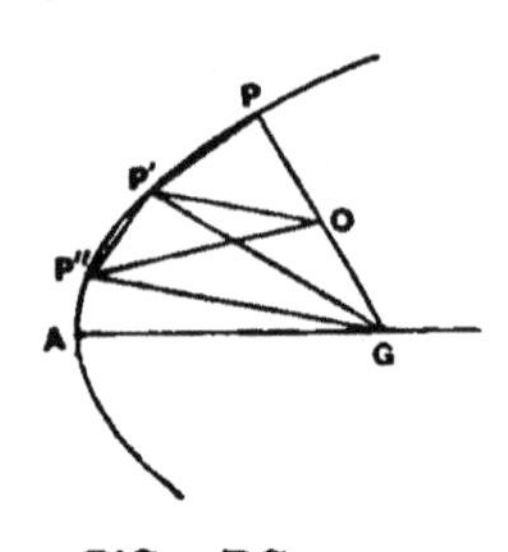

Since $$P'G > PG,$$
$$\angle GPP' > \angle GP'P.$$

Therefore, *a fortiori*,

$$\angle OPP' > \angle OP'P,$$

or $$OP' > OP.$$

Similarly $OP'' > OP'$ [&c. as in Prop. 84].

[There follow three propositions establishing for the three curves, by *reductio ad absurdum*, the converse of the propositions 85 and 86 just given. It is also proved that the normal makes with the axis towards the nearer vertex an acute angle.]

Proposition 88.

[V. 16, 17, 18.]

If E' be a point on the minor axis of an ellipse at a distance from B equal to half the parameter of BB' $\left[\text{or } \frac{CA^2}{CB}\right]$, *then $E'B$ is the* **maximum** *straight line from E to the curve; and, if P be any other point on it, $E'P$ diminishes as P moves further from B on either side.*

Also $E'B^2 - E'P^2 = Bn^2 . \frac{p_b - BB'}{BB'} \left[= Bn^2 . \frac{CA^2 - CB^2}{CB^2}\right]$.

Apollonius proves this separately for the cases (1) where $\frac{p_b}{2} < BB'$, (2) where $\frac{p_b}{2} = BB'$, and (3) where $\frac{p_b}{2} > BB'$.

The method of proof is the same for all three cases, and only the first case of the three is given here.

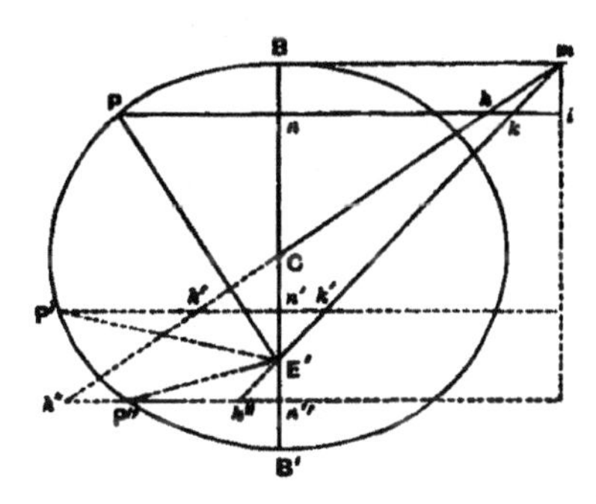

By Prop. 81 (which is applicable to either axis) we have, if $Bm = \frac{p_b}{2} = BE'$, and Pn meets Cm, $E'm$ in h, k respectively,

$$Pn^2 = 2\,(\text{quadrilateral } mBnh).$$

Also $$nE'^2 = 2\triangle nkE'.$$

$$\therefore\ PE'^2 = 2\triangle mBE' - 2\triangle mhk.$$

But $$BE'^2 = 2\triangle mBE'.$$

$$\therefore\ BE'^2 - PE'^2 = 2\triangle mhk$$

$$= mi\,.\,(hi - ki) = mi\,.\,(hi - mi)$$

$$= mi^2\,.\,\frac{mB - CB}{CB}$$

$$= Bn^2\,.\,\frac{p_b - BB'}{BB'},$$

whence the proposition follows.

Proposition 89.

[V. 19.]

If BE' be measured along the minor axis of an ellipse equal to half the parameter $\left[\text{or } \frac{CA^2}{CB}\right]$ *and any point O be taken on the minor axis such that $BO > BE'$, then OB is the* **maximum** *straight line from O to the curve; and, if P be any other point on it, OP diminishes continually as P moves in either direction from B to B'.*

The proof follows the method of Props. 84, 87.

Proposition 90.

[V. 20, 21, 22.]

If g be a point on the minor axis of an ellipse such that $Bg > BC$ *and* $Bg < \frac{1}{2}p_b \left[\text{or } \frac{CA^2}{CB}\right]$, *and if Cn be measured towards B so that*

$$Cn : ng = BB' : p_b \; [= CB^2 : CA^2],$$

then the perpendicular through n to BB′ will meet the curve in two points P such that Pg is the **maximum** *straight line from g to the curve.*

Also, if P′ be any other point on the curve, P′g diminishes as P′ moves further from P on either side to B or B′, and

$$Pg^2 - P'g^2 = nn'^2 . \frac{p_b - BB'}{BB'} \left[= nn'^2 . \frac{CA^2 - CB^2}{CB^2}\right].$$

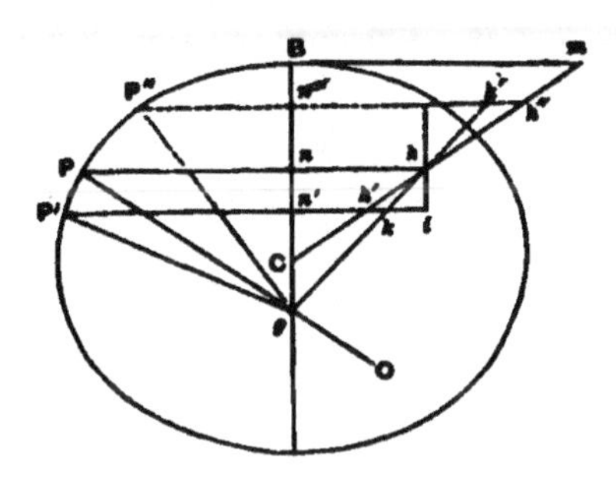

Draw Bm perpendicular to BB' and equal to half its parameter p_b. Join Cm meeting Pn in h and $P'n'$ in h', and join gh meeting $P'n'$ in k.

Then since, by hypothesis,

$$Cn : ng = BB' : p_b = BC : Bm,$$

and $\qquad Cn : nh = BC : Bm$, by similar triangles,

it follows that $ng = nh$. Also $gn' = n'k$, and $hi = ik$, where hi is perpendicular to $P'n'$.

Now $\quad Pn^2 = 2$ (quadrilateral $mBnh$),

$$ng^2 = 2\triangle hng;$$

$$\therefore\ Pg^2 = 2\,(mBnh + \triangle hng).$$

Similarly $\quad P'g^2 = 2\,(mBn'h' + \triangle kn'g).$

By subtraction,

$$Pg^2 - P'g^2 = 2\triangle hh'k$$

$$= hi\,.\,(h'i - ki)$$

$$= hi\,.\,(h'i - hi)$$

$$= hi^2\,.\left(\frac{Bm - BC}{BC}\right)$$

$$= nn'^2\,.\,\frac{p_b - BB'}{BB'};$$

whence it follows that Pg is the *maximum* straight line from g to the curve, and the difference between Pg^2 and $P'g^2$ is the area described.

COR. 1. It follows from the same method of proof as that used in Props. 84, 87, 89 that, if O be any point on Pg produced beyond the minor axis, PO is the *maximum* straight line that can be drawn from O to the same part of the ellipse in which Pg is a maximum, i.e. to the semi-ellipse BPB', and if OP' be drawn to any other point on the semi-ellipse, OP' diminishes as P' moves from P to B or B'.

COR. 2. In the particular case where g coincides with the centre C, the maximum straight line from C to the ellipse is perpendicular to BB', viz. CA or CA'. Also, if g be not the centre, the angle PgB must be acute if Pg is a *maximum*; and, if Pg is a maximum [or a normal],

$$Cn : ng = CB^2 : CA^2.$$

[This corollary is proved separately by *reductio ad absurdum*.]

Proposition 91.

[V. 23.]

If g be on the minor axis of an ellipse, and gP is a maximum straight line from g to the curve, and if gP meet the major axis in G, GP is a **minimum** *straight line from G to the curve.*

[In other words, the minimum from G and the maximum from g determine one and the same normal.]

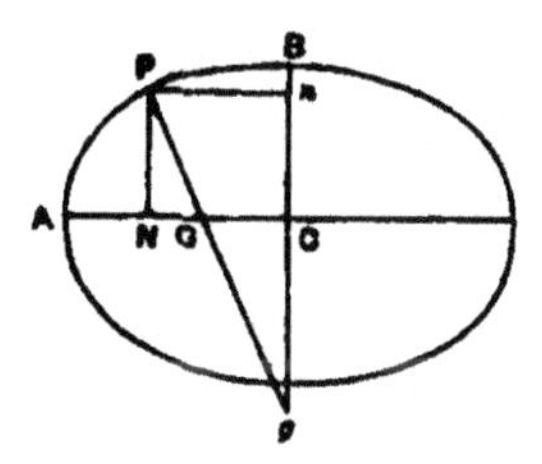

We have $Cn : ng = BB' : p_b$ [Prop. 90]

$[= CB^2 : CA^2]$

$= p_a : AA'$.

Also $Cn : ng = PN : ng$

$= NG : Pn$, by similar triangles

$= NG : CN$.

$\therefore NG : CN = p_a : AA'$,

or PG is the normal determined as the *minimum* straight line from G. [Prop. 86]

Proposition 92.

[V. 24, 25, 26.]

Only one normal can be drawn from any one point of a conic, whether such normal be regarded as the minimum straight line from the point in which it meets AA', or as the maximum straight line from the point in which (in the case of an ellipse) it meets the minor axis.

This is at once proved by *reductio ad absurdum* on assuming that PG, PH (meeting the axis AA' in G, H) are minimum straight lines from G and H to the curve, and on a similar assumption for the minor axis of an ellipse.

Proposition 93.

[V. 27, 28, 29, 30.]

The normal at any point P on a conic, whether regarded as a minimum straight line from its intersection with the axis AA' or as a maximum from its intersection with BB' (in the case of an ellipse), is perpendicular to the tangent at P.

Let the tangent at P meet the axis of the parabola, or the axis AA' of a hyperbola or an ellipse, in T. Then we have to prove that TPG is a right angle.

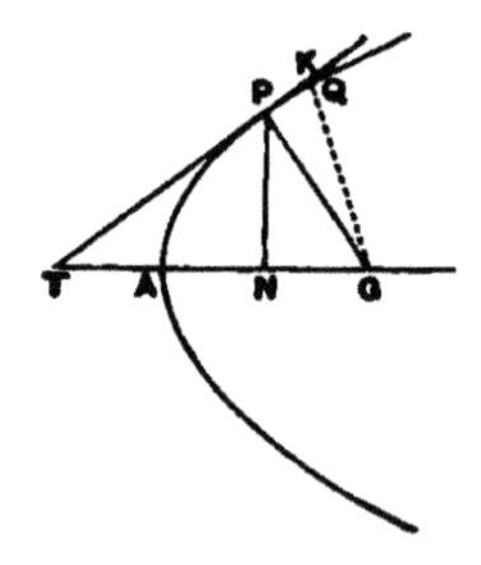

(1) For the *parabola* we have

$$AT = AN, \text{ and } NG = \frac{p_a}{2};$$

$$\therefore NG : p_a = AN : NT,$$

so that

$$TN \cdot NG = p_a \cdot AN$$
$$= PN^2.$$

And the angle at N is a right angle;

$$\therefore \angle TPG \text{ is a right angle.}$$

(2) For the *hyperbola* or *ellipse*

$PN^2 : CN.NT$

$= p_a : AA'$ [Prop. 14]

$= NG : CN$, by the property of the minimum, [Prop. 86]

$= TN.NG : CN.NT$.

$\therefore PN^2 = TN.NG$, while the angle at N is right;

$\therefore \angle TPG$ is a right angle.

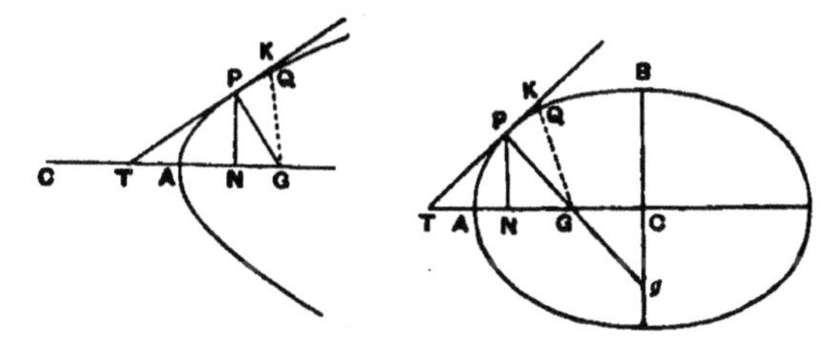

(3) If Pg be the maximum straight line from g on the minor axis of an ellipse, and if Pg meet AA' in G, PG is a minimum from G, and the result follows as in (2).

[Apollonius gives an alternative proof applicable to all three conics. If GP is not perpendicular to the tangent, let GK be perpendicular to it.

Then $\angle GKP > \angle GPK$, and therefore $GP > GK$.

Hence *a fortiori* $GP > GQ$, where Q is the point in which GK cuts the conic; and this is impossible because GP is a *minimum*. Therefore &c.]

Proposition 94.

[V. 31, 33, 34.]

(1) *In general, if O be any point within a conic and OP be a maximum or a minimum straight line from O to the conic, a straight line PT drawn at right angles to PO will touch the conic at P.*

(2) *If O' be any point on OP produced outside the conic, then, of all straight lines drawn from O' to meet the conic in one point but not produced so as to meet it in a second point, $O'P$ will be the minimum; and of the rest that which is nearer to it will be less than that which is more remote.*

(1) *First*, let OP be a *maximum*. Then, if TP does not touch the conic, let it cut it again at Q, and draw OK to meet PQ in K and the curve in R.

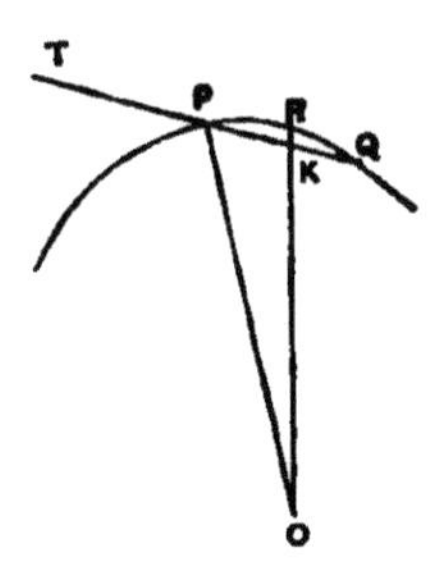

Then, since the angle OPK is right, $\angle OPK > \angle OKP$.

Therefore $OK > OP$, and *a fortiori* $OR > OP$: which is impossible, since OP is a maximum.

Therefore TP must touch the conic at P.

Secondly, let OP be a *minimum*. If possible, let TP cut the curve again in Q. From any point between TP and the curve draw a straight line to P and draw ORK perpendicular to this

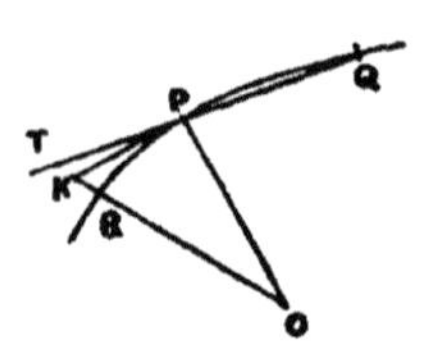

line meeting it at K and the curve in R. Then the angle OKP is a right angle. Therefore $OP > OK$, and *a fortiori* $OP > OR$: which is impossible, since OP is a minimum. Therefore TP must touch the curve.

(2) Let O' be any point on OP produced. Draw the tangent at P, as PK, which is therefore at right angles to OP. Then draw $O'Q$, $O'R$ to meet the curve in one point only, and let $O'Q$ meet PK in K.

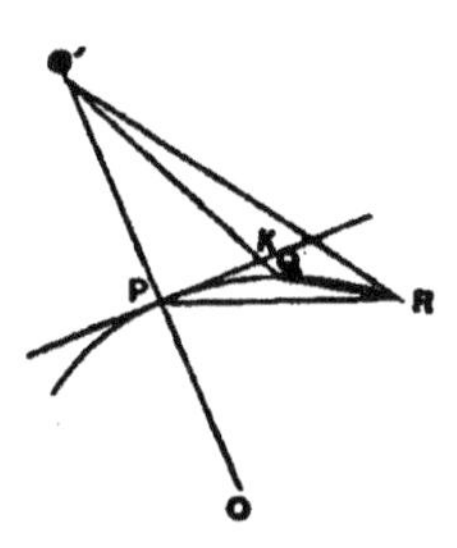

Then $O'K > O'P$. Therefore *a fortiori* $O'Q > O'P$, and $O'P$ is a minimum.

Join RP, RQ. Then the angle $O'QR$ is obtuse, and therefore the angle $O'RQ$ is acute. Therefore $O'R > O'Q$, and so on.

Proposition 95.

[V. 35, 36, 37, 38, 39, 40.]

(1) *If the normal at P meet the axis of a parabola or the axis AA' of a hyperbola or ellipse in G, the angle PGA increases as P or G moves further and further from A, but in the hyperbola the angle PGA will always be less than the complement of half the angle between the asymptotes.*

(2) *Two normals at points on the same side of the axis AA' will meet on the opposite side of that axis.*

(3) *Two normals at points on the same quadrant of an ellipse, as AB, will meet at a point within the angle ACB'.*

(1) Suppose P' is further from the vertex than P. Then, since PG, $P'G'$ are minimum straight lines from G, G' to the curve, we have

(*a*) For the *parabola*

$$NG = \frac{p_a}{2} = N'G',$$

and $$P'N' > PN;$$

$$\therefore \angle P'G'A > \angle PGA.$$

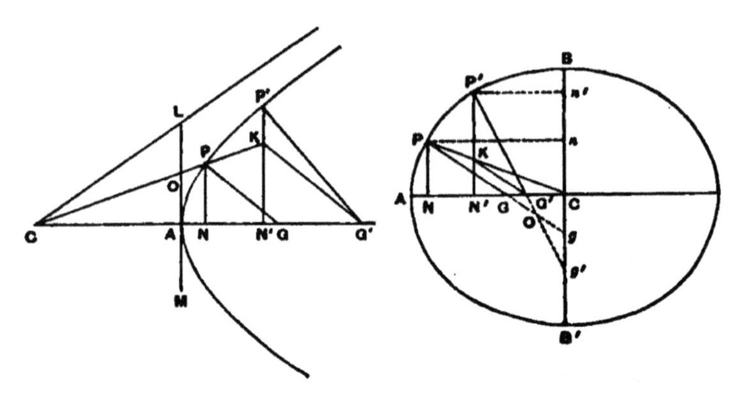

(*b*) For the *hyperbola* and *ellipse*, joining CP and producing it if necessary to meet $P'N'$ in K, and joining KG', we have

$$N'G' : CN' = p_a : AA' \quad \text{[Prop. 86]}$$
$$= NG : CN;$$
$$\therefore N'G' : NG = CN' : CN$$
$$= KN' : PN, \text{ by similar triangles.}$$

Therefore the triangles PNG, $KN'G'$ are similar, and

$$\angle KG'N' = \angle PGN.$$

Therefore $$\angle P'G'N' > \angle PGN.$$

(*c*) In the *hyperbola*, let AL be drawn perpendicular to AA' to meet the asymptote in L and CP in O. Also let AM be equal to $\frac{p_a}{2}$.

Now $AA' : p_a = CA : AM = CN : NG$,

and $OA : CA = PN : CN$, by similar triangles;

therefore, *ex aequali*, $OA : AM = PN : NG$.

Hence $AL : AM > PN : NG$.

But $$AL : AM = CA : AL; \qquad \text{[Prop. 28]}$$

$$\therefore CA : AL > PN : NG;$$

$$\therefore \angle PGN \text{ is less than } \angle CLA.$$

(2) It follows at once from (1) that two normals at points on one side of AA' will meet on the other side of AA'.

(3) Regard the two normals as the *maximum* straight lines from g, g', the points where they meet the minor axis of the ellipse.

Then $$Cn' : n'g' = BB' : p_b \qquad \text{[Prop. 90]}$$

$$= Cn : ng;$$

$$\therefore Cn' : Cg' = Cn : Cg.$$

But $$Cn' > Cn; \quad \therefore Cg' > Cg,$$

whence it follows that Pg, $P'g'$ must cross at a point O before cutting the minor axis. Therefore O lies on the side of BB' towards A.

And, by (2) above, O lies below AC; therefore O lies within the $\angle ACB'$.

Proposition 96.

[V. 41, 42, 43.]

(1) *In a parabola or an ellipse any normal PG will meet the curve again.*

(2) *In the hyperbola (a), if AA' be not greater than p_a, no normal can meet the curve in a second point on the same branch; but (b), if $AA' > p_a$, some normals will meet the same branch again and others not.*

(1) For the *ellipse* the proposition is sufficiently obvious, and in the *parabola*, since PG meets a diameter (the axis), it will meet another diameter, viz. that through the point of contact of the tangent parallel to PG, i.e. the diameter bisecting it. Therefore it will meet the curve again.

(2) (*a*) Let CL, CL' be the asymptotes, and let the tangent at A meet them in L, L'. Take AM equal to $\frac{p_a}{2}$. Let PG be any normal and PN the ordinate.

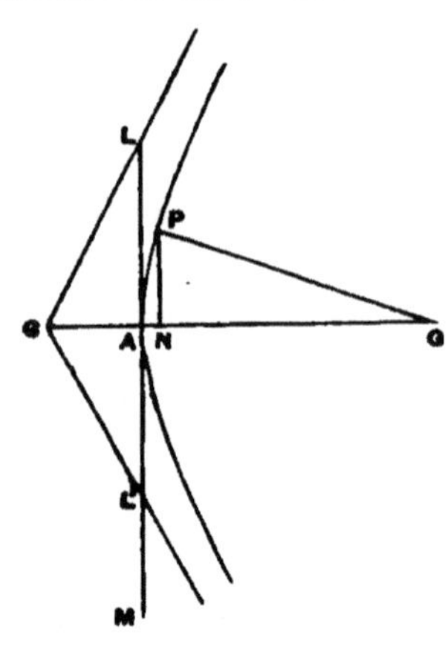

Then, by hypothesis, $CA \not> AM$,

and $$CA : AM = CA^2 : AL^2; \qquad \text{[Prop. 28]}$$

$$\therefore CA \not> AL;$$

hence the angle CLA is not greater than ACL or ACL'.

But $$\angle CLA > \angle PGN; \qquad \text{[Prop. 95]}$$

$$\therefore \angle ACL' > \angle PGN.$$

It follows that the angle ACL' together with the angle adjacent to PGN will be greater than two right angles.

Therefore PG will not meet CL' towards L' and therefore will not meet the branch of the hyperbola again.

(*b*) Suppose $CA > AM$ or $\frac{p_a}{2}$. Then

$$LA : AM > LA : AC.$$

Take a point K on AL such that

$$KA : AM = LA : AC.$$

Join CK, and produce it to meet the hyperbola in P, and let PN be the ordinate, and PG the normal, at P.

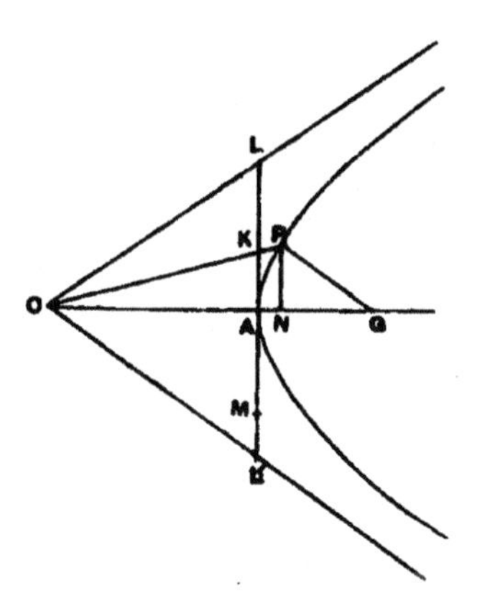

PG is then the *minimum* from G to the curve, and

$$NG : CN = p_a : AA'$$
$$= AM : AC.$$

Also $\qquad CN : PN = AC : AK$, by similar triangles.

Therefore, *ex aequali*, $\quad NG : PN = AM : AK$

$$= CA : AL, \text{ from above.}$$

Hence $\qquad \angle ACL' = \angle ACL = \angle PGN$;

$\therefore$ PG, CL' are parallel and do not meet.

But the normals at points between A and P make with the axis angles less than the angle PGN, and normals at points beyond P make with the axis angles greater than PGN.

Therefore normals at points between A and P will not meet the asymptote CL', or the branch of the hyperbola, again; but normals beyond P will meet the branch again.

Proposition 97.

[V. 44, 45, 46, 47, 48.]

If P_1G_1, P_2G_2 be normals at points on one side of the axis of a conic meeting in O, and if O be joined to any other point P on the conic (it being further supposed in the case of the ellipse that all three lines OP_1, OP_2, OP cut the same half of the axis), then

(1) *OP cannot be a normal to the curve;*

(2) *if OP meet the axis in K, and PG be the normal at P,*

$$AG < AK \text{ when } P \text{ is intermediate between } P_1 \text{ and } P_2,$$

and $$AG > AK \text{ when } P \text{ does not lie between } P_1 \text{ and } P_2.$$

I. First let the conic be a PARABOLA.

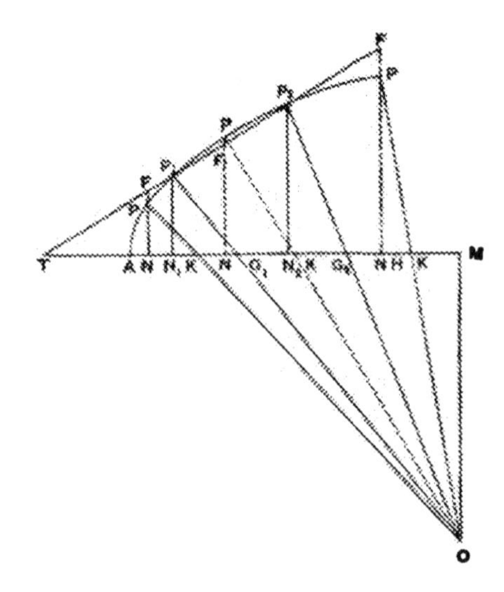

Let P_1P_2 meet the axis in T, and draw the ordinates P_1N_1, P_2N_2.

Draw OM perpendicular to the axis, and measure MH towards the vertex equal to $\frac{p_a}{2}$.

Then $$MH = N_2G_2,$$
and $$N_2H = G_2M.$$

Therefore $$MH : HN_2 = N_2G_2 : G_2M$$
$$= P_2N_2 : MO, \text{ by similar triangles.}$$

Therefore $$HM . MO = P_2N_2 . N_2H$$
Similarly $$HM . MO = P_1N_1 . N_1H \quad \text{.................. (A).}$$

Therefore $$HN_1 : HN_2 = P_2N_2 : P_1N_1$$
$$= TN_2 : TN_1,$$
whence $$N_1N_2 : HN_1 = N_1N_2 : TN_2;$$
$$\therefore TN_2 = HN_1$$
and $$TN_1 = HN_2 \quad \text{.......................... (B).}$$

If P be a variable point and PN the ordinate*, we have now three cases:

$$TN < TN_1 \text{ or } HN_2 \text{..............................(1),}$$
$$TN > TN_1 \text{ or } HN_2, \text{ but } < TN_2 \text{ or } HN_1 \text{......(2),}$$
$$TN > TN_2 \text{ or } HN_1 \text{..............................(3).}$$

Thus, denoting the several cases by the numbers (1), (2), (3), we have

$$N_2N : TN > N_2N : HN_2 \text{.................(1),}$$
$$< N_2N : HN_2 \text{.................(2),}$$
$$< N_2N : HN_2 \text{.................(3),}$$

and we derive respectively

$$TN_2 : TN > HN : HN_2 \text{.................(1),}$$
$$< HN : HN_2 \text{.................(2),}$$
$$> HN : HN_2 \text{.................(3).}$$

* It will be observed that there are three sets of points P, N, K, in the figure denoted by the same letters. This is done in order to exhibit the three different cases; and it is only necessary to bear in mind that attention must be confined to one at a time as indicated in the course of the proof.

If NP meet P_1P_2 in F, we have, by similar triangles,

$$P_2N_2 : FN > HN : HN_2 \text{......(1) and (3)},$$
$$< HN : HN_2 \text{..................(2)}.$$

But in (1) and (3) $FN > PN$, and in (2) $FN < PN$.

Therefore, *a fortiori* in all the cases,

$$P_2N_2 : PN > HN : HN_2 \text{...... (1) and (3)},$$
$$< HN : HN_2 \text{.................(2)}.$$

Thus $P_2N_2 . N_2H > PN . NH$......................(1) and (3),
$$< PN . NH \text{..............................(2)}.$$

Hence $\left.\begin{array}{l} HM . MO > PN . NH ...\text{(1) and (3)} \\ \quad\quad\quad\; < PN . NH .... \text{(2)}\end{array}\right\}$, by (A) above.

Therefore $MO : PN > NH : HM$.................. (1) and (3),
$$< NH : HM \text{............................ (2)}$$

and $MO : PN = MK : NK$.

Therefore $MK : NK > NH : HM$ .... (1) and (3),
$$< NH : HM \text{..................(2)},$$

whence we obtain $MN : NK > MN : HM$ .... (1) and (3),
$$< MN : HM \text{..................(2)},$$

so that HM or $NG > NK$ in (1) and (3),
and $< NK$ in (2).

Thus the proposition is proved.

II. Let the conic be a HYPERBOLA or an ELLIPSE.

Let the normals at P_1, P_2 meet at O, and draw OM perpendicular to the axis. Divide CM in H (internally for the hyperbola and externally for the ellipse) so that

$$CH : HM = AA' : p_a \text{ [or } CA^2 : CB^2\text{]},$$

and let OM be similarly divided at L. Draw HVR parallel to OM and LVE, ORF parallel to CM.

Suppose P_2P_1 produced to meet EL in T, and let P_1N_1, P_2N_2 meet it in U_1, U_2.

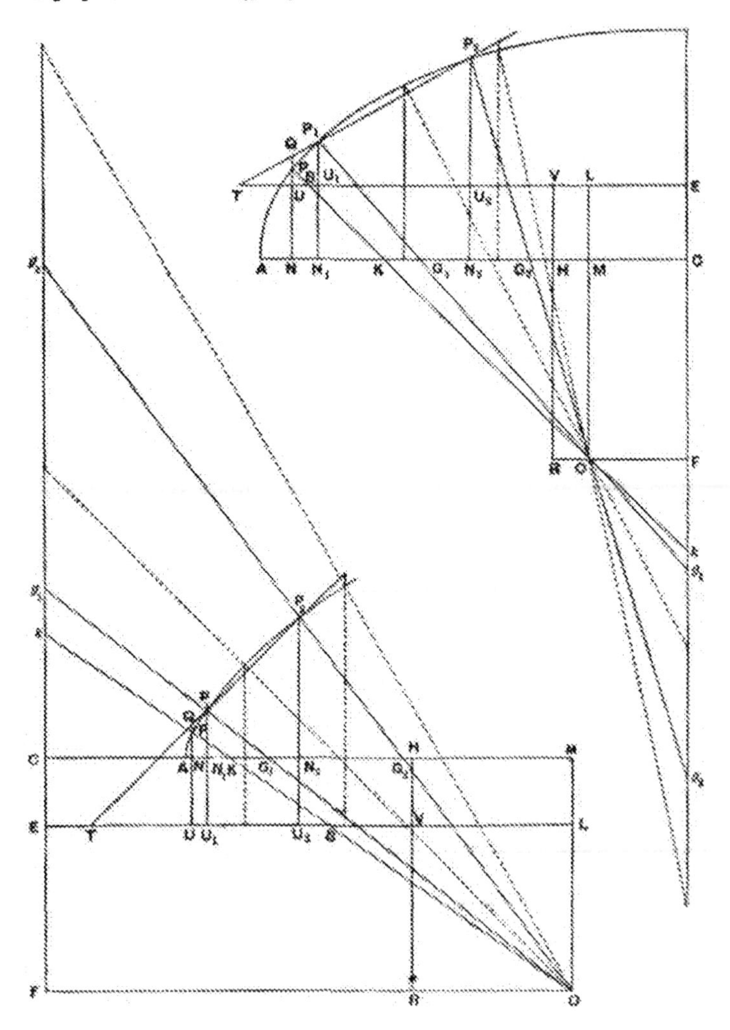

Take any other point P on the curve. Join OP meeting the axes in K, k, and let PN meet P_1P_2 in Q and EL in U.

Now $CN_2 : N_2G_2 = AA' : p_a = CH : HM$.

Therefore, *componendo* for the hyperbola and *dividendo* for the ellipse,

$$CM : CH = CG_2 : CN_2$$
$$= CG_2 \sim CM : CN_2 \sim CH$$
$$= MG_2 : HN_2$$
$$= MG_2 : VU_2 \quad \text{.................... (A).}$$

Next

$$FE : EC = AA' : p_a = CN_2 : N_2G_2,$$

so that $FC : CE = CG_2 : N_2G_2$.

Thus $FC : N_2U_2 = CG_2 : N_2G_2$

$= Cg_2 : P_2N_2$, by similar triangles,

$= FC \pm Cg_2 : N_2U_2 \pm P_2N_2$

$= Fg_2 : P_2U_2$ .... (B).

Again

$$FC . CM : EC . CH = (FC : CE) . (CM : CH)$$
$$= (Fg_2 : P_2U_2) . (MG_2 : VU_2),$$

from (A) and (B),

and $FC . CM = Fg_2 . MG_2$, $\because Fg_2 : CM = FC : MG_2$.

$\therefore EC . CH = P_2U_2 . U_2V$,

or $CE . EV = P_2U_2 . U_2V$

$= P_1U_1 . U_1V$, in like manner;

$\therefore U_1V : U_2V = P_2U_2 : P_1U_1$

$= TU_2 : TU_1$, by similar triangles,

whence $U_1U_2 : U_1V = U_1U_2 : TU_2$;

$$\left.\begin{aligned} \therefore TU_2 &= VU_1 \\ TU_1 &= VU_2 \end{aligned}\right\} \quad \text{.......................... (C).}$$

and

Now suppose (1) that $AN < AN_1$;

then $U_2V > TU$, from (C) above;

$\therefore UU_2 : TU > UU_2 : U_2V$;

hence $TU_2 : TU > UV : U_2V$;

$\therefore P_2U_2 : QU > UV : U_2V$,

by similar triangles.

Therefore $P_2U_2 . U_2V > QU . UV$,

and *a fortiori* $> PU . UV$.

But $P_2U_2 . U_2V = CE . EV$, from above,

$$= LO . OR, \because CE : LO = OR : EV;$$

$$\therefore LO . OR > PU . UV.$$

Suppose (2) that $AN > AN_1$ but $< AN_2$.

Then $$TU_1 < UV;$$

$$\therefore U_1U : TU_1 > U_1U : UV,$$

whence $$TU : TU_1 > U_1V : UV;$$

$$\therefore QU : P_1U_1 > U_1V : UV,$$

by similar triangles.

Therefore (*a fortiori*) $PU . UV > P_1U_1 . U_1V$

$$> LO . OR.$$

Lastly (3) let AN be $> AN_2$.

Then $$TU_1 > UV;$$

$$\therefore U_1U : TU_1 < U_1U : UV,$$

whence $$TU : TU_1 < U_1V : UV,$$

or $$QU : P_1U_1 < U_1V : UV;$$

$$\therefore P_1U_1 . U_1V > QU . UV,$$

and *a fortiori* $$> PU . UV;$$

$$\therefore LO . OR > PU . UV,$$

as in (1) above.

Thus we have for cases (1) and (3)

$$LO . OR > PU . UV,$$

and for (2) $$LO . OR < PU . UV.$$

That is, we shall have, supposing the upper symbol to refer to (1) and (3) and the lower to (2),

$$LO : PU \gtrless UV : OR,$$

i.e. $$LS : SU \gtrless UV : LV;$$

$$\therefore LU : US \gtrless LU : LV,$$

and $$LV \gtrless US.$$

It follows that

$$FO : LV \lesseqgtr FO : SU, \text{ or } Fk : PU,$$

or

$$CM : MH \lesseqgtr Fk : PU;$$

$$\therefore FC : CE \lesseqgtr Fk : PU$$

$$\lesseqgtr Fk \mp FC : PU \mp CE$$

$$\lesseqgtr Ck : PN$$

$$\lesseqgtr CK : NK.$$

Therefore, *componendo* or *dividendo*,

$$FE : EC \lesseqgtr CN : NK,$$

or

$$CN : NK \gtreqless FE : EC,$$

i.e.

$$\gtreqless AA' : p_a.$$

But

$$CN : NG = AA' : p_a;$$

$$\therefore NK \lesseqgtr NG;$$

i.e. when P is not between P_1 and P_2 $NK < NG$, and when P lies between P_1 and P_2, $NK > NG$, whence the proposition follows.

COR. 1. In the particular case of a quadrant of an ellipse where P_2 coincides with B, i.e. where O coincides with g_1, it follows that no other normal besides P_1g_1, Bg_1 can be drawn through g_1 to the quadrant, and, if P be a point between A and P_1, while Pg_1 meets the axis in K, $NG > NK$.

But if P lie between P_1 and B, $NG < NK$.

[This is separately proved by Apollonius from the property in Prop. 95 (3).]

COR. 2. *Three normals at points on one quadrant of an ellipse cannot meet at one point.*

This follows at once from the preceding propositions.

COR. 3. *Four normals at points on one semi-ellipse bounded by the major axis cannot meet at one point.*

For, if four such normals cut the major axis and meet in one point, the centre must (1) separate one normal from the three others, or (2) must separate two from the other two, or (3) must lie on one of them.

In cases (1) and (3) a contradiction of the preceding proposition is involved, and in case (2) a contradiction of Prop. 95 (3) which requires two points of intersection, one on each side of the minor axis.

Proposition 98.

[V. 49, 50.]

In any conic, if M be any point on the axis such that AM is not greater than half the latus rectum, and if O be any point on the perpendicular to the axis through M, then no straight line drawn to any point on the curve on the side of the axis opposite to O and meeting the axis between A and M can be a normal.

Let OP be drawn to the curve meeting the axis in K, and let PN be the ordinate at P.

We have in the *parabola*, since $AM \ngtr \frac{p_a}{2}$,

$$NM < \frac{p_a}{2}, \quad \text{i.e.} < NG.$$

Therefore, *a fortiori*, $NK < NG$.

For the *hyperbola* and *ellipse* $AA' : p_a$ is not greater than $CA : AM$,

and $$CN : NM > CA : AM;$$

$$\therefore CN : NM > AA' : p_a$$
$$> CN : NG;$$
$$\therefore NM < NG,$$

and *a fortiori* $$NK < NG.$$

Therefore OP is not a normal.

PROPOSITIONS LEADING IMMEDIATELY TO THE DETERMINATION OF THE *EVOLUTE*.

Proposition 99.

[V. 51, 52.]

If AM measured along the axis be greater than $\frac{p_a}{2}$ (but in the case of the ellipse less than AC), and if MO be drawn perpendicular to the axis, then a certain length [y] can be assigned such that

(*a*) *if $OM > y$, no normal can be drawn through O which cuts the axis; but, if OP be any straight line drawn to the curve cutting the axis in K, $NK < NG$, where PN is the ordinate and PG the normal at P;*

(*b*) *if $OM = y$, only one normal can be so drawn through O, and, if OP be any other straight line drawn to the curve and meeting the axis in K, $NK < NG$, as before;*

(*c*) *if $OM < y$, two normals can be so drawn through O, and, if OP be any other straight line drawn to the curve, NK is less or greater than NG according as OP is not, or is, intermediate between the two normals.*

I. Suppose the conic is a PARABOLA.

Measure MH towards the vertex equal to $\frac{p_a}{2}$, and divide AH at N_1 so that $HN_1 = 2N_1A$.

Take a length y such that

$$y : P_1N_1 = N_1H : HM,$$

where P_1N_1 is the ordinate passing through N_1.

(a) Suppose $OM > y$.

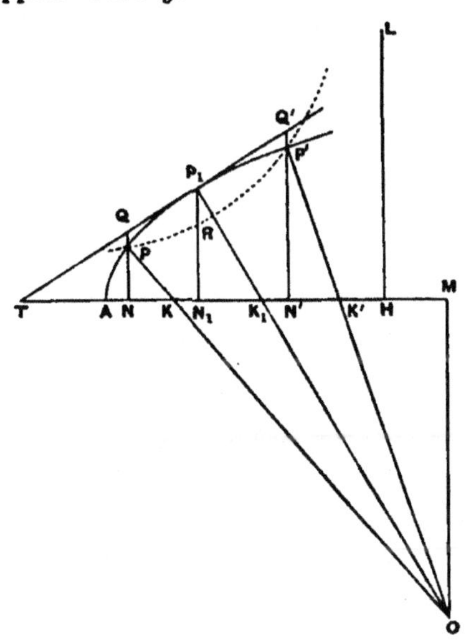

Join OP_1 meeting the axis in K_1.

Then $$y : P_1N_1 = N_1H : HM;$$

$$\therefore\ OM : P_1N_1 > N_1H : HM,$$

or $$MK_1 : K_1N_1 > N_1H : HM;$$

hence $$MN_1 : N_1K_1 > MN_1 : HM,$$

so that $$N_1K_1 < HM,$$

i.e. $$N_1K_1 < \frac{p_a}{2}.$$

Therefore OP_1 is not a normal, and $N_1K_1 < N_1G_1$.

Next let P be any other point. Join OP meeting the axis in K, and let the ordinate PN meet the tangent at P_1 in Q.

Then, if $AN < AN_1$, we have, since $N_1T = 2AN_1 = N_1H$,

$$N_1H > NT;$$

$$\therefore N_1N : NT > N_1N : HN_1;$$

thus $TN_1 : TN > HN : HN_1$,

or $P_1N_1 : QN > HN : HN_1$,

and *a fortiori*

$$P_1N_1 : PN > HN : HN_1,$$

or $P_1N_1 . N_1H > PN . NH$;

If $$AN > AN_1,$$

$$N_1T > NH;$$

$$\therefore N_1N : NH > N_1N : N_1T,$$

whence

$$HN_1 : HN > TN : TN_1$$

$$> QN : P_1N_1$$

$$> PN : P_1N_1,$$

a fortiori

$$\therefore P_1N_1 . N_1H > PN . NH.$$

But $$OM . MH > P_1N_1 . N_1H, \text{ by hypothesis};$$

$$\therefore OM . MH > PN . NH,$$

or $$OM : PN > NH : HM,$$

i.e. $$MK : KN > NH : HM,$$

by similar triangles.

Therefore, *componendo*, $MN : NK > MN : HM$,

whence $$NK < HM \text{ or } \frac{p_a}{2}.$$

Therefore OP is not a normal, and $NK < NG$.

(*b*) Suppose $OM = y$, and we have in this case

$$MN_1 : N_1K_1 = MN_1 : HM,$$

or $$N_1K_1 = HM = \frac{p_a}{2} = N_1G_1,$$

and P_1O is a normal.

If P is any other point, we have, as before,

$$P_1N_1 . N_1H > PN . NH,$$

and $P_1N_1 . N_1H$ is in this case equal to $OM . MH$.

Therefore $$OM . MH > PN . NH,$$

and it follows as before that OP is not normal, and $NK < NG$.

(*c*) Lastly, if $OM < y$,

$$OM : P_1N_1 < N_1H : HM,$$

or $$OM . MH < P_1N_1 . N_1H.$$

Let N_1R be measured along N_1P_1 so that

$$OM . MH = RN_1 . N_1H.$$

Thus R lies within the curve.

Let HL be drawn perpendicular to the axis, and with AH, HL as asymptotes draw a hyperbola passing through R. This hyperbola will therefore cut the parabola in two points, say P, P'.

Now, by the property of the hyperbola,

$$PN.NH = RN_1.N_1H$$

$$= OM.MH, \text{ from above;}$$

$$\therefore\ OM : PN = NH : HM,$$

or $$MK : KN = NH : HM,$$

and, *componendo*, $$MN : NK = MN : HM;$$

$$\therefore\ NK = HM = \frac{p_a}{2} = NG,$$

and PO is normal.

Similarly $P'O$ is normal.

Thus we have two normals meeting in O, and the rest of the proposition follows from Prop. 97.

[It is clear that in the second case where $OM = y$, O is the intersection of two consecutive normals, i.e. is the centre of curvature at the point P_1.

If then x, y be the coordinates of O, so that $AM = x$, and if $4a = p_a$,

$$HM = 2a,$$

$$N_1H = \tfrac{2}{3}(x - 2a),$$

$$AN_1 = \tfrac{1}{3}(x - 2a).$$

Also $$y^2 : P_1N_1^2 = N_1H^2 : HM^2,$$

or $$y^2 : 4a.AN_1 = N_1H^2 : 4a^2;$$

$$\therefore\ ay^2 = AN_1.N_1H^2$$

$$= \tfrac{4}{27}(x - 2a)^3,$$

or $$27ay^2 = 4(x - 2a)^3,$$

which is the Cartesian equation of the **evolute** of a parabola.]

II. Let the curve be a HYPERBOLA or an ELLIPSE.

We have $AM > \frac{p_a}{2}$, so that $CA : AM < AA' : p_a$.

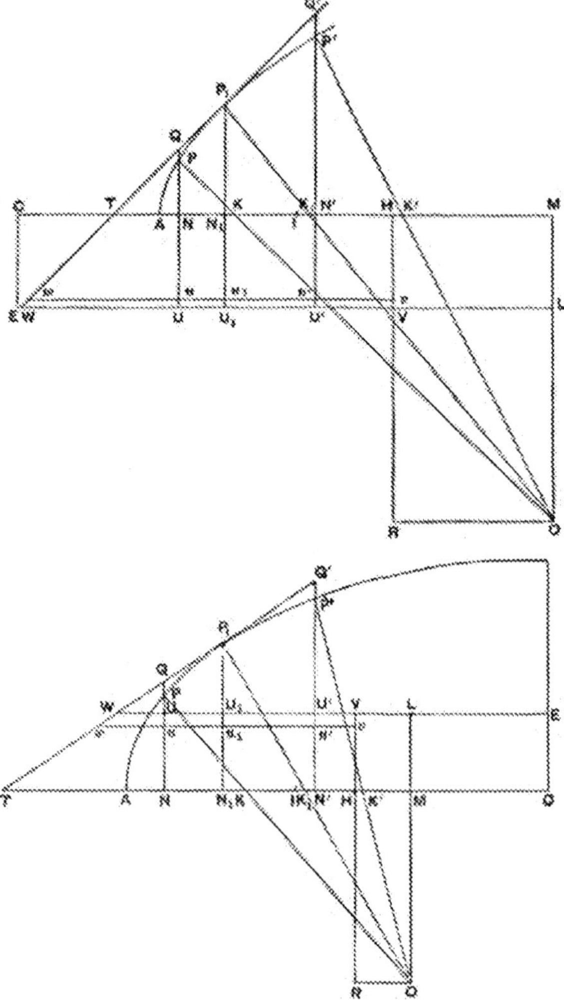

Therefore, if H be taken on AM such that $CH : HM = AA' : p_a$, H will fall between A and M.

Take two mean proportionals CN_1, CI between CA and CH*, and let P_1N_1 be the ordinate through N_1.

Take a point L on OM (in the hyperbola) or on OM produced (in the ellipse) such that $OL : LM = AA' : p_a$. Draw LVE, OR both parallel to the axis, and CE, HVR both perpendicular to the axis. Let the tangent at P_1 meet the axis in T and EL in W, and let P_1N_1 meet EL in U_1. Join OP_1, meeting the axis in K_1.

Let now y be such a length that

$$y : P_1N_1 = (CM : MH).(HN_1 : N_1C).$$

(*a*) Suppose first that $OM > y$;

$$\therefore\ OM : P_1N_1 > y : P_1N_1.$$

But

$$OM : P_1N_1 = (OM : ML).(ML : P_1N_1)$$
$$= (OM : ML).(N_1U_1 : P_1N_1),$$

and

$$y : P_1N_1 = (CM : MH).(HN_1 : N_1C)$$
$$= (OM : ML).(HN_1 : N_1C);$$

$$\therefore\ N_1U_1 : P_1N_1 > HN_1 : N_1C \text{ .... (1)},$$

or

$$P_1N_1 . N_1H < CN_1 . N_1U_1.$$

Adding or subtracting the rectangle $U_1N_1 . N_1H$, we have

$$P_1U_1 . U_1V < CH . HV$$
$$< LO . OR, \because\ CH : HM = OL : LM.$$

But, for a normal at P_1, we must have [from the proof of Prop. 97]

$$P_1U_1 . U_1V = LO . OR.$$

Therefore P_1O is not a normal, and [as in the proof of Prop. 97]

$$N_1K_1 < N_1G_1.$$

* For Apollonius' method of finding two mean proportionals see the Introduction.

Next let P be any other point than P_1, and let U, N, K have the same relation to P that U_1, N_1, K_1 have to P_1.

Also, since $U_1N_1 : N_1P_1 > HN_1 : N_1C$ by (1) above, let u_1 be taken on U_1N_1 such that

$$u_1N_1 : N_1P_1 = HN_1 : N_1C \qquad\qquad (2),$$

and draw wuu_1v parallel to WUU_1V.

Now $CN_1 . CT = CA^2$, so that $CN_1 : CA = CA : CT$;

$\therefore$ CT is a third proportional to CN_1, CA.

But CN_1 is a third proportional to CH, CI,

and $$CN_1 : CA = CI : CN_1 = CH : CI;$$

$$\therefore\ CH : CN_1 = CN_1 : CT$$
$$= CH \sim CN_1 : CN_1 \sim CT$$
$$= HN_1 : N_1T.$$

And $$CH : CN_1 = P_1u_1 : P_1N_1,$$

since $u_1N_1 : N_1P_1 = HN_1 : N_1C$, from (2) above;

$$\therefore\ HN_1 : N_1T = P_1u_1 : P_1N_1$$
$$= u_1w : N_1T;$$

thus $$u_1w = HN_1 = u_1v.$$

If $AN < AN_1$,

$wu < u_1v$,

and $u_1u : uw > u_1u : u_1v$,

whence $u_1w : uw > uv : u_1v$.

$\therefore$ $P_1u_1 : Qu > uv : u_1v$

(where PN meets P_1T in Q);

thus $P_1u_1 . u_1v > Qu . uv$

$> Pu . uv$,

a fortiori.

But, since

$HN_1 : N_1C = u_1N_1 : P_1N_1$,

$P_1N_1 . N_1H = CN_1 . N_1u_1$,

and, adding or subtracting the rectangle $u_1N_1 . N_1H$,

If $AN > AN_1$,

$wu_1 > uv$;

$\therefore$ $uu_1 : uv > uu_1 : wu_1$,

whence

$vu_1 : vu > wu : wu_1$

$> Qu : P_1u_1$;

thus $P_1u_1 . u_1v > Qu . uv$

$> Pu . uv$,

a fortiori,

and the proof proceeds as in the first column, leading to the same result,

$$PU . UV < LO . OR.$$

$$P_1u_1 . u_1v = CH . Hv;$$

$$\therefore CH . Hv > Pu . uv,$$

and, adding or subtracting the rectangle $uU . UV$,

$$PU . UV < CH . Hv + uU . UV$$

for the hyperbola,

or

$$PU . UV < CH . Hv - uU . UV$$

for the ellipse,

$\therefore$ in either case, *a fortiori*,

$$PU . UV < CH . HV,$$

or $\quad PU . UV < LO . OR.$

Therefore, as in the proof of Prop. 97, PO is not a normal, but $NK < NG$.

(*b*) Next suppose $OM = y$, so that $OM : P_1N_1 = y : P_1N_1$, and we obtain in this case

$$U_1N_1 : N_1P_1 = HN_1 : N_1C;$$

$$\therefore CN_1 . N_1U_1 = P_1N_1 . N_1H.$$

Adding or subtracting $U_1N_1 . N_1H$, we have

$$P_1U_1 . U_1V = CH . HV = LO . OR,$$

and this [Prop. 97] is the property of the normal at P_1.

Therefore one normal can be drawn from O.

If P be any other point on the curve, it will be shown as before that $U_1W = U_1V$, because in this case the lines WV, uv coincide; also

$UU_1 : UW > UU_1 : U_1V$ in the case where $UW < U_1V$,

and

$UU_1 : UV > UU_1 : U_1W$ in the case where $U_1W > UV$,

whence, exactly as before, we derive that

$$P_1U_1 . U_1V > QU . UV$$

$$> PU . UV, \text{ a fortiori},$$

and thence that $PU . UV < LO . OR$.

Therefore PO is not a normal, and $NK < NG$.

(*c*) Lastly, if $OM < y$, we shall have in this case

$$N_1U_1 : P_1N_1 < HN_1 : N_1C,$$

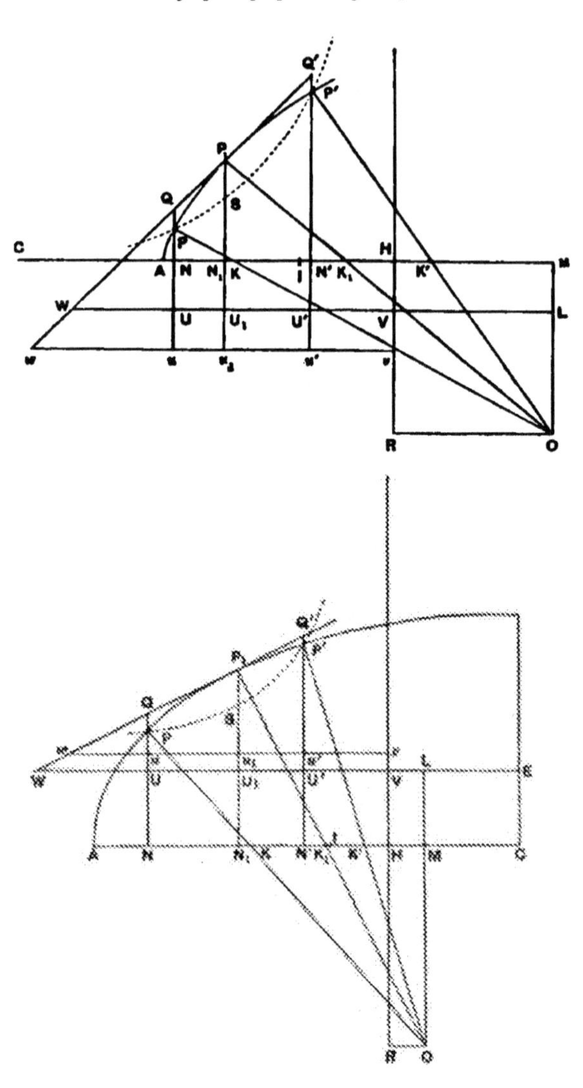

and we shall derive

$$LO \, . \, OR < P_1U_1 \, . \, U_1V.$$

Let S be taken on P_1N_1 such that $LO \, . \, OR = SU_1 \, . \, U_1V$, and through S describe a hyperbola whose asymptotes are VW and VH produced. This hyperbola will therefore meet the conic in two points P, P', and by the property of the hyperbola

$$PU \, . \, UV = P'U' \, . \, U'V = SU_1 \, . \, U_1V = LO \, . \, OR,$$

so that PO, $P'O$ are both normals.

The rest of the proposition follows at once from Prop. 97.

[It is clear that in case (*b*) O is the point of intersection of two consecutive normals, or the centre of the circle of curvature at P.

To find the Cartesian equation of the **evolute** we have

$$\left. \begin{aligned} x &= CM, \\ \frac{CH}{HM} &= \frac{a^2}{b^2}, \text{ or } \frac{CH}{x - CH} = \frac{a^2}{b^2} \end{aligned} \right\} \quad \ldots\ldots\ldots\ldots\ldots(1).$$

Also
$$\frac{y}{P_1N_1} = \frac{CM}{MH} \cdot \frac{HN_1}{N_1C} \quad \ldots\ldots\ldots\ldots\ldots\ldots (2),$$

and
$$\frac{CN_1^2}{a^2} \mp \frac{P_1N_1^2}{b^2} = 1 \quad \ldots\ldots\ldots\ldots\ldots\ldots(3),$$

where the upper sign refers to the hyperbola.

And, lastly, $a : CN_1 = CN_1 : CI = CI : CH$ ....(4).

From (4) $\qquad CN_1^2 = a \, . \, CI,$

and
$$CN_1 = \frac{a \, . \, CH}{CI};$$

$$\therefore \; CN_1^3 = a^2 \, . \, CH \quad \ldots\ldots\ldots\ldots\ldots\ldots(5).$$

Now, from (2),

$$\begin{aligned} \frac{y}{P_1N_1} &= \frac{CM}{MH} \cdot \frac{HN_1}{N_1C} \\ &= \frac{a^2 \pm b^2}{b^2} \cdot \frac{CH \sim CN_1}{CN_1}, \text{ by aid of (1)}, \\ &= \frac{a^2 \pm b^2}{b^2} \cdot \frac{CN_1^2 \sim a^2}{a^2}, \text{ by (5)}, \\ &= \frac{a^2 \pm b^2}{b^2} \cdot \frac{P_1N_1^2}{b^2}, \text{ by (3)}. \end{aligned}$$

Thus $$P_1N_1^2 = \frac{b^4 y}{a^2 \pm b^2},$$

whence $$P_1N_1^2 = b^2 \cdot \left(\frac{by}{a^2 \pm b^2}\right)^2 \quad\text{.........................(6).}$$

But, from (1), $CH = \dfrac{a^2 x}{a^2 \pm b^2}$.

Therefore, by (5), $CN_1^2 = \dfrac{a^4 x}{a^2 \pm b^2}$,

whence $$CN_1^2 = a^2 \cdot \left(\frac{ax}{a^2 \pm b^2}\right)^2 \quad\text{.........................(7).}$$

Thus, from (6) and (7), by the aid of (3),

$$\left(\frac{ax}{a^2 \pm b^2}\right)^{\frac{2}{3}} \mp \left(\frac{by}{a^2 \pm b^2}\right)^{\frac{2}{3}} = 1,$$

or $$(ax)^{\frac{2}{3}} \mp (by)^{\frac{2}{3}} = (a^2 \pm b^2)^{\frac{2}{3}}.]$$

Proposition 100.

[V. 53, 54.]

If O be a point on the minor axis of an ellipse, then

(a) if $OB : BC \not< AA' : p_a$, and P be any point on either of the quadrants BA, BA' except the point B, and if OP meet the major axis in K,

PO cannot be a normal, but $NK < NG$;

(b) if $OB : BC < AA' : p_a$, one normal only besides OB can be drawn to either of the two quadrants as OP, and, if P' be any other point, $N'K'$ is less or greater than $N'G'$ according as P' is further from, or nearer to, the minor axis than P.

[This proposition follows at once as a particular case of the preceding, but Apollonius proves it separately thus.]

(a) We have $OB : BC < On : nC$;

$\therefore On : nC$, or $CN : NK > AA' : p_a$,

whence $CN : NK > CN : NG$,

and $NK < NG$.

(*b*) Suppose now that

$$O'B : BC < AA' : p_a.$$

Take a point n on $O'B$ such that

$$O'n : nC = AA' : p_a.$$

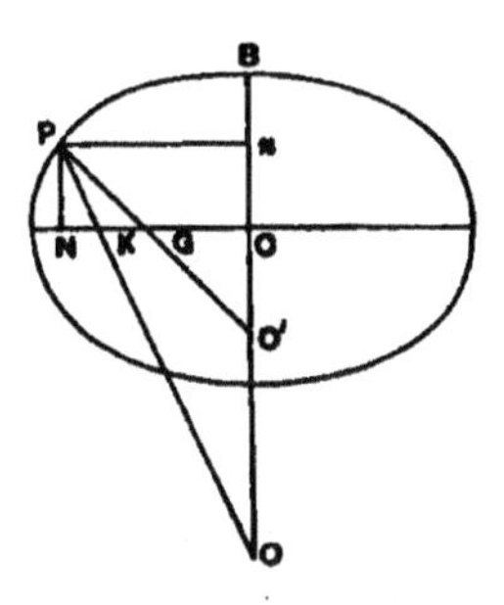

Therefore $\qquad CN : NK_1 = AA' : p_a,$

where N is the foot of the ordinate of P, the point in which nP drawn parallel to the major axis meets the ellipse, and K_1 is the point in which $O'P$ meets the major axis;

$$\therefore\; NK_1 = NG, \text{ and } PO' \text{ is a normal.}$$

PO', BO' are then two normals through O', and the rest of the proposition follows from Prop. 97.

CONSTRUCTION OF NORMALS.

Proposition 101.

[V. 55, 56, 57.]

If O is any point below the axis AA' of an ellipse, and $AM > AC$ (where M is the foot of the perpendicular from O on the axis), then one normal to the ellipse can always be drawn through O cutting the axis between A and C, but never more than one such normal.

Produce OM to L and CM to H so that

$$OL : LM = CH : HM = AA' : p_a,$$

and draw LI, IH parallel and perpendicular to the axis respectively. Then with IL, IH as asymptotes describe a [rectangular] hyperbola passing through O.

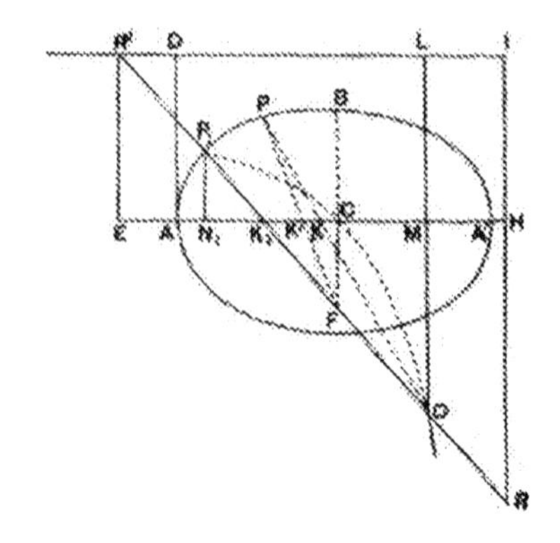

This will meet the ellipse in some point P_1. For, drawing AD, the tangent at A, to meet IL produced in D, we have

$$AH : HM > CH : HM$$

$$> AA' : p_a$$

$$> OL : LM;$$

$$\therefore AH . LM > OL . HM,$$

or $$AD . DI > OL . LI.$$

Thus, from the property of the hyperbola, it must meet AD between A and D, and therefore must meet the ellipse in some point P_1.

Produce OP_1 both ways to meet the asymptotes in R, R', and draw $R'E$ perpendicular to the axis.

Therefore $OR = P_1R'$, and consequently $EN_1 = MH$.

Now $$AA' : p_a = OL : LM$$

$$= ME : EK_1, \text{ by similar triangles.}$$

Also $$AA' : p_a = CH : HM;$$

$$\therefore AA' : p_a = ME - CH : EK_1 - MH$$

$$= CN_1 : N_1K_1,$$

since $$EN_1 = MH.$$

Therefore $N_1K_1 = N_1G_1$, and P_1O is a normal.

Let P be any other point such that OP meets AC in K.

Produce BC to meet OP_1 in F, and join FP, meeting the axis in K'.

Then, since two normals [at P_1, B] meet in F, FP is not a normal, but $NK' > NG$. Therefore, *a fortiori*, $NK > NG$. And, if P is between A and P_1, $NK < NG$. [Prop. 97, Cor. 1.]

Proposition 102.

[V. 58, 59, 60, 61.]

If O be any point outside a conic, but not on the axis whose extremity is A, we can draw a normal to the curve through O.

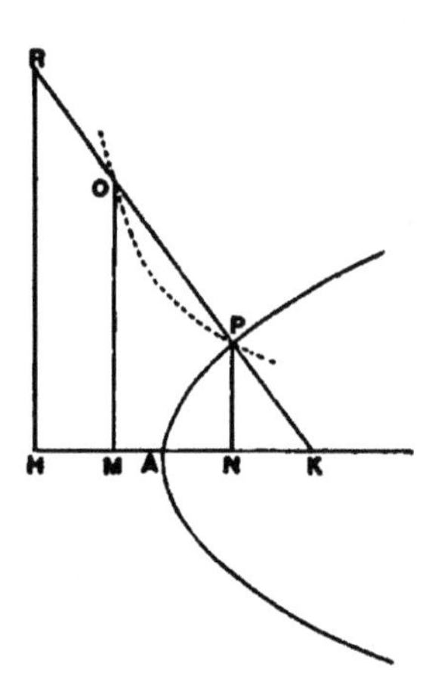

For the *parabola* we have only to measure MH in the direction of the axis produced outside the curve, and of length equal to $\frac{p_a}{2}$, to draw HR perpendicular to the axis on the same side as O, and, with HR, HA as asymptotes, to describe a [rectangular] hyperbola through O. This will meet the curve in a point P, and, if OP be joined and produced to meet the axis in K and HR in R, we have at once $HM = NK$.

Therefore $$NK = \frac{p_a}{2},$$

and PK is a normal.

In the *hyperbola* or *ellipse* take H on CM or on CM produced, and L on OM or OM produced, so that

$$CH : HM = OL : LM = AA' : p_a.$$

Then draw HIR perpendicular to the axis, and ILR' through L parallel to the axis.

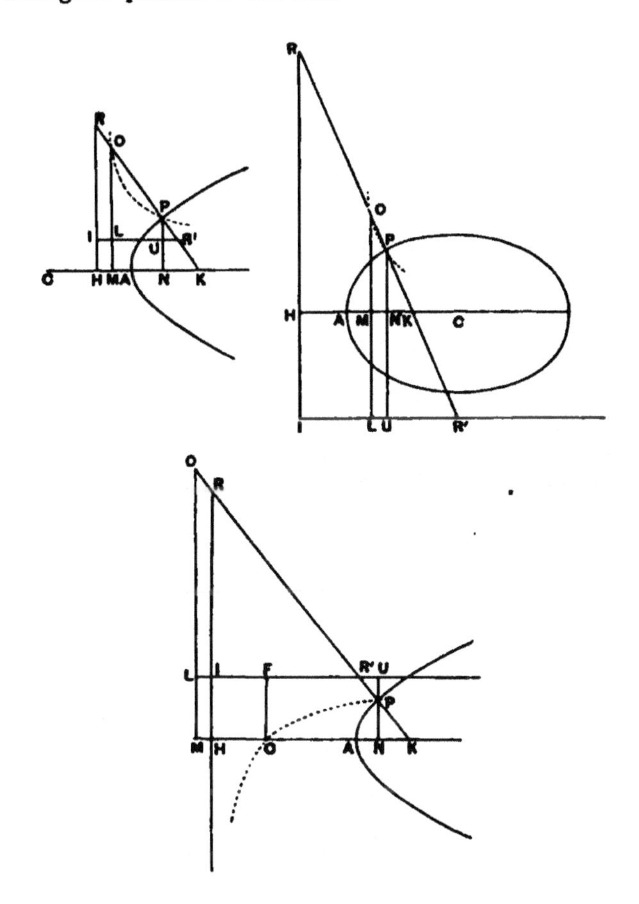

(1) If M falls on the side of C towards A, draw with asymptotes IR, IL, and through O, a [rectangular] hyperbola cutting the curve in P.

(2) If M falls on the side of C further from A in the *hyperbola*, draw a [rectangular] hyperbola with IH, IR' as asymptotes and through C, the centre, cutting the curve in P.

Then OP will be a normal.

For we have (1) $MK : HN = MK : LR'$,

since $OR = PR'$, and therefore $IL = UR'$.

Therefore $MK : HN = MO : OL$, by similar triangles,

$= MC : CH$,

$\because CH : HM = OL : LM$.

Therefore, alternately,

$$MK : MC = NH : HC \quad \ldots\ldots\ldots\ldots\ldots \text{(A)}.$$

In case (2) $OL : LM = CH : HM$,

or $OL . LI = CH . HI$,

[so that O, C are on opposite branches of the same rectangular hyperbola].

Therefore $PU : OL = LI : IU$,

or, by similar triangles,

$$UR' : R'L = LI : IU,$$

whence $R'L = IU = HN$;

$\therefore MK : HN = MK : R'L$

$= MO : OL$

$= MC : CH$,

and $MK : MC = NH : HC$, as before (A).

Thus, in either case, we derive

$$CK : CM = CN : CH,$$

and hence, alternately,

$$CN : CK = CH : CM,$$

so that $CN : NK = CH : HM$

$= AA' : p_a$;

$\therefore NK = NG$,

and OP is the normal at P.

(3) For the *hyperbola*, in the particular case where M coincides with C, or O is on the conjugate axis, we need only divide OC in L, so that

$$OL : LC = AA' : p_a,$$

and then draw LP parallel to AA' to meet the hyperbola in P. P is then the foot of the normal through O, for

$$\begin{aligned} AA' : p_a &= OL : LC \\ &= OP : PK \\ &= CN : NK, \end{aligned}$$

and $$NK = NG.$$

[The particular case is that in which the hyperbola used in the construction reduces to two straight lines.]

Proposition 103.

[V. 62, 63.]

If O be an internal point, we can draw through O a normal to the conic.

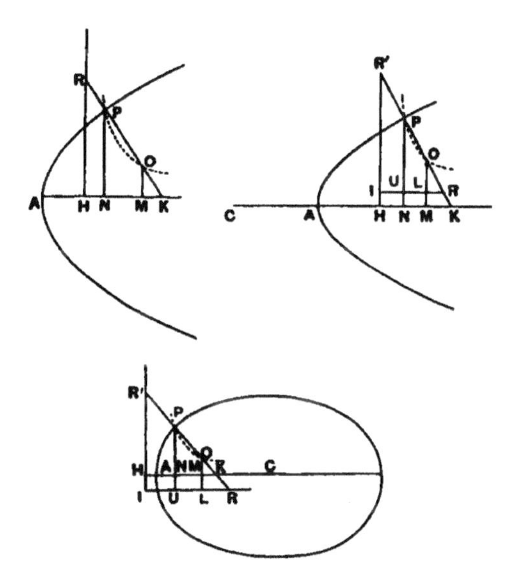

The construction and proof proceed as in the preceding proposition, *mutatis mutandis.*

The case of the *parabola* is obvious; and for the *hyperbola* or *ellipse*

$$MK : HN = OM : OL$$

$$= CM : CH.$$

$$\therefore CM : CH = CM \pm MK : CH \pm HN$$

$$= CK : CN;$$

$$\therefore NK : CN = HM : CH$$

$$= p_a : AA';$$

$$\therefore NK = NG,$$

and PO is a normal.

OTHER PROPOSITIONS RESPECTING MAXIMA AND MINIMA.

Proposition 104.

[V. 64, 65, 66, 67.]

If O be a point below the axis of any conic such that either no *normal, or only* one *normal, can be drawn to the curve through O which cuts the axis (between A and C in the case of the ellipse), then OA is the* **least** *of the lines OP cutting the axis, and that which is nearer to OA is less than that which is more remote.*

If OM be perpendicular to the axis, we must have

$$AM > \frac{p_a}{2},$$

and also OM must be either greater than or equal to y, where

(a) in the case of the *parabola*

$$y : P_1N_1 = N_1H : HM;$$

(b) in the case of the *hyperbola* or *ellipse*

$$y : P_1N_1 = (CM : MH) . (HN_1 : N_1C),$$

with the notation of Prop. 99.

In the case where $OM > y$, we have proved in Prop. 99 for all three curves that, for any straight line OP drawn from O to the curve and cutting the axis in K, $NK < NG$;

but, in the case where $OM = y$, $NK < NG$ for any point P between A and P_1 except P_1 itself, for which $N_1K_1 = N_1G_1$.

Also for any point P more remote from A than P_1 it is still true that $NK < NG$.

I. Consider now the case of any of the three conics where, for all points P, $NK < NG$.

Let P be any point other than A. Draw the tangents AY, PT. Then the angle OAY is obtuse. Therefore the perpendicular at A to AO, as AL, falls within the curve. Also, since $NK < NG$, and PG is perpendicular to PT, the angle OPT is acute.

(1) Suppose, if possible, $OP = OA$.

With OP as radius and O as centre describe a circle. Since the angle OPT is acute, this circle will cut the tangent PT,

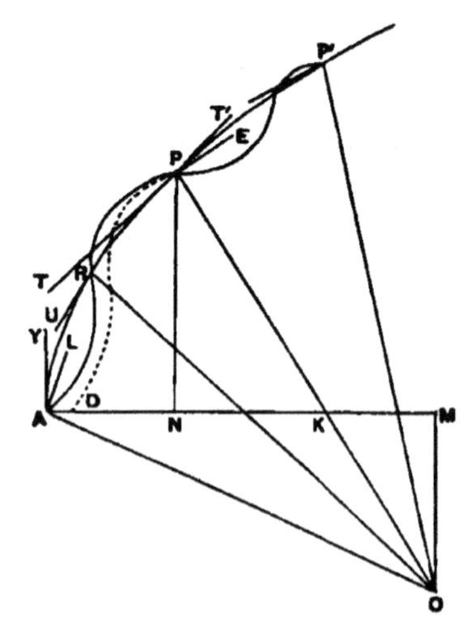

but AL will lie wholly without it. It follows that the circle must cut the conic in some intermediate point as R. If RU be the tangent to the conic at R, the angle ORU is acute. Therefore RU must meet the circle. But it falls wholly outside it: which is absurd.

Therefore OP is *not* equal to OA.

(2) Suppose, if possible, $OP < OA$.

In this case the circle drawn with O as centre and OP as radius must cut AM in some point, D. And an absurdity is proved in the same manner as before.

Therefore OP is neither equal to OA nor less than OA, i.e. $OA < OP$.

It remains to be proved that, if P' be a point beyond P, $OP < OP'$.

If the tangent TP be produced to T', the angle OPT' is obtuse because the angle OPT is acute. Therefore the perpendicular from P to OP, viz. PE, falls within the curve, and the same proof as was used for A, P will apply to P, P'.

Therefore $OA < OP$, $OP < OP'$, &c.

II. Where only *one* normal, OP_1, cutting the axis can be drawn from O, the above proof applies to all points P between A and P_1 (excluding P_1 itself) and also applies to the comparison between two points P each of which is more remote from A than P_1.

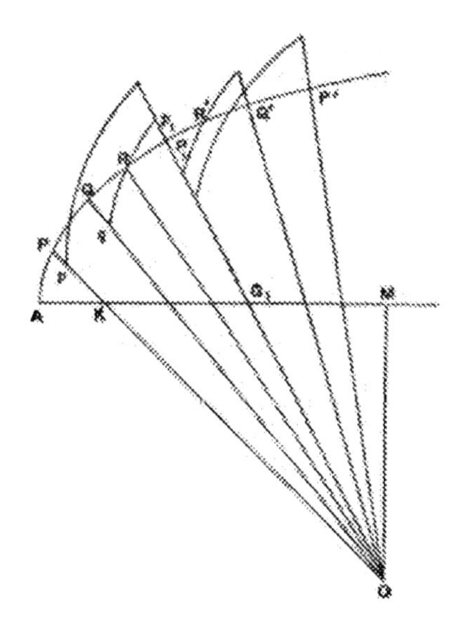

It only remains therefore to prove that

(*a*) $OP_1 >$ any straight line OP between OA and OP_1,

(*b*) $OP_1 <$ any straight line OP' beyond OP_1.

(*a*) Suppose first, if possible, that $OP = OP_1$, and let Q be any point between them, so that, by the preceding proof, $OQ > OP$. Measure along OQ a length Oq such that Oq is greater than OP_1 and less than OQ. With O as centre and Oq as radius describe a circle meeting OP_1 produced in p_1. This circle must then meet the conic in an intermediate point R.

Thus, by the preceding proof, OQ is less than OR, and therefore is less than Oq: which is absurd.

Therefore OP is not equal to OP_1.

Again suppose, if possible, that $OP > OP_1$. Then, by taking on OP_1 a length Op_1 greater than OP_1 and less than OP, an absurdity is proved in the same manner.

Therefore, since OP is neither equal to nor greater than OP_1,

$$OP < OP_1.$$

(*b*) If OP' lies more remote from OA than OP_1, an exactly similar proof will show that $OP_1 < OP'$.

Thus the proposition is completely established.

Proposition 105. (Lemma.)

[V. 68, 69, 70, 71.]

If two tangents at points Q, Q' on one side of the axis of a conic meet in T, and if Q be nearer to the axis than Q', then $TQ < TQ'$.

The proposition is proved at once for the parabola and hyperbola and for the case where Q, Q' are on one quadrant of an ellipse: for the angle TVQ' is greater than the angle TVQ, and $QV = VQ'$.

Therefore the base TQ is less than the base TQ'.

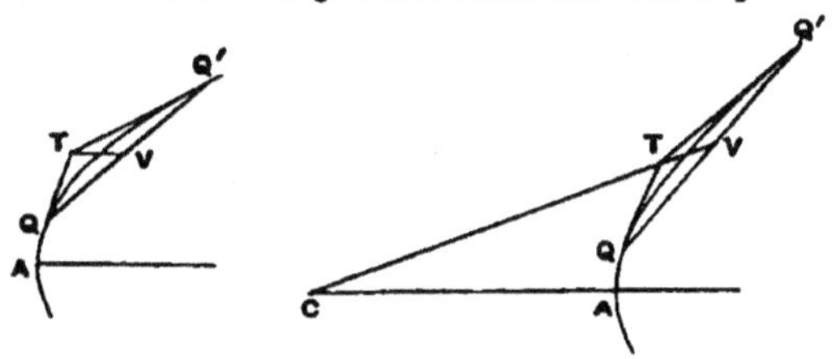

In the case where Q, Q' are on different quadrants of an ellipse, produce the ordinate $Q'N'$ to meet the ellipse again in q'. Join $q'C$ and produce it to meet the ellipse in R. Then $Q'N' = N'q'$, and $q'C = CR$, so that $Q'R$ is parallel to the axis. Let RM be the ordinate of R.

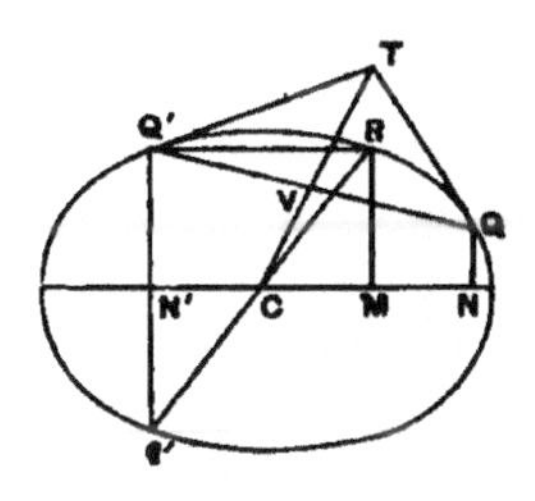

Now $\qquad RM > QN$;

$\therefore$ [Prop. 86, Cor.] $\qquad CQ > CR$,

i.e. $\qquad > CQ'$;

$\therefore \angle CVQ > \angle CVQ'$,

and, as before, $\qquad TQ < TQ'$.

Proposition 106.

[V. 72.]

If from a point O below the axis of a parabola or hyperbola it is possible to draw two normals OP_1, OP_2 cutting the axis (P_1 being nearer to the vertex A than P_2), and if further P be any other point on the curve and OP be joined, then

(1) *if P lies between A and P_2, OP_1 is the* **greatest** *of all the lines OP, and that which is nearer to OP_1 on each side is greater than that which is more remote;*

(2) *if P lies between P_1 and P_2, or beyond P_2, OP_2 is the* **least** *of all the lines OP, and the nearer to OP_2 is less than the more remote.*

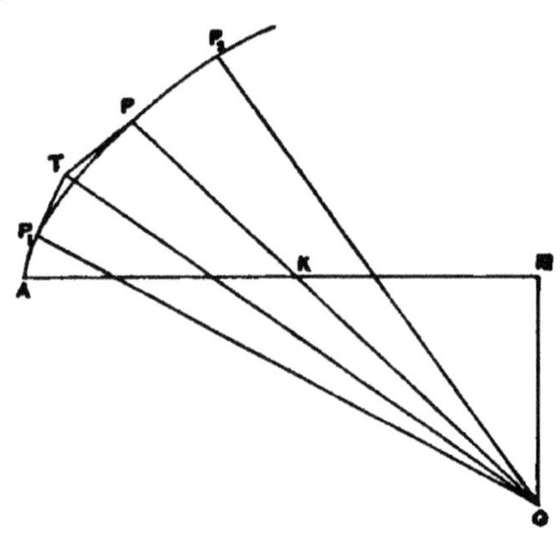

By Prop. 99, if P is between A and P_1, OP is not a normal, but $NK < NG$. Therefore, by the same proof as that employed in Prop. 104, we find that OP increases continually as P moves from A towards P_1.

We have therefore to prove that OP diminishes continually as P moves from P_1 to P_2. Let P be any point between P_1 and P_2, and let the tangents at P_1, P meet in T. Join OT.

Then, by Prop. 105, $\qquad TP_1 < TP.$

Also $$TP_1^2 + OP_1^2 > TP^2 + OP^2,$$

since $AK > AG$, and consequently the angle OPT is obtuse.

Therefore $$OP < OP_1.$$

Similarly it can be proved that, if P' is a point between P and P_2, $OP' < OP$.

That OP increases continually as P moves from P_2 further away from A and P_1 is proved by the method of Prop. 104.

Thus the proposition is established.

Proposition 107.

[V. 73.]

If O be a point below the major axis of an ellipse such that it is possible to draw through O one *normal only to the whole of the semi-ellipse ABA', then, if OP_1 be that normal and P_1 is on the quadrant AB, OP_1 will be the* **greatest** *of all the straight lines drawn from O to the semi-ellipse, and that which is nearer to OP_1 will be greater than that which is more remote. Also OA' will be the* **least** *of all the straight lines drawn from O to the semi-ellipse.*

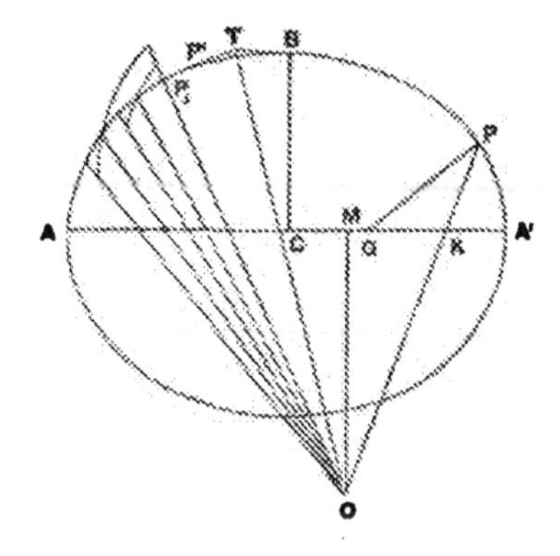

It follows from Props. 99 and 101 that, if OM be perpendicular to the axis, M must lie between C and A', and that OM must be greater than the length y determined as in Prop. 99.

Thus for all points P between A' and B, since K is nearer to A' than G is, it is proved by the method of Prop. 104 that OA' is the *least* of all such lines OP, and OP increases continually as P passes from A' to B.

For any point P' between B and P_1 we use the method of Prop. 106, drawing the tangents at P' and B, meeting in T.

Thus we derive at once that $OB < OP'$, and similarly that OP' increases continually as P' passes from B to P_1.

For the part of the curve between P_1 and A we employ the method of *reductio ad absurdum* used in the second part of Prop. 104.

Proposition 108.

[V. 74.]

If O be a point below the major axis of an ellipse such that two *normals only can be drawn through it to the whole semi-ellipse ABA', then that normal, OP_2, which cuts the minor axis is the* **greatest** *of all straight lines from O to the semi-ellipse, and that which is nearer to it is greater than that which is more remote. Also OA, joining O to the nearer vertex A, is the* **least** *of all such straight lines.*

It follows from Prop. 99 that, if O be nearer to A than to A', then P_1, the point at which O is the centre of curvature, is on the quadrant AB, and that OP_1 is one of the only two possible normals, while P_2, the extremity of the other, is on the quadrant BA'; also $OM = y$ determined as in Prop. 99.

In this case, since only one normal can be drawn to the quadrant AB, we prove that OP increases as P moves from A to P_1 by the method of Prop. 104, as also that OP increases as P moves from P_1 to B.

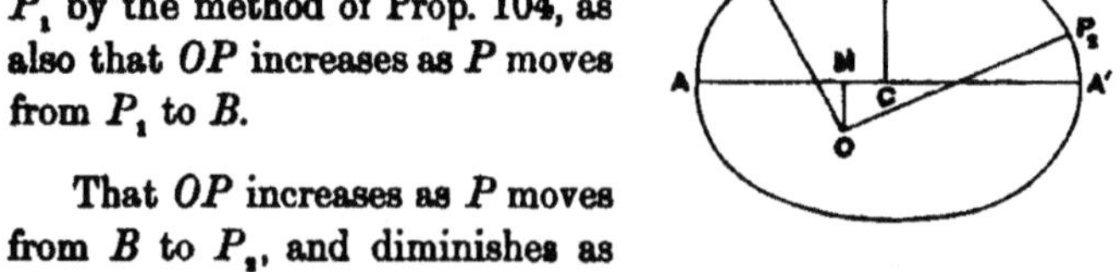

That OP increases as P moves from B to P_2, and diminishes as it passes from P_2 to A', is established by the method employed in the last proposition.

Proposition 109.

[V. 75, 76, 77.]

If O be a point below the major axis of an ellipse such that three *normals can be drawn to the semi-ellipse ABA' at points P_1, P_2, P_3, where P_1, P_2 are on the quadrant AB and P_3 on the quadrant BA', then (if P_1 be nearest to the vertex A),*

(1) *OP_3 is the* **greatest** *of all lines drawn from O to points on the semi-ellipse between A' and P_2, and the nearer to OP_3 on either side is greater than the more remote;*

(2) *OP_1 is the* **greatest** *of all lines from O to points on the semi-ellipse from A to P_2, and the nearer to OP_1 on either side is greater than the more remote.*

(3) *of the two* **maxima**, *$OP_3 > OP_1$.*

Part (2) of this proposition is established by the method of Prop. 106.

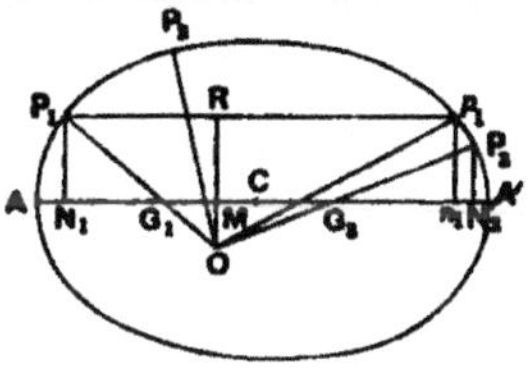

Part (1) is proved by the method of Prop. 107.

It remains to prove (3).

We have

$$CN_1 : N_1G_1 = AA' : p_a = CN_3 : N_3G_3;$$

$$\therefore\ MN_1 : N_1G_1 < CN_3 : N_3G_3$$

$$< MN_3 : N_3G_3,\ \textit{a fortiori},$$

whence $$MG_1 : N_1G_1 < MG_3 : N_3G_3;$$

and, by similar triangles,

$$OM : P_1N_1 < OM : P_3N_3,$$

or $$P_1N_1 > P_3N_3.$$

If then P_1p_1 be parallel to the axis, meeting the curve in p_1, we have at once, on producing OM to R,

$$p_1R > P_1R,$$

so that $$Op_1 > OP_1;$$

$$\therefore\ \textit{a fortiori}\ OP_3 > OP_1.$$

As particular cases of the foregoing propositions we have

(1) If O be on the minor axis, and *no* normal except OB can be drawn to the ellipse, OB is *greater* than any other straight line from O to the curve, and the nearer to it is greater than the more remote.

(2) If O be on the minor axis, and *one* normal (besides OB) can be drawn to either quadrant as OP_1, then OP_1 is the *greatest* of all straight lines from O to the curve, and the nearer to it is greater than the more remote.

EQUAL AND SIMILAR CONICS.

DEFINITIONS.

1. Conic sections are said to be **equal** when one can be applied to the other in such a way that they everywhere coincide and nowhere cut one another. When this is not the case they are **unequal**.

2. Conics are said to be **similar** if, the same number of ordinates being drawn to the axis at proportional distances from the vertex, all the ordinates are respectively proportional to the corresponding abscissae. Otherwise they are **dissimilar**.

3. The straight line subtending a segment of a circle or a conic is called the **base** of the segment.

4. The **diameter** of the segment is the straight line which bisects all chords in it parallel to the base, and the point where the diameter meets the segment is the **vertex** of the segment.

5. **Equal** segments are such that one can be applied to the other in such a way that they everywhere coincide and nowhere cut one another. Otherwise they are **unequal**.

6. Segments are **similar** in which the angles between the respective bases and diameters are equal, and in which, parallels to the base being drawn from points on each segment to meet the diameter at points proportionally distant from the vertex, each parallel is respectively proportional to the corresponding abscissa in each.

Proposition 110.

[VI. 1, 2.]

(1) *In two parabolas, if the ordinates to a diameter in each are inclined to the respective diameters at equal angles, and if the corresponding parameters are equal, the two parabolas are equal.*

(2) *If the ordinates to a diameter in each of two hyperbolas or two ellipses are equally inclined to the respective diameters, and if the diameters as well as the corresponding parameters are equal respectively, the two conics are equal, and conversely.*

This proposition is at once established by means of the fundamental properties

(1) $QV^2 = PL \, . \, PV$ for the *parabola*, and

(2) $QV^2 = PV \, . \, VR$ for the *hyperbola* or *ellipse*
proved in Props. 1—3.

Proposition 111.

[VI. 3.]

Since an ellipse is limited, while a parabola and a hyperbola proceed to infinity, an ellipse cannot be equal to either of the other curves. Also a parabola cannot be equal to a hyperbola.

For, if a parabola be equal to a hyperbola, they can be applied to one another so as to coincide throughout. If then equal abscissae AN, AN' be taken along the axes in each we have for the parabola

$$AN : AN' = PN^2 : P'N'^2.$$

Therefore the same holds for the hyperbola: which is impossible, because

$$PN^2 : P'N'^2 = AN \, . \, A'N : AN' \, . \, A'N'.$$

Therefore a parabola and hyperbola cannot be equal.

[Here follow six easy propositions, chiefly depending upon the symmetrical form of a conic, which need not be reproduced.]

Proposition 112.

[VI. 11, 12, 13.]

(1) *All parabolas are similar.*

(2) *Hyperbolas, or ellipses, are similar to one another when the "figure" on a diameter of one is similar to the "figure" on a diameter of the other and the ordinates to the diameters in each make equal angles with the diameters respectively.*

(1) The result is derived at once from the property

$$PN^2 = p_a . AN.$$

(2) Suppose the diameters to be *axes* in the first place (conjugate axes for *hyperbolas*, and both major or both minor axes for *ellipses*) so that the ordinates are at right angles to the diameters in both.

Then the ratio $p_a : AA'$ is the same in both curves. Therefore, using capital letters for one conic and small letters for the other, and making $AN : an$ equal to $AA' : aa'$, we have at the same time

$$PN^2 : AN . NA' = pn^2 : an . na'.$$

But $$AN . NA' : AN^2 = an . na' : an^2,$$

because $$A'N : AN = a'n : an;$$

$$\therefore PN^2 : AN^2 = pn^2 : an^2,$$

or $$PN : AN = pn : an,$$

and the condition of similarity is satisfied (Def. 2).

Again, let PP', pp' be diameters in two hyperbolas or two ellipses, such that the corresponding ordinates make equal angles with the diameters, and the ratios of each diameter to its parameter are equal.

Draw tangents at P, p meeting the axes in T, t respectively. Then the angles CPT, cpt are equal. Draw AH, ah perpendicular to the axes and meeting CP, cp in H, h; and on CH, ch as diameters describe circles, which therefore pass respectively through A, a. Draw QAR, qar through A, a parallel respectively to the tangents at P, p and meeting the circles just described in R, r.

Let V, v be the middle points of AQ, aq, so that V, v lie on CP, cp respectively.

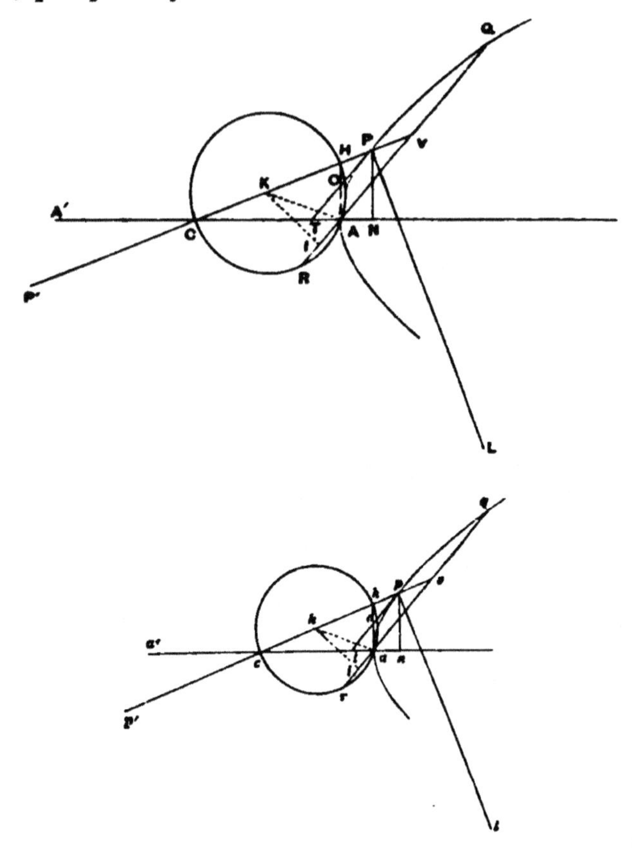

Then, since the "figures" on PP', pp' are similar,

$$AV^2 : CV \,.\, VH = av^2 : cv \,.\, vh, \qquad \text{[Prop. 14]}$$

or $$AV^2 : AV \,.\, VR = av^2 : av \,.\, vr,$$

whence $$AV : VR = av : vr \ldots\ldots\ldots\ldots\ldots\ldots\ldots(\alpha),$$

and, since the angle AVC is equal to the angle avc, it follows that the angles at C, c are equal.

[For, if K, k be the centres of the circles, and I, i the middle points of AR, ar, we derive from (α)

$$VA : AI = va : ai;$$

and, since $$\angle KVI = \angle kvi,$$

the triangles KVI, kvi are similar.

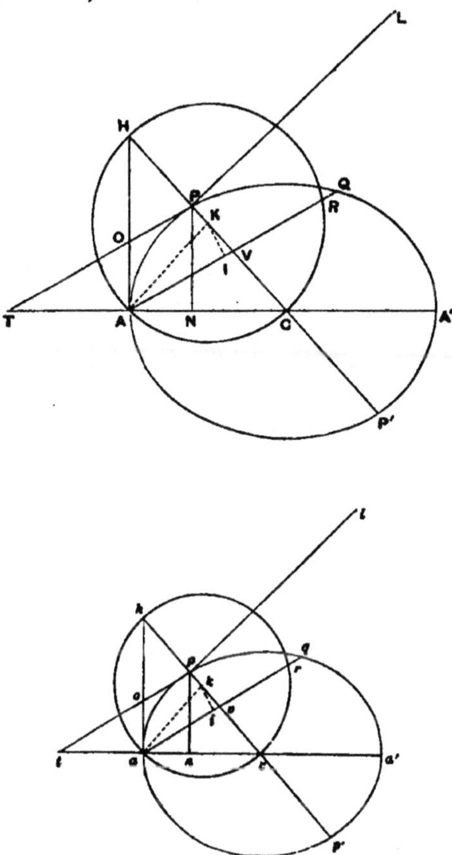

Therefore, since VI, vi are divided at A, a in the same ratio' the triangles KVA, kva are similar;

$$\therefore \angle AKV = \angle akv;$$

hence the halves of these angles, or of their supplements, are equal, or

$$\angle KCA = \angle kca.]$$

Therefore, since the angles at P, p are also equal, the triangles CPT, cpt are similar.

Draw PN, pn perpendicular to the axes, and it will follow that

$$PN^2 : CN . NT = pn^2 : cn . nt,$$

whence the ratio of AA' to its parameter and that of aa' to its parameter are equal. [Prop. 14]

Therefore (by the previous case) the conics are similar.

Proposition 113.

[VI. 14, 15.]

A parabola is neither similar to a hyperbola nor to an ellipse; and a hyperbola is not similar to an ellipse.

[Proved by *reductio ad absurdum* from the ordinate properties.]

Proposition 114.

[VI. 17, 18.]

(1) *If PT, pt be tangents to two similar conics meeting the axes in T, t respectively and making equal angles with them; if, further, PV, pv be measured along the diameters through P, p so that*

$$PV : PT = pv : pt,$$

and if QQ', qq' be the chords through V, v parallel to PT, pt respectively: then the segments QPQ', qpq' are similar and similarly situated.

(2) *And,* conversely, *if the segments are similar and similarly situated, $PV : PT = pv : pt$, and the tangents are equally inclined to the axes.*

I. Let the conics be *parabolas*.

Draw the tangents at A, a meeting the diameters through P, p in H, h, and let PL, pl be such lengths that

$$PL : 2PT = OP : PH$$
and
$$pl : 2pt = op : ph,$$

where O, o are the points of intersection of AH, PT and ah, pt.

Therefore PL, pl are the parameters of the ordinates to the diameters PV, pv. [Prop. 22]

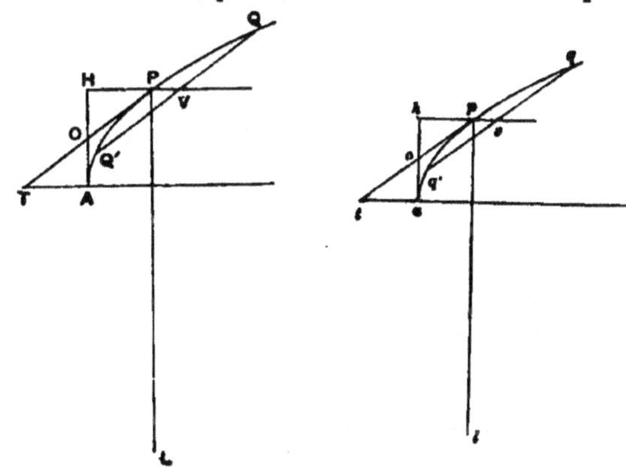

Hence $$QV^2 = PL . PV,$$
$$qv^2 = pl . pv.$$

(1) Now, since $\angle PTA = \angle pta$,
$$\angle OPH = \angle oph,$$
and the triangles OPH, oph are similar.

Therefore $$OP : PH = op : ph,$$
so that $$PL : PT = pl : pt.$$

But, by hypothesis,
$$PV : PT = pv : pt;$$
$$\therefore PL : PV = pl : pv,$$
and, since QV is a mean proportional between PV, PL, and qv between pv, pl,
$$QV : PV = qv : pv.$$

Similarly, if V', v' be points on PV, pv such that

$$PV : PV' = pv : pv',$$

and therefore $$PL : PV' = pl : pv',$$

it follows that the ordinates passing through V', v' are in the same ratio to their respective abscissae.

Therefore the segments are similar. (Def. 6.)

(2) If the segments are similar and similarly situated, we have to prove that

$$\angle PTA = \angle pta,$$

and $$PV : PT = pv : pt.$$

Now the tangents at P, p are parallel to QQ', qq' respectively, and the angles at V, v are equal.

Therefore the angles PTA, pta are equal.

Also, by similar segments,

$$QV : PV = qv : pv,$$

while $$PL : QV = QV : PV, \text{ and } pl : qv = qv : pv;$$

$$\therefore PL : PV = pl : pv.$$

But $$\left.\begin{aligned} PL : 2PT &= OP : PH \\ pl : 2pt &= op : ph \end{aligned}\right\},$$

and $$OP : PH = op : ph,$$

by similar triangles.

Therefore $$PV : PT = pv : pt.$$

II. If the curves be *hyperbolas* or *ellipses*, suppose a similar construction made, and let the ordinates PN, pn be drawn to the major or conjugate axes. We can use the figures of Prop. 112, only remembering that the chords are here QQ', qq', and do not pass through A, a.

(1) Since the conics are similar, the ratio of the axis to its parameter is the same for both.

Therefore $PN^2 : CN . NT = pn^2 : cn . nt.$ [Prop. 14]

Also the angles PTN, ptn are equal,

therefore $PN : NT = pn : nt.$

Hence $PN : CN = pn : cn,$

and $\angle PCN = \angle pcn.$

Therefore also $\angle CPT = \angle cpt.$

It follows that the triangles OPH, oph are similar.

Therefore $OP : PH = op : ph.$

But $$\left.\begin{array}{l} OP : PH = PL : 2PT \\ op : ph = pl : 2pt \end{array}\right\},$$

whence $PL : PT = pl : pt.$

Also, by similar triangles,

$$PT : CP = pt : cp;$$

$$\therefore PL : CP = pl : cp,$$

or $PL : PP' = pl : pp'$ ....(A).

Therefore the "figures" on the diameters PP', pp' are similar.

Again, we made $PV : PT = pv : pt,$

so that $PL : PV = pl : pv$......................(B).

We derive, by the method employed in Prop. 112, that

$$QV : PV = qv : pv,$$

and that, if PV, pv be proportionally divided in the points V', v', the ordinates through these points are in the same ratios.

Also the angles at V, v are equal.

Therefore the segments are similar.

(2) If the segments are similar, the ordinates are in the ratio of their abscissae, and we have

$$\left.\begin{array}{r} QV : PV = qv : pv \\ PV : PV' = pv : pv' \\ PV' : Q'V' = pv' : q'v' \end{array}\right\}.$$

Then $$QV^2 : Q'V'^2 = qv^2 : q'v'^2;$$

$$\therefore PV . VP' : PV' . V'P' = pv . vp' : pv' . v'p',$$

and $$PV : PV' = pv : pv',$$

so that $$P'V : P'V' = p'v : p'v'.$$

From these equations it follows that

$$\left.\begin{aligned} PV' : VV' &= pv' : vv' \\ P'V' : VV' &= p'v' : vv' \end{aligned}\right\},$$

and

whence $$P'V' : PV' = p'v' : pv';$$

$$\therefore P'V' . V'P : PV'^2 = p'v' . v'p : pv'^2.$$

But $$PV'^2 : Q'V'^2 = pv'^2 : q'v'^2;$$

$$\therefore P'V' . V'P : Q'V'^2 = p'v' . v'p : q'v'^2.$$

But these ratios are those of PP', pp' to their respective parameters.

Therefore the "figures" on PP', pp' are similar; and, since the angles at V, v are equal, the conics are similar.

Again, since the conics are similar, the "figures" on the axes are similar.

Therefore $$PN^2 : CN . NT = pn^2 : cn . nt,$$

and the angles at N, n are right, while the angle CPT is equal to the angle cpt.

Therefore the triangles CPT, cpt are similar, and the angle CTP is equal to the angle ctp.

Now, since $$PV . VP' : QV^2 = pv . vp' : qv^2,$$

and $$QV^2 : PV^2 = qv^2 : pv^2;$$

it follows that $$PV : P'V = pv : p'v,$$

whence $$PP' : PV = pp' : pv.$$

But, by the similar triangles CPT, cpt,

$$CP : PT = cp : pt,$$

or $$PP' : PT = pp' : pt;$$

$$\therefore PV : PT = pv : pt,$$

and the proposition is proved.

Proposition 115.

[VI. 21, 22.]

If two ordinates be drawn to the axes of two parabolas, or the major or conjugate axes of two similar ellipses or two similar hyperbolas, as $PN, P'N'$ *and* $pn, p'n'$, *such that the ratios* $AN : an$ *and* $AN' : an'$ *are each equal to the ratio of the respective latera recta, then the segments* PP', pp' *will be similar; also* PP' *will not be similar to any segment in the other conic which is cut off by two ordinates other than* $pn, p'n'$, *and vice versa.*

[The method of proof adopted follows the lines of the previous propositions, and accordingly it is unnecessary to reproduce it.]

Proposition 116.

[VI. 26, 27.]

If any cone be cut by two parallel planes making hyperbolic or elliptic sections, the sections will be similar but not equal.

On referring to the figures of Props. 2 and 3, it will be seen at once that, if another plane parallel to the plane of section be drawn, it will cut the plane of the axial triangle in a straight line $p'pm$ parallel to $P'PM$ and the base in a line dme parallel to DME; also $p'pm$ will be the diameter of the resulting hyperbola or ellipse, and the ordinates to it will be parallel to dme, *i.e.* to DME.

Therefore the ordinates to the diameters are equally inclined to those diameters in both curves.

Also, if PL, pl are the corresponding parameters,

$$PL : PP' = BF . FC : AF^2 = pl : pp'.$$

Hence the rectangles $PL \cdot PP'$ and $pl \cdot pp'$ are similar.

It follows that the conics are similar. [Prop. 112]

And they cannot be equal, since $PL \cdot PP'$ cannot be equal to $pl \cdot pp'$. [Cf. Prop. 110 (2)]

[A similar proposition holds for the *parabola*, since, by Prop. 1, $PL : PA$ is a constant ratio. Therefore two parallel parabolic sections have different parameters.]

Proposition 117.

[VI. 28.]

In a given right cone to find a parabolic section equal to a given parabola.

Let the given parabola be that of which am is the axis and al the latus rectum. Let the given right cone be OBC, where O is the apex and BC the circular base, and let OBC be a triangle through the axis meeting the base in BC.

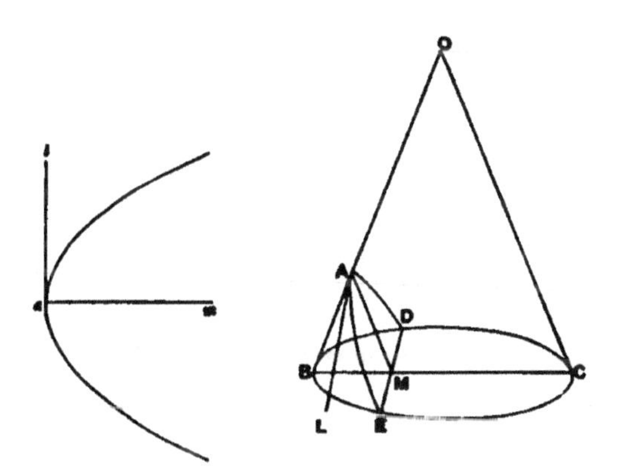

Measure OA along OB such that

$$al : OA = BC^2 : BO \,.\, OC.$$

Draw AM parallel to OC meeting BC in M, and through AM draw a plane at right angles to the plane OBC and cutting the circular base in DME.

Then DE is perpendicular to AM, and the section DAE is a parabola whose axis is AM.

Also [Prop. 1], if AL is the latus rectum,

$$AL : AO = BC^2 : BO . OC,$$

whence $AL = al$, and the parabola is equal to the given one [Prop. 110].

No other parabola with vertex on OB can be found which is equal to the given parabola except DAE. For, if another such parabola were possible, its plane must be perpendicular to the plane OBC and its axis must be parallel to OC. If A' were the supposed vertex and $A'L'$ the latus rectum, we should have $A'L' : A'O = BC^2 : BO . OC = AL : AO$. Thus, if A' does not coincide with A, $A'L'$ cannot be equal to AL or al, and the parabola cannot be equal to the given one.

Proposition 118.

[VI. 29.]

In a given right cone to find a section equal to a given hyperbola. (A necessary condition of possibility is that the ratio of the square on the axis of the cone to the square on the radius of the base must not be greater than the ratio of the transverse axis of the given hyperbola to its parameter.)

Let the given hyperbola be that of which aa', al are the transverse axis and parameter respectively.

I. Suppose $OI^2 : BI^2 < aa' : al$, where I is the centre of the base of the given cone.

Let a circle be circumscribed about the axial triangle OBC, and produce OI to meet the circle again in D.

Then $OI : ID = OI^2 : BI^2$,

so that $OI : ID < aa' : al$.

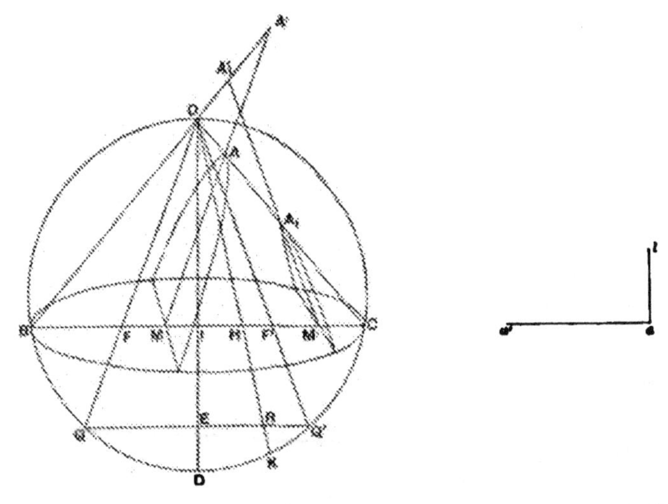

Take E on ID such that $OI : IE = aa' : al$, and through E draw the chord QQ' parallel to BC.

Suppose now that AA', A_1A_1' are placed in the angle formed by OC and BO produced, such that $AA' = A_1A_1' = aa'$, and AA', A_1A_1' are respectively parallel to OQ, OQ', meeting BO in M, M'.

Through $A'AM$, $A_1'A_1M'$ draw planes perpendicular to the plane of the triangle OBC making hyperbolic sections, of which $A'AM$, $A_1'A_1M'$ will therefore be the transverse axes.

Suppose OQ, OQ' to meet BC in F, F'.

Then $aa' : al = OI : IE$

$$= OF : FQ \text{ or } OF' : F'Q'$$

$$= OF^2 : OF.FQ \text{ or } OF'^2 : OF'.F'Q'$$

$$= OF^2 : BF.FC \text{ or } OF'^2 : BF'.F'C$$

$$= AA' : AL \text{ or } A_1A_1' : A_1L_1,$$

where AL, A_1L_1 are the parameters of AA', A_1A_1' in the sections respectively.

It follows, since $AA' = A_1A' = aa'$,

that $AL = A_1L_1 = al$.

Hence the two hyperbolic sections are each equal to the given hyperbola.

There are no other equal sections having their vertices on OC.

For (1), if such a section were possible and OH were parallel to the axis of such a section, OH could not be coincident either with OQ or OQ'. This is proved after the manner of the preceding proposition for the parabola.

If then (2) OH meet BC in H, QQ' in R, and the circle again in K, we should have, if the section were possible,

$$\begin{aligned} aa' : al &= OH^2 : BH \cdot HC \\ &= OH^2 : OH \cdot HK \\ &= OH : HK; \end{aligned}$$

which is impossible, since

$$aa' : al = OI : IE = OH : HR.$$

II. If $OI^2 : BI^2 = aa' : al$, we shall have $OI : ID = aa' : al$, and OQ, OQ' will both coincide with OD.

In this case there will be only one section equal to the given hyperbola whose vertex is on OC, and the axis of this section will be perpendicular to BC.

III. If $OI^2 : BI^2 > aa' : al$, no section can be found in the right cone which is equal to the given hyperbola.

For, if possible, let there be such a section, and let ON be drawn parallel to its axis meeting BC in N.

Then we must have $aa' : al = ON^2 : BN \cdot NC$,

so that $OI^2 : BI \cdot IC > ON^2 : BN \cdot NC$.

But $ON^2 > OI^2$, while $BI \cdot IC > BN \cdot NC$: which is absurd.

Proposition 119.

[VI. 30.]

In a given right cone to find a section equal to a given ellipse.

In this case we describe the circle about OBC and suppose F, F' taken on BC produced in both directions such that, if OF, OF' meet the circle in Q, Q',

$$OF : FQ = OF' : F'Q' = aa' : al.$$

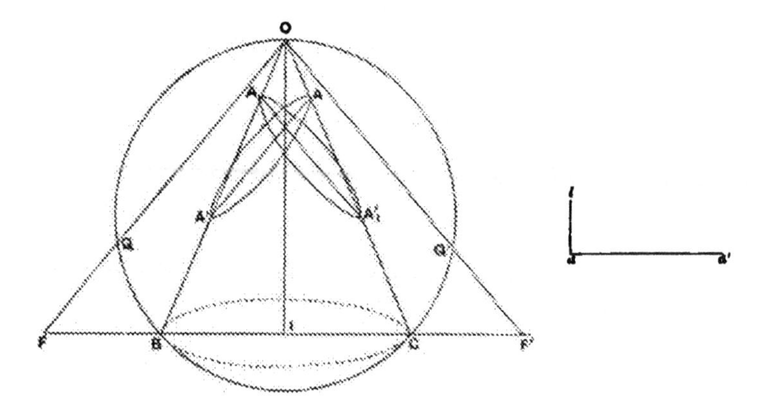

Then we place straight lines AA', A_1A_1' in the angle BOC so that they are each equal to aa', while AA' is parallel to OQ and A_1A_1' to OQ'.

Next suppose planes drawn through AA', A_1A_1' each perpendicular to the plane of OBC, and these planes determine two sections each of which is equal to the given ellipse.

The proof follows the method of the preceding proposition.

Proposition 120.

[VI. 31.]

To find a right cone similar to a given one and containing a given parabola as a section of it.

Let OBC be an axial section of the given right cone, and let the given parabola be that of which AN is the axis and AL the latus rectum. Erect a plane passing through AN and perpendicular to the plane of the parabola, and in this plane make the angle NAM equal to the angle OBC.

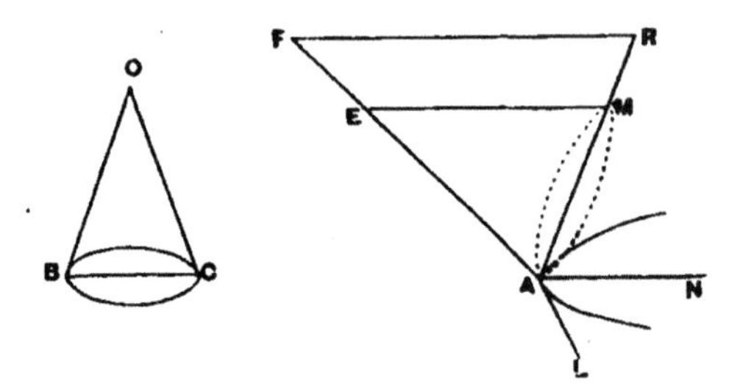

Let AM be taken of such a length that $AL : AM = BC : BO$, and on AM as base, in the plane MAN, describe the triangle EAM similar to the triangle OBC. Then suppose a cone described with vertex E and base the circle on AM as diameter in a plane perpendicular to the plane EAM.

The cone EAM will be the cone required.

For $\quad \angle MAN = \angle OBC = \angle EAM = \angle EMA$;

therefore EM is parallel to AN, the axis of the parabola.

Thus the plane of the given parabola cuts the cone in a section which is also a parabola.

Now
$$AL : AM = BC : BO$$
$$= AM : AE,$$
or
$$AM^2 = EA . AL;$$
$$\therefore\ AM^2 : AE . EM = AL : EM$$
$$= AL : EA.$$

Hence AL is the latus rectum of the parabolic section of the cone made by the plane of the given parabola. It is also the latus rectum of the given parabola.

Therefore the given parabola is itself the parabolic section, and EAM is the cone required.

There can be no other right cone similar to the given one, having its vertex on the same side of the given parabola, and containing that parabola as a section.

For, if another such cone be possible, with vertex F, draw through the axis of this cone a plane cutting the plane of the given parabola at right angles. The planes must then intersect in AN, the axis of the parabola, and therefore F must lie in the plane of EAN.

Again, if AF, FR are the sides of the axial triangle of the cone, FR must be parallel to AN, or to EM, and

$$\angle AFR = \angle BOC = \angle AEM,$$

so that F must lie on AE or AE produced. Let AM meet FR in R.

Then, if AL' be the latus rectum of the parabolic section of the cone FAR made by the plane of the given parabola,

$$\begin{aligned} AL' : AF &= AR^2 : AF.FR \\ &= AM^2 : AE.EM \\ &= AL : AE. \end{aligned}$$

Therefore AL', AL cannot be equal; or the given parabola is not a section of the cone FAR.

Proposition 121.

[VI. 32.]

To find a right cone similar to a given one and containing a given hyperbola as a section of it. (If OBC be the given cone and D the centre of its base BC, and if AA', AL be the axis and parameter of the given hyperbola, a necessary condition of possibility is that the ratio $OD^2 : DB^2$ must not be greater than the ratio $AA' : AL$.)

Let a plane be drawn through the axis of the given hyperbola and perpendicular to its plane; and on $A'A$, in the plane so described, describe a segment of a circle containing an

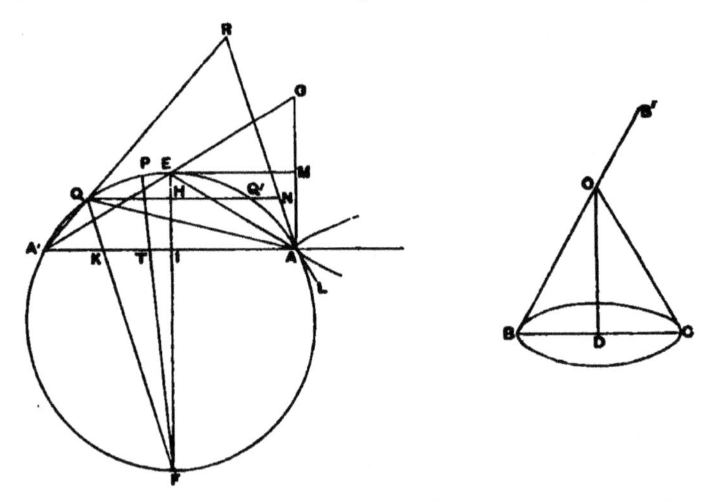

angle equal to the exterior angle $B'OC$ at the vertex of the given cone. Complete the circle, and let EF be the diameter of it bisecting AA' at right angles in I. Join $A'E$, AE, and draw AG parallel to EF meeting $A'E$ produced in G.

Then, since EF bisects the angle $A'EA$, the angle EGA is equal to the angle EAG. And the angle AEG is equal to the angle BOC, so that the triangles EAG, OBC are similar.

Draw EM perpendicular to AG.

Then $$OD^2 : DB^2 = EM^2 : MA^2$$
$$= IA^2 : EI^2$$
$$= FI : IE.$$

I. Suppose that

$$OD^2 : DB^2 < AA' : AL,$$

so that $$FI : IE < AA' : AL.$$

Take a point H on EI such that $FI : IH = AA' : AL$, and through H draw the chord QQ' of the circle parallel to AA'. Join $A'Q$, AQ, and in the plane of the circle draw AR making with AQ an angle equal to the angle OBC. Let AR meet $A'Q$ produced in R, and QQ' produced in N.

Join FQ meeting AA' in K.

Then, since the angle QAR is equal to the angle OBC, and

$$\angle FQA = \tfrac{1}{2}\angle A'QA = \tfrac{1}{2}\angle B'OC,$$

AR is parallel to FQ.

Also the triangle QAR is similar to the triangle OBC.

Suppose a cone formed with vertex Q and base the circle described on AR as diameter in a plane perpendicular to that of the circle FQA.

This cone will be such that the given hyperbola is a section of it.

We have, by construction,

$$AA' : AL = FI : IH$$
$$= FK : KQ, \text{ by parallels,}$$
$$= FK . KQ : KQ^2$$
$$= A'K . KA : KQ^2.$$

But, by the parallelogram $QKAN$,

$$A'K : KQ = QN : NR,$$

and $$KA : KQ = QN : NA,$$

whence $$A'K . KA : KQ^2 = QN^2 : AN . NR.$$

It follows that

$$AA' : AL = QN^2 : AN . NR.$$

Therefore [Prop. 2] AL is the parameter of the hyperbolic section of the cone QAR made by the plane of the given hyperbola. The two hyperbolas accordingly have the same axis and parameter, whence they coincide [Prop. 110 (2)]; and the cone QAR has the required property.

Another such cone is found by taking the point Q' instead of Q and proceeding as before.

No other right cone except these two can be found which is similar to the given one, has its apex on the same side of the plane of the given hyperbola, and contains that hyperbola as a section.

For, if such a cone be possible with apex P, draw through its axis a plane cutting the plane of the given hyperbola at right angles. The plane thus described must then pass through the axis of the given hyperbola, whence P must lie in the plane of the circle FQA. And, since the cone is similar to the given cone, P must lie on the arc $A'QA$.

Then, by the converse of the preceding proof, we must have (if FP meet $A'A$ in T)

$$AA' : AL = FT : TP;$$

$$\therefore\ FT : TP = FI : IH,$$

which is impossible.

II. Suppose that

$$OD^2 : DB^2 = AA' : AL,$$

so that $$FI : IE = AA' : AL.$$

In this case Q, Q' coalesce with E, and the cone with apex E and base the circle on AG as diameter perpendicular to the plane of FQA is the cone required.

III. If $OD^2 : DB^2 > AA' : AL$, no right cone having the desired properties can be drawn.

For, if possible, let P be the apex of such a cone, and we shall have, as before,

$$FT : TP = AA' : AL.$$

But $AA' : AL < OD^2 : DB^2$, or $FI : IE$.

Hence $FT : TP < FI : IE$, which is absurd.

Therefore, etc.

Proposition 122.

[VI. 33.]

To find a right cone similar to a given one and containing a given ellipse as a section of it.

As before, take a plane through AA' perpendicular to the plane of the given ellipse; and in the plane so drawn describe on AA' as base a segment of a circle containing an angle equal to the angle BOC, the vertical angle of the given cone. Bisect the arc of the segment in F.

Draw two lines FK, FK' to meet AA' produced both ways and such that, if they respectively meet the segment in Q, Q',

$$FK : KQ = FK' : K'Q' = AA' : AL.$$

Draw QN parallel to AA', and AN parallel to QF, meeting in N. Join AQ, $A'Q$, and let $A'Q$ meet AN in R.

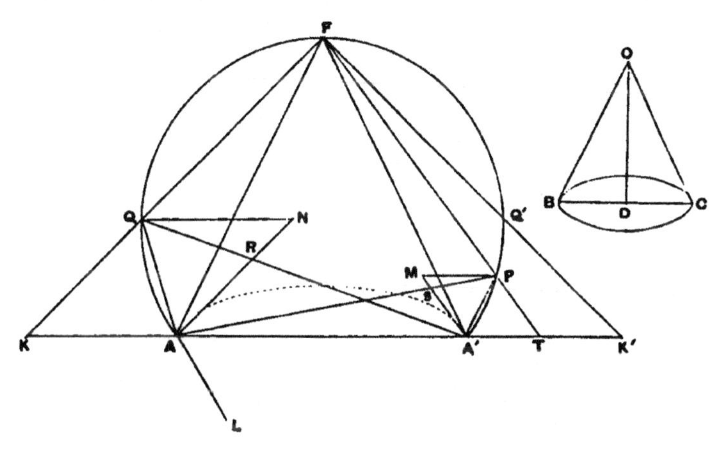

Conceive a cone drawn with Q as apex and as base the circle on AR as diameter and in a plane at right angles to that of AFA'.

This cone will be such that the given ellipse is one of its sections.

For, since FQ, AR are parallel,

$$\angle FQR = \angle ARQ.$$

$$\therefore \angle ARQ = \angle FAA'$$

$$= \angle OBC.$$

And $$\angle AQR = \angle AFA'$$

$$= \angle BOC.$$

Therefore the triangles QAR, OBC are similar, and likewise the cones QAR, OBC.

Now $AA' : AL = FK : KQ$, by construction,

$$= FK . KQ : KQ^2$$

$$= A'K . KA : KQ^2$$

$$= (A'K : KQ) . (KA : KQ)$$

$$= (QN : NR) . (QN : NA), \text{ by parallels,}$$

$$= QN^2 : AN . NR.$$

Therefore [Prop. 3] AL is the latus rectum of the elliptic section of the cone QAR made by the plane of the given ellipse. And AL is the latus rectum of the given ellipse. Therefore that ellipse is itself the elliptic section.

In like manner another similar right cone can be found with apex Q' such that the given ellipse is a section.

No other right cone besides these two can be found satisfying the given conditions and having its apex on the same side of the plane of the given ellipse. For, as in the preceding proposition, its apex P, if any, must lie on the arc AFA'. Draw PM parallel to $A'A$, and $A'M$ parallel to FP, meeting in M. Join AP, $A'P$, and let AP meet $A'M$ in S.

The triangle $PA'S$ will then be similar to OBC, and we shall have $PM^2 : A'M . MS = AT . TA' : TP^2 = FT . TP : TP^2$, in the same way as before.

We must therefore have

$$AA' : AL = FT : TP;$$

and this is impossible, because

$$AA' : AL = FK : KQ.$$

VALUES OF CERTAIN FUNCTIONS OF THE LENGTHS OF CONJUGATE DIAMETERS.

Proposition 123 (Lemma).

[VII. 1.]

In a parabola, if PN be an ordinate and AH be measured along the axis away from N and equal to the latus rectum,*

$$AP^2 = AN . NH. \qquad [= AN (AN + p_a)]$$

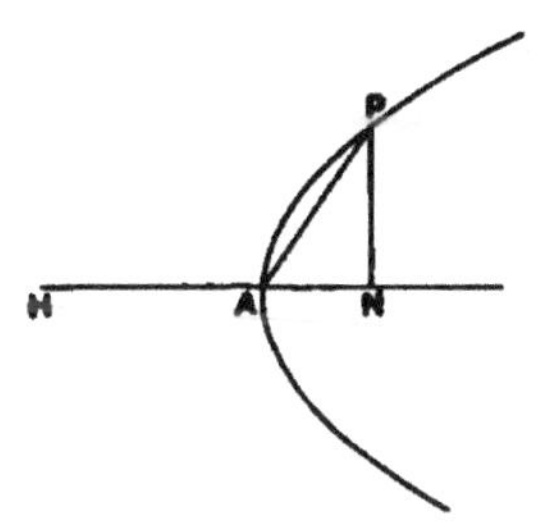

This is proved at once from the property $PN^2 = p_a . AN$, by adding AN^2 to each side.

Proposition 124 (Lemma).

[VII. 2, 3.]

If AA′ be divided at H, internally for the hyperbola, and externally for the ellipse, so that $AH : HA' = p_a : AA'$, then, if PN be any ordinate,

$$AP^2 : AN . NH = AA' : A'H.$$

* Though Book VII. is mainly concerned with conjugate diameters of a central conic, one or two propositions for the parabola are inserted, no doubt in order to show, in connection with particular propositions about a central conic, any obviously corresponding properties of the parabola.

Produce AN to K, so that

$$AN . NK = PN^2;$$

thus $$AN . NK : AN . A'N$$
$$= PN^2 : AN . A'N$$
$$= p_a : AA' \quad \text{[Prop. 8]}$$
$$= AH : A'H, \text{ by construction,}$$

or $$NK : A'N = AH : A'H.$$

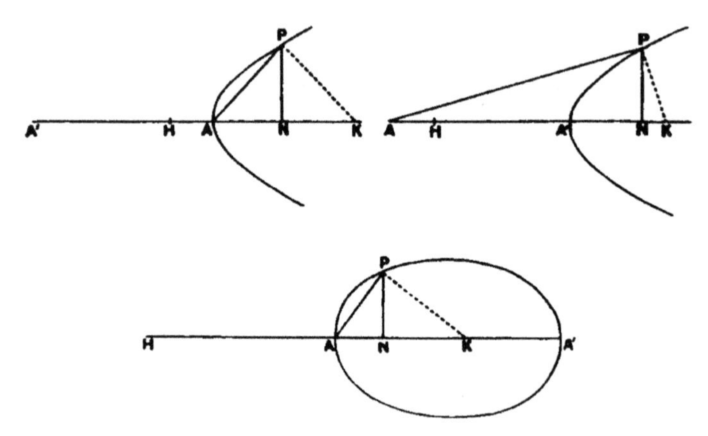

It follows that

$$A'N \pm NK : A'N = A'H \pm AH : A'H$$

(where the upper sign applies to the hyperbola).

Hence $$A'K : A'N = AA' : A'H;$$

$$\therefore \; A'K \pm AA' : A'N \pm A'H = AA' : A'H,$$

or $$AK : NH = AA' : A'H.$$

Thus $AN . AK : AN . NH = AA' : A'H$.

But $AN . AK = AP^2$, since $AN . NK = PN^2$.

Therefore $AP^2 : AN . NH = AA' : A'H$.

The same proposition is true if AA' is the *minor* axis of an ellipse and p_a the corresponding parameter.

Proposition 125 (Lemma).

[VII. 4.]

If in a hyperbola or an ellipse the tangent at P meet the axis AA′ in T, and if CD be the semi-diameter parallel to PT, then

$$PT^2 : CD^2 = NT : CN.$$

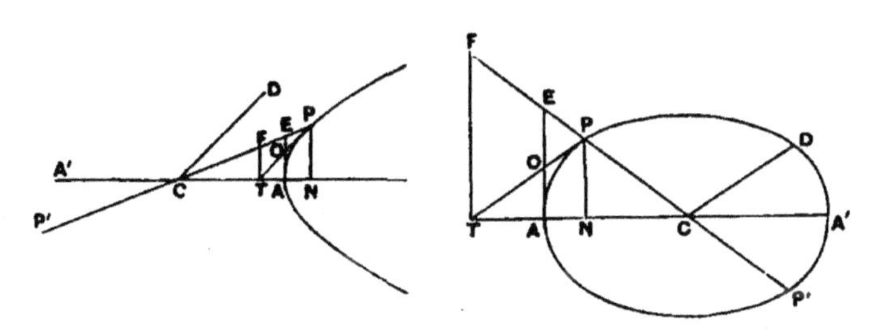

Draw AE, TF at right angles to CA to meet CP, and let AE meet PT in O.

Then, if p be the parameter of the ordinates to PP', we have

$$\frac{p}{2} : PT = OP : PE. \qquad \text{[Prop. 23]}$$

Also, since CD is parallel to PT, it is conjugate to CP.

Therefore $$\frac{p}{2} \cdot CP = CD^2 \text{ ....(1).}$$

Now $$OP : PE = TP : PF;$$

$$\therefore \frac{p}{2} : PT = PT : PF,$$

or $$\frac{p}{2} \cdot PF = PT^2 \text{ ....(2).}$$

From (1) and (2) we have

$$PT^2 : CD^2 = PF : CP$$
$$= NT : CN.$$

Proposition 126 (Lemma).

[VII. 5.]

In a parabola, if p be the parameter of the ordinates to the diameter through P, and PN the principal ordinate, and if AL be the latus rectum,

$$p = AL + 4AN.$$

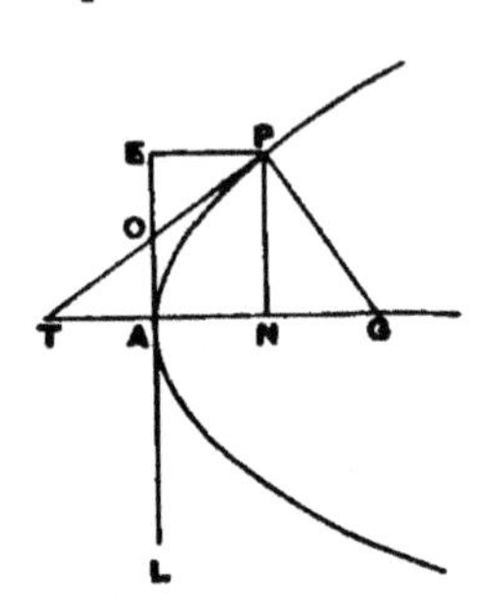

Let the tangent at A meet PT in O and the diameter through P in E, and let PG, at right angles to PT, meet the axis in G.

Then, since the triangles PTG, EPO are similar,

$$GT : TP = OP : PE,$$

$$\therefore\ GT = \frac{p}{2} \quad \text{..............(1).} \qquad \text{[Prop. 22]}$$

Again, since TPG is a right angle,

$$TN . NG = PN^2$$
$$= LA . AN,$$

by the property of the parabola.

But $\quad TN = 2AN. \qquad$ [Prop. 12]

Therefore $\quad AL = 2NG \quad$ ....(2);

thus $\quad AL + 4AN = 2(TN + NG)$

$$= 2TG$$

$= p$, from (1) above.

[*Note.* The property of the normal (NG = half the latus rectum) is incidentally proved here by regarding it as the perpendicular through P to the tangent at that point. Cf. Prop. 85 where the normal is regarded as the minimum straight line from G to the curve.]

DEF. If AA' be divided, internally for the hyperbola, and externally for the ellipse, in each of two points H, H' such that

$$A'H : AH = AH' : A'H' = AA' : p_a,$$

where p_a is the parameter of the ordinates to AA', then AH, $A'H'$ (corresponding to p_a in the proportion) are called **homologues**.

In this definition AA' may be either the *major* or the *minor* axis of an ellipse.

Proposition 127.

[VII. 6, 7.]

If AH, $A'H'$ be the "homologues" in a hyperbola or an ellipse, and PP', DD' any two conjugate diameters, and if AQ be drawn parallel to DD' meeting the curve in Q, and QM be perpendicular to AA', then

$$PP'^2 : DD'^2 = MH' : MH.$$

Join $A'Q$, and let the tangent at P meet AA' in T.

Then, since $A'C = CA$, and $QV = VA$ (where CP meets QA in V), $A'Q$ is parallel to CV.

Now $\quad PT^2 : CD^2 = NT : CN \quad$ [Prop. 125]

$\quad\quad = AM : A'M$, by similar triangles.

And, also by similar triangles,

$$CP^2 : PT^2 = A'Q^2 : AQ^2,$$

whence, *ex aequali*,

$$\begin{aligned} CP^2 : CD^2 &= (AM : A'M) . (A'Q^2 : AQ^2) \\ &= (AM : A'M) \times (A'Q^2 : A'M . MH') \\ &\quad \times (A'M . MH' : AM . MH) \times (AM . MH : AQ^2). \end{aligned}$$

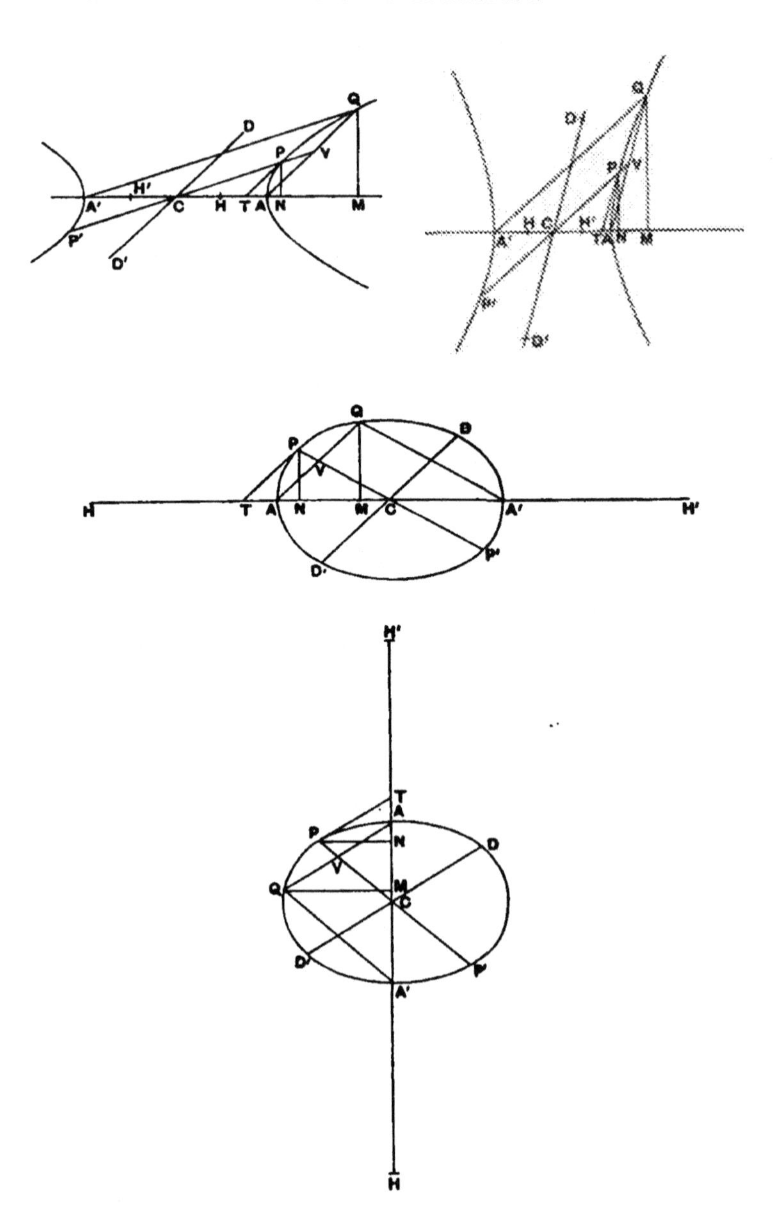
Q
D
P
V
H'
A'
C
H
T
A
N
M
P'
D'
Q
B
P
V
H
T
A
N
M
C
A'
H'
P'
D'
H'
T
A
P
N
D
V
Q
M
C
D'
P'
A'
H

But, by Prop. 124,

$$A'Q^2 : A'M \, . \, MH' = AA' : AH',$$

and $$AM \, . \, MH : AQ^2 = A'H : AA' = AH' : AA'.$$

Also $A'M \, . \, MH' : AM \, . \, MH = (A'M : AM) \, . \, (MH' : MH)$.

It follows that

$$CP^2 : CD^2 = MH' : MH,$$

or $$PP'^2 : DD'^2 = MH' : MH.$$

This result may of course be written in the form

$$PP' : p = MH' : MH,$$

where p is the parameter of the ordinates to PP'.

Proposition 128.

[VII. 8, 9, 10, 11.]

In the figures of the last proposition the following relations hold for both the hyperbola and the ellipse:

(1) $AA'^2 : (PP' \pm DD')^2 = A'H \, . \, MH' : (MH' \pm \sqrt{MH \, . \, MH'})^2$,

(2) $AA'^2 : PP' \, . \, DD' = A'H : \sqrt{MH \, . \, MH'}$,

(3) $AA'^2 : (PP'^2 \pm DD'^2) = A'H : MH \pm MH'$.

(1) We have

$$AA'^2 : PP'^2 = CA^2 : CP^2;$$

$$\therefore \; AA'^2 : PP'^2 = CN \, . \, CT : CP^2 \qquad \text{[Prop. 14]}$$

$$= A'M \, . \, A'A : A'Q^2,$$

by similar triangles.

Now $$A'Q^2 : A'M \, . \, MH' = AA' : AH' \qquad \text{[Prop. 124]}$$

$$= AA' : A'H$$

$$= A'M \, . \, A'A : A'M \, . \, A'H,$$

whence, alternately,

$$A'M \, . \, A'A : A'Q^2 = A'M \, . \, A'H : A'M \, . \, MH'.$$

Therefore, from above,

$$AA'^2 : PP'^2 = A'H : MH' \quad \ldots\ldots\ldots\ldots\ldots (\alpha),$$

$$= A'H \, . \, MH' : MH'^2.$$

Again, $PP'^2 : DD'^2 = MH' : MH$ ...(β), [Prop. 127]

$$= MH'^2 : MH \cdot MH';$$

$$\therefore PP' : DD' = MH' : \sqrt{MH \cdot MH'} \text{ ......} (\gamma).$$

Hence $PP' : PP' \pm DD' = MH' : MH' \pm \sqrt{MH \cdot MH'}$,

and $PP'^2 : (PP' \pm DD')^2 = MH'^2 : (MH' \pm \sqrt{MH \cdot MH'})^2$.

Therefore by (α) above, *ex aequali*,

$$AA'^2 : (PP' \pm DD')^2 = A'H \cdot MH' : (MH' \pm \sqrt{MH \cdot MH'})^2.$$

(2) We derive from (γ) above

$$PP'^2 : PP' \cdot DD' = MH' : \sqrt{MH \cdot MH'}.$$

Therefore by (α), *ex aequali*,

$$AA'^2 : PP' \cdot DD' = A'H : \sqrt{MH \cdot MH'}.$$

(3) From (β),

$$PP'^2 : (PP'^2 \pm DD'^2) = MH' : MH \pm MH'.$$

Therefore by (α), *ex aequali*,

$$AA'^2 : (PP'^2 \pm DD'^2) = A'H : MH \pm MH'.$$

Proposition 129.

[VII. 12, 13, 29, 30.]

In every ellipse the sum, and in every hyperbola the difference, of the squares on any two conjugate diameters is equal to the sum or difference respectively of the squares on the axes.

Using the figures and construction of the preceding two propositions, we have

$$AA'^2 : BB'^2 = AA' : p_a$$

$$= A'H : AH, \text{ by construction,}$$

$$= A'H : A'H'.$$

Therefore

$$AA'^2 : AA'^2 \pm BB'^2 = A'H : A'H \pm A'H'$$

(where the upper sign belongs to the *ellipse*),

or $AA'^2 : AA'^2 \pm BB'^2 = A'H : HH'$ .... (α).

Again, by (α) in Prop. 128 (1),

$$AA'^2 : PP'^2 = A'H : MH',$$

and, by means of (β) in the same proposition,

$$PP'^2 : (PP'^2 \pm DD'^2) = MH' : MH \pm MH'$$
$$= MH' : HH'.$$

From the last two relations we obtain

$$AA'^2 : (PP'^2 \pm DD'^2) = A'H : HH'.$$

Comparing this with (α) above, we have at once

$$(PP'^2 \pm DD'^2) = (AA'^2 \pm BB'^2).$$

Proposition 130.

[VII. 14, 15, 16, 17, 18, 19, 20.]

The following results can be derived from the preceding propositions, viz.

(1) *For the ellipse,*

$$AA'^2 : PP'^2 \sim DD'^2 = A'H : 2CM;$$

and for both the ellipse and hyperbola, if p denote the parameter of the ordinates to PP',

(2) $$AA'^2 : p^2 = A'H . MH' : MH'^2,$$

(3) $$AA'^2 : (PP' \pm p)^2 = A'H . MH' : (MH \pm MH')^2,$$

(4) $$AA'^2 : PP' . p = A'H : MH, \text{ and}$$

(5) $$AA'^2 : PP'^2 \pm p^2 = A'H . MH' : MH'^2 \pm MH^2.$$

(1) We have

$$AA'^2 : PP'^2 = A'H : MH', \quad \text{[Prop. 128 (1), } (\alpha)]$$

and $$PP'^2 : PP'^2 \sim DD'^2 = MH' : MH' \sim MH \quad [\textit{ibid.}, (\beta)]$$
$$= MH' : 2CM \text{ in the } \textit{ellipse}.$$

Therefore for the *ellipse*

$$AA'^2 : PP'^2 \sim DD'^2 = A'H : 2CM.$$

(2) For either curve

$$AA''^2 : PP''^2 = A'H : MH', \text{ as before,}$$
$$= A'H \cdot MH' : MH'^2,$$

and, by Prop. 127,

$$PP''^2 : p^2 = MH'^2 : MH^2;$$
$$\therefore AA''^2 : p^2 = A'H \cdot MH' : MH^2.$$

(3) By Prop. 127,

$$PP' : p = MH' : MH;$$
$$\therefore PP''^2 : (PP' \pm p)^2 = MH'^2 : (MH \pm MH')^2.$$

And $$AA''^2 : PP''^2 = A'H \cdot MH' : MH'^2, \text{ as before;}$$
$$\therefore AA''^2 : (PP' \pm p)^2 = A'H \cdot MH' : (MH \pm MH')^2.$$

(4) $$AA''^2 : PP''^2 = A'H : MH', \text{ as before,}$$

and $$PP''^2 : PP' \cdot p = PP' : p$$
$$= MH' : MH; \quad \text{[Prop. 127]}$$
$$\therefore AA''^2 : PP' \cdot p = A'H : MH.$$

(5) $$AA''^2 : PP''^2 = A'H \cdot MH' : MH'^2, \text{ as before,}$$

and $$PP''^2 : PP''^2 \pm p^2 = MH'^2 : MH'^2 \pm MH^2,$$

by means of Prop. 127;

$$\therefore AA''^2 : PP''^2 \pm p^2 = A'H \cdot MH' : MH'^2 \pm MH^2.$$

Proposition 131.

[VII. 21, 22, 23.]

In a hyperbola, if $AA' \underset{\text{or}}{\gtrless} BB'$, then, if PP', DD' be any other two conjugate diameters, $PP' \underset{\text{or}}{\gtrless} DD'$ respectively; and the ratio $PP' : DD'$ continually $\left\{\begin{matrix}\text{decreases}\\ \text{or increases}\end{matrix}\right\}$ as P moves further from A on either side.

Also, if $AA' = BB'$, $PP' = DD'$.

(1) Of the figures of Prop. 127, the first corresponds to the case where $AA' > BB'$, and the second to the case where $AA' < BB'$.

Taking then the $\left\{\begin{matrix}\text{first}\\ \text{second}\end{matrix}\right\}$ figure respectively, it follows from

$$PP'^2 : DD'^2 = MH' : MH \qquad \text{[Prop. 127]}$$

that $$PP' \underset{\text{or}<}{>} DD'.$$

Also $AA'^2 : BB'^2 = AA' : p_a = A'H : AH$, by construction,

$$= AH' : AH,$$

and $$AH' : AH \underset{\text{or}<}{>} MH' : MH,$$

while $MH' : MH \left\{\begin{matrix}\text{diminishes}\\ \text{or increases}\end{matrix}\right\}$ continually as M moves further from A, *i.e.* as Q, or P, moves further from A along the curve.

Therefore $$AA'^2 : BB'^2 \underset{\text{or}<}{>} PP'^2 : DD'^2,$$

and the latter ratio $\left\{\begin{matrix}\text{diminishes}\\ \text{or increases}\end{matrix}\right\}$ as P moves further from A.

And the same is true of the ratios

$$AA' : BB' \text{ and } PP' : DD'.$$

(2) If $AA' = BB'$, then $AA' = p_a$, and both H and H' coincide with C.

In this case therefore

$$AH = AH' = AC,$$
$$MH = MH' = CM,$$

and $PP' = DD'$ always.

Proposition 132.

[VII. 24.]

In an ellipse, if AA' be the major, and BB' the minor, axis, and if PP', DD' be any other two conjugate diameters, then

$$AA' : BB' > PP' : DD',$$

and the latter ratio diminishes continually as P moves from A to B.

We have $CA^2 : CB^2 = AN . NA' : PN^2$;

$$\therefore AN . NA' > PN^2,$$

and, adding CN^2 to each,

$$CA^2 > CP^2,$$

or $$AA' > PP' \quad \text{.......................(1).}$$

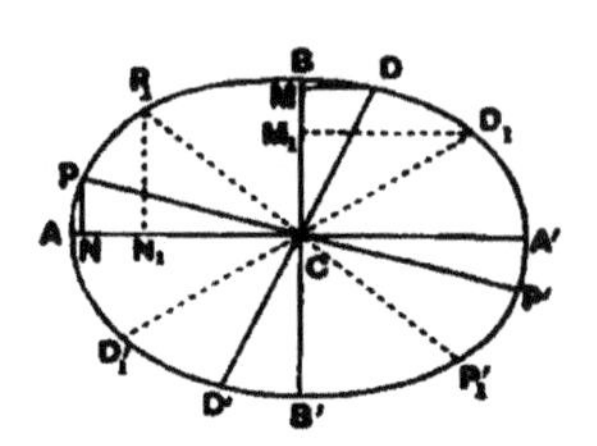

Also $CB^2 : CA^2 = BM . MB' : DM^2$,

where DM is the ordinate to BB'.

Therefore $BM . MB' < DM^2$,

and, adding CM^2, $CB^2 < CD^2$;

$$\therefore BB' < DD' \quad \text{.....................(2).}$$

Again, if P_1P_1', D_1D_1' be another pair of conjugates, P_1 being further from A than P, D_1 will be further from B than D.

And $AN . NA' : AN_1 . N_1A' = PN^2 : P_1N_1^2$.

But $AN_1 . N_1A' > AN . NA'$;

$$\therefore P_1N_1^2 > PN^2,$$

and $AN_1 . N_1A' - AN . NA' > P_1N_1^2 - PN^2$.

But, as above, $AN_1 . N_1A' > P_1N_1^2$,

and $AN_1 . N_1A' - AN . NA' = CN^2 - CN_1^2$;

$$\therefore CN^2 - CN_1^2 > P_1N_1^2 - PN^2;$$

thus $$CP^2 > CP_1^2,$$

or $$PP' > P_1P_1' \quad \text{.......................(3).}$$

In an exactly similar manner we prove that

$$DD' < D_1D_1' \quad \text{.....................(4).}$$

We have therefore, by (1) and (2),

$$AA' : BB' > PP' : DD',$$

and, by (3) and (4), $PP' : DD' > P_1P_1' : D_1D_1'$.

COR. It is at once clear, if p_a, p, p_1 are the parameters corresponding to AA', PP', P_1P_1', that

$$p_a < p, \quad p < p_1, \text{ etc.}$$

Proposition 133.

[VII. 25, 26.]

(1) *In a hyperbola or an ellipse*

$$AA' + BB' < PP' + DD',$$

where PP', DD' are any conjugate diameters other than the axes.

(2) *In the hyperbola $PP' + DD'$ increases continually as P moves further from A, while in the ellipse it increases as P moves from A until PP', DD' take the position of the equal conjugate diameters, when it is a* **maximum**.

(1) For the *hyperbola*

$$AA'^2 \sim BB'^2 = PP'^2 \sim DD'^2 \qquad \text{[Prop. 129]}$$

or $(AA' + BB') . (AA' \sim BB') = (PP' + DD') . (PP' \sim DD')$,

and, by the aid of Prop. 131,

$$AA' \sim BB' > PP' \sim DD';$$

$$\therefore AA' + BB' < PP' + DD'.$$

Similarly it is proved that $PP' + DD'$ increases as P moves further from A.

In the case where $AA' = BB'$, $PP' = DD'$, and $PP' > AA'$; and the proposition still holds.

(2) For the *ellipse*

$$AA' : BB' > PP' : DD';$$

$$\therefore (AA'^2 + BB'^2) : (AA' + BB')^2 > (PP'^2 + DD'^2) : (PP' + DD')^2.*$$

But $$AA'^2 + BB'^2 = PP'^2 + DD'^2; \qquad \text{[Prop. 129]}$$

$$\therefore AA' + BB' < PP' + DD'.$$

* Apollonius draws this inference directly, and gives no intermediate steps.

Similarly it may be proved that $PP' + DD'$ increases as P moves from A until PP', DD' take the position of the equal conjugate diameters, when it begins to diminish again.

Proposition 134.

[VII. 27.]

In every ellipse or hyperbola having unequal axes

$$AA' \sim BB' > PP' \sim DD',$$

where PP', DD' are any other conjugate diameters. Also, as P moves from A, $PP' \sim DD'$ diminishes, in the hyperbola continually, and in the ellipse until PP', DD' take up the position of the equal conjugate diameters.

For the *ellipse* the proposition is clear from what was proved in Prop. 132.

For the *hyperbola*

$$AA'^2 \sim BB'^2 = PP'^2 \sim DD'^2,$$

and

$$PP' > AA'.$$

It follows that

$$AA' \sim BB' > PP' \sim DD',$$

and the latter diminishes continually as P moves further from A.

[This proposition should more properly have come before Prop. 133, because it is really used (so far as regards the hyperbola) in the proof of that proposition.]

Proposition 135.

[VII. 28.]

In every hyperbola or ellipse

$$AA' . BB' < PP' . DD',$$

and $PP' . DD'$ increases as P moves away from A, in the hyperbola continually, and in the ellipse until PP', DD' coincide with the equal conjugate diameters.

We have $\qquad AA' + BB' < PP' + DD', \qquad$ [Prop. 133]

so that $\qquad \therefore (AA' + BB')^2 < (PP' + DD')^2.$

And, for the *ellipse*,

$$AA'^2 + BB'^2 = PP'^2 + DD'^2. \qquad \text{[Prop. 129]}$$

Therefore, by subtraction,

$$AA'.BB' < PP'.DD',$$

and in like manner it will be shown that $PP.DD'$ increases until PP', DD' coincide with the equal conjugate diameters.

For the *hyperbola* [proof omitted in Apollonius] $PP' > AA'$, $DD' > BB'$, and PP', DD' both increase continually as P moves away from A. Hence the proposition is obvious.

Proposition 136.

[VII. 31.]

If PP', DD' be two conjugate diameters in an ellipse or in conjugate hyperbolas, and if tangents be drawn at the four extremities forming a parallelogram $LL'MM'$, then

the parallelogram $LL'MM'$ = rect. $AA'.BB'$.

Let the tangents at P, D meet the axis AA' in T, T' respectively. Let PN be an ordinate to AA', and take a length PO such that

$$PO^2 = CN.NT.$$

Now $$CA^2 : CB^2 = CN.NT : PN^2 \qquad \text{[Prop. 14]}$$

$$= PO^2 : PN^2,$$

or $$CA : CB = PO : PN;$$

$$\therefore CA^2 : CA.CB = PO.CT : CT.PN.$$

Hence, alternately,

$$CA^2 : PO.CT = CA.CB : CT.PN,$$

or $$CT.CN : PO.CT = CA.CB : CT.PN \quad \text{......(1)}.$$

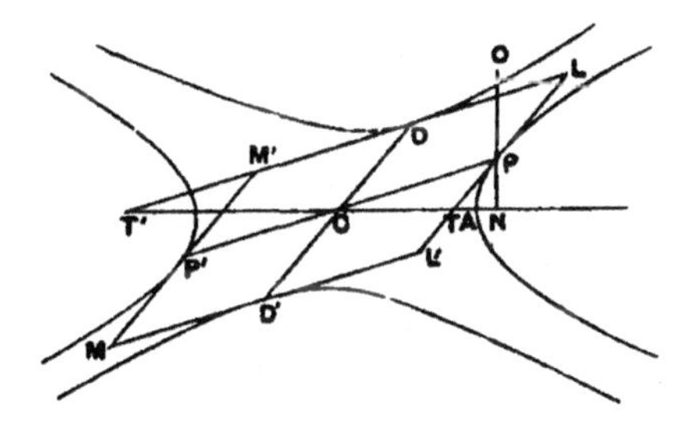

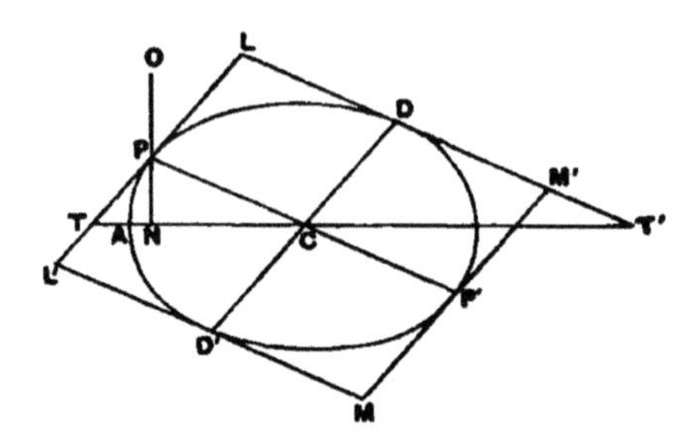

Again, $PT^2 : CD^2 = NT : CN$, [Prop. 125]

so that $2\triangle CPT : 2\triangle T'DC = NT : CN$.

But the parallelogram (CL) is a mean proportional between $2\triangle CPT$ and $2\triangle T'DC$,

for
$$2\triangle CPT : (CL) = PT : CD$$
$$= CP : DT'$$
$$= (CL) : 2\triangle T'DC.$$

Also PO is a mean proportional between CN and NT.

Therefore
$$2\triangle CPT : (CL) = PO : CN = PO . CT : CT . CN$$
$$= CT . PN : CA . CB, \text{ from (1) above.}$$

And $$2\triangle CPT = CT . PN;$$
$$\therefore (CL) = CA . CB,$$

or, quadrupling each side,
$$\square LL'MM' = AA' . BB'.$$

Proposition 137.

[VII. 33, 34, 35.]

Supposing p_a to be the parameter corresponding to the axis AA' in a hyperbola, and p to be the parameter corresponding to a diameter PP',

(1) *if AA' be not less than p_a, then $p_a < p$, and p increases continually as P moves further from A;*

(2) *if AA' be less than p_a but not less than $\frac{p_a}{2}$, then $p_a < p$, and p increases as P moves away from A;*

(3) *if $AA' < \frac{p_a}{2}$, there can be found a diameter P_0P_0' on either side of the axis such that $p_0 = 2P_0P_0'$. Also p_0 is less than any other parameter p, and p increases as P moves further from P_0 in either direction.*

(1) (*a*) If $AA' = p_a$, we have [Prop. 131 (2)]

$$PP' = p = DD',$$

and PP', and therefore p, increases continually as P moves away from A.

(*b*) If $AA' > p_a$, $AA' > BB'$, and, as in Prop. 131 (1), $PP' : DD'$, and therefore $PP' : p$, diminishes continually as P moves away from A. But PP' increases. Therefore p increases all the more.

(2) Suppose $AA' < p_a$ but $\nless \frac{p_a}{2}$.

Let P be any point on the branch with vertex A; draw $A'Q$ parallel to CP meeting the same branch in Q, and draw the ordinate QM.

Divide $A'A$ at H, H' so that

$$A'H : HA = AH' : H'A' = AA' : p_a,$$

as in the preceding propositions.

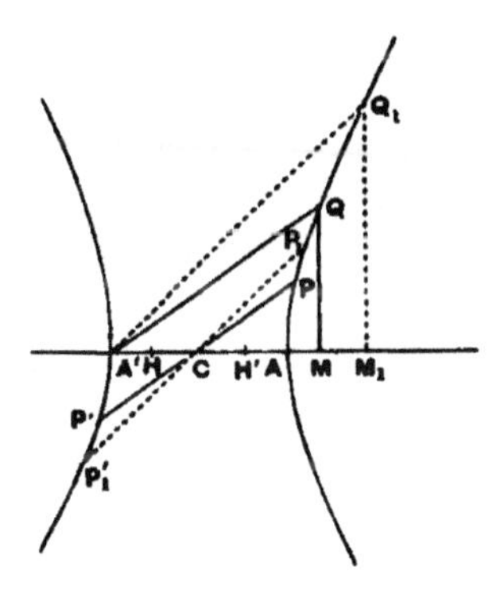

Therefore $AA''^2 : p_a^2 = A'H . AH' : AH^2$ ....(α).

We have now $AH > AH'$ but $\not> 2AH'$.

And $MH + HA > 2AH$;

$$\therefore\ MH + HA : AH > AH : AH',$$

or $(MH + HA)\,AH' > AH^2$ ....(β).

It follows that

$(MH + HA)\,AM : (MH + HA)\,AH'$, or $AM : AH'$,

$$< (MH + HA)\,AM : AH^2.$$

Therefore, *componendo*,

$$MH' : AH' < (MH + HA)\,AM + AH^2 : AH^2$$

$$< MH^2 : AH^2 \text{........................}(\gamma),$$

whence $A'H . MH' : A'H . AH' < MH^2 : AH^2$,

or, alternately,

$$A'H . MH' : MH^2 < A'H . AH' : AH^2.$$

But, by Prop. 130 (2), and by the result (α) above, these ratios are respectively equal to $AA''^2 : p^2$, and $AA''^2 : p_a^2$.

Therefore $AA''^2 : p^2 < AA''^2 : p_a^2$,

or $p_a < p$.

Again, if P_1 be a point further from A than P is, and if $A'Q_1$ is parallel to CP_1, and M_1 is the foot of the ordinate QM_1, then, since $AH \not> 2AH'$,

$$MH < 2MH';$$

also $M_1H + HM > 2MH$.

Thus $(M_1H + HM)\,MH' > MH^2$.

This is a similar relation to that in (β) above except that M is substituted for A, and M_1 for M.

We thus derive, by the same proof, the corresponding result to (γ) above, or

$$M_1H' : MH' < M_1H^2 : MH^2,$$

whence $A'H . M_1H' : M_1H^2 < A'H . MH' : MH^2$,

or $AA''^2 : p_1^2 < AA''^2 : p^2$,

so that $p < p_1$, and the proposition is proved.

(3) Now let AA' be less than $\frac{p_a}{2}$.

Take a point M_0 such that $HH' = H'M_0$, and let Q_0, P_0 be related to M_0 in the same way that Q, P are to M.

Then $$P_0P_0' : p_0 = M_0H' : M_0H. \quad \text{[Prop. 127]}$$

It follows, since $HH' = H'M_0$, that

$$p_0 = 2P_0P_0'.$$

Next, let P be a point on the curve between P_0 and A, and Q, M corresponding points.

Then $$M_0H' . H'M < HH'^2,$$

since $$MH' < M_0H'.$$

Add to each side the rectangle $(MH + HH')\,MH'$, and we have

$$(M_0H + HM)\,MH' < MH^2.$$

This again corresponds to the relation (β) above, with M substituted for A, M_0 for M, and $<$ instead of $>$.

The result corresponding to (γ) above is

$$M_0H' : MH' > M_0H^2 : MH^2;$$

$$\therefore\ A'H . M_0H' : M_0H^2 > A'H . MH' : MH^2,$$

or $$AA'^2 : p_0^2 > AA'^2 : p^2.$$

Therefore $$p > p_0.$$

And in like manner we prove that p increases as P moves from P_0 to A.

Lastly, let P be more remote from A than P_0 is.

In this case $$H'M > H'M_0,$$

and we have $$MH' . H'M_0 > HH'^2,$$

and, by the last preceding proof, interchanging M and M_0 and substituting the opposite sign of relation,

$$AA'^2 : p^2 < AA'^2 : p_0^2,$$

and $$p > p_0.$$

In the same way we prove that p increases as P moves further away from P and A.

Hence the proposition is established.

Proposition 138.

[VII. 36.]

In a hyperbola with unequal axes, if p_a be the parameter corresponding to AA' and p that corresponding to PP',

$$AA' \sim p_a > PP' \sim p,$$

and $PP' \sim p$ diminishes continually as P moves away from A.

With the same notation as in the preceding propositions,

$$A'H : HA = AH' : H'A' = AA' : p_a,$$

whence $$AA'^2 : (AA' \sim p_a)^2 = A'H . AH' : HH'^2.$$

Also [Prop. 130 (3)]

$$AA'^2 : (PP' \sim p)^2 = A'H . MH' : HH'^2.$$

But $$A'H . MH' > A'H . AH';$$

$$\therefore\ AA'^2 : (PP' \sim p)^2 > AA'^2 : (AA' \sim p_a)^2.$$

Hence $$AA' \sim p_a > PP' \sim p.$$

Similarly, if P_1, M_1 be further from A than P, M are, we have

$$A'H . M_1H' > A'H . MH',$$

and it follows that

$$PP' \sim p > P_1P_1' \sim p_1,$$

and so on.

Proposition 139.

[VII. 37.]

In an ellipse, if P_0P_0', D_0D_0' be the equal conjugate diameters and PP', DD' any other conjugate diameters, and if p_0, p, p_a, p_b be the parameters corresponding to P_0P_0', PP', AA', BB' respectively, then

(1) *$AA' \sim p_a$ is the* **maximum** *value of $PP' \sim p$ for all points P between A and P_0, and $PP' \sim p$ diminishes continually as P moves from A to P_0,*

(2) *$BB' \sim p_b$ is the* **maximum** *value of $PP' \sim p$ for all points P between B and P_0, and $PP' \sim p$ diminishes continually as P passes from B to P_0,*

(3) $BB' \sim p_b > AA' \sim p_a$.

The results (1) and (2) follow at once from Prop. 132.

(3) Since $p_b : BB' = AA' : p_a$, and $p_b > AA'$, it follows at once that $BB' \sim p_b > AA' \sim p_a$.

Proposition 140.

[VII. 38, 39, 40.]

(1) *In a hyperbola, if AA' be not less than $\frac{1}{3} p_a$,*

$$PP' + p > AA' + p_a,$$

where PP' is any other diameter and p the corresponding parameter; and $PP' + p$ will be the smaller the nearer P approaches to A.

(2) *If $AA' < \frac{1}{3} p_a$, there is on each side of the axis a diameter, as P_0P_0', such that $P_0P_0' = \frac{1}{3} p_0$; and $P_0P_0' + p_0$ is less than $PP' + p$, where PP' is any other diameter on the same side of the axis. Also $PP' + p$ increases as P moves away from P_0.*

(1) The construction being the same as before, we suppose

(*a*) $$AA' \nless p_a.$$

In this case [Prop. 137 (1)] PP' increases as P moves from A, and p along with it.

Therefore $PP' + p$ also increases continually.

(*b*) Suppose $AA' < p_a$ but $\nless \frac{1}{3} p_a$;

$$\therefore AH' \nless \tfrac{1}{3} AH;$$

thus $$AH' \nless \tfrac{1}{4}(AH + AH'),$$

and $$(AH + AH') . 4AH' \nless (AH + AH')^2.$$

Hence $4(AH+AH')AM : 4(AH+AH')AH'$, or $AM : AH'$,

$$\ngtr 4(AH + AH')AM : (AH + AH')^2;$$

and, *componendo*,

$$MH' : AH' \ngtr 4(AH + AH')AM + (AH + AH')^2 : (AH + AH')^2.$$

Now

$$(MH + MH')^2 - (AH + AH')^2 = 2AM(MH + MH' + AH + AH')$$
$$> 4AM(AH + AH');$$

$$\therefore\ 4AM(AH + AH') + (AH + AH')^2 < (MH + MH')^2.$$

It follows that

$$MH' : AH' < (MH + MH')^2 : (AH + AH')^2,$$

or $\quad A'H \,.\, MH' : (MH + MH')^2 < A'H \,.\, AH' : (AH + AH')^2;$

$\therefore\ AA'^2 : (PP' + p)^2 < AA'^2 : (AA' + p_a)^2$ [by Prop. 130 (3)].

Hence $\qquad AA' + p_a < PP' + p.$

Again, since $\qquad AH' \nless \frac{1}{4}(AH + AH'),$

$$MH' > \frac{1}{4}(MH + MH');$$

$$\therefore\ 4(MH + MH')\,MH' > (MH + MH')^2.$$

And, if P_1 be another point further from A than P is, and Q_1, M_1 points corresponding to Q, M, we have, by the same proof as before (substituting M for A, and M_1 for M),

$$A'H \,.\, M_1H' : (M_1H + M_1H')^2 < A'H \,.\, MH' : (MH + MH')^2.$$

We derive $\qquad PP' + p < P_1P_1' + p_1;$

and the proposition is established.

(2) We have $AH' < \frac{1}{3}AH$, so that $AH' < \frac{1}{4}HH'$.

Make $H'M_0$ equal to $\frac{1}{4}HH'$, so that $M_0H' = \frac{1}{3}M_0H$.

Then $\qquad P_0P_0' : p_0 = M_0H' : M_0H = 1 : 3,$

and $\qquad P_0P_0' = \dfrac{p_0}{3}.$

Next, since $\qquad M_0H' = \frac{1}{3}M_0H,$

$$M_0H' = \frac{1}{4}(M_0H + M_0H').$$

Now suppose P to be a point between A and P_0, so that

$$M_0H' > MH';$$

$$\therefore\ (M_0H + M_0H')^2 > (M_0H + MH') \,.\, 4M_0H'.$$

Subtracting from each side the rectangle $(M_0H + MH').4MM_0$,

$$(MH + MH')^2 > (M_0H + MH').4MH';$$

$$\therefore\ (M_0H + MH').4MM_0 : (M_0H + MH').4MH', \text{ or } MM_0 : MH',$$

$$> (M_0H + MH').4MM_0 : (MH + MH')^2.$$

Therefore, *componendo*,

$$M_0H' : MH' > (M_0H + MH').4MM_0 + (MH + MH')^2 : (MH + MH')^2$$

$$> (M_0H + M_0H')^2 : (MH + MH')^2.$$

Hence

$$A'H.M_0H' : (M_0H + M_0H')^2 > A'H.MH' : (MH + MH')^2.$$

Therefore [Prop. 130 (3)]

$$AA'^2 : (P_0P_0' + p_0)^2 > AA'^2 : (PP' + p)^2,$$

and

$$PP' + p > P_0P_0' + p_0.$$

Again, if P_1 be a point between P and A, we have

$$(MH + MH')^2 > (MH + M_1H').4MH',$$

and we prove exactly as before that

$$P_1P_1' + p_1 > PP' + p,$$

and so on.

Lastly, if $MH > M_0H$, we shall have

$$(MH + M_0H').4M_0H' > (M_0H + M_0H')^2.$$

If to both sides of this inequality there be added the rectangle $(MH + M_0H').4MM_0$, they become respectively

$$(MH + M_0H').4MH' \text{ and } (MH + MH')^2,$$

and the method of proof used above gives

$$P_0P_0' + p_0 < PP' + p,$$

and so on.

Hence the proposition is established.

Proposition 141.

[VII. 41.]

In any ellipse, if PP' be any diameter and p its parameter, $PP'+p>AA'+p_a$, and $PP'+p$ is the less the nearer P is to A. Also $BB'+p_b>PP'+p$.

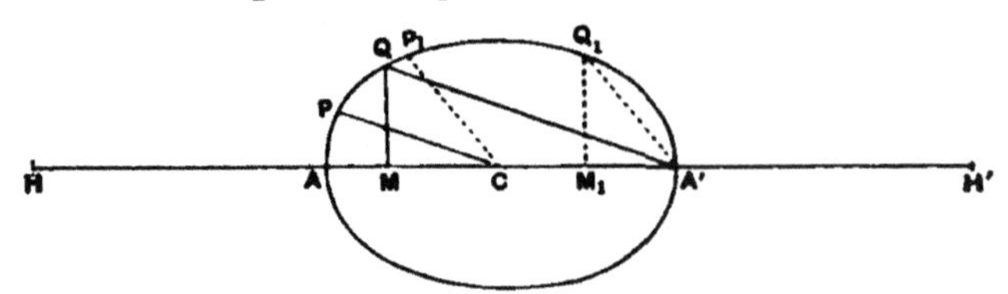

With the same construction as before,

$$A'H : HA = AH' : H'A'$$
$$= AA' : p_a$$
$$= p_b : BB'.$$

Then $\quad AA'^2 : (AA'+p_a)^2 = A'H^2 : HH'^2$

$$= A'H \cdot AH' : HH'^2 \quad\ldots\ldots\ldots(\alpha).$$

Also $\quad AA'^2 : BB'^2 = AA' : p_a = A'H : A'H'$

$$= A'H \cdot A'H' : A'H'^2$$

and $\quad BB'^2 : (BB'+p_b)^2 = A'H'^2 : HH'^2$

Therefore, *ex aequali*,

$$AA'^2 : (BB'+p_b)^2 = A'H \cdot A'H' : HH'^2 \quad\ldots\ldots\ldots(\beta).$$

From (α) and (β), since $AH'>A'H'$,

$$AA'+p_a<BB'+p_b.$$

Again $AA'^2 : (PP'+p)^2 = A'H \cdot MH' : HH'^2$, [Prop. 130 (3)]

and $\quad AA'^2 : (P_1P_1'+p_1)^2 = A'H \cdot M_1H' : HH'^2$,

where P_1 is between P and B, from which it follows, since

$$AH'>MH'>M_1H'>A'H',$$

that $\quad AA'+p_a<PP'+p,$

$$PP'+p<P_1P_1'+p_1,$$
$$P_1P_1'+p_1<BB'+p_b,$$

and the proposition follows.

Proposition 142.

[VII. 42.]

In a hyperbola, if PP' be any diameter with parameter p,

$$AA'.p_a < PP'.p,$$

and $PP'.p$ increases as P moves away from A.

We have $$A'H : HA = AA'^2 : AA'.p_a,$$

and $$A'H : MH = AA'^2 : PP'.p, \quad \text{[Prop. 130 (4)]}$$

while $$AH < MH;$$

$$\therefore AA'.p_a < PP'.p,$$

and, since MH increases as P moves from A, so does $PP'.p$.

Proposition 143.

[VII. 43.]

In an ellipse $AA'.p_a < PP'.p$, where PP' is any diameter, and $PP'.p$ increases as P moves away from A, reaching a **maximum** *when P coincides with B or B'.*

The result is derived at once, like the last proposition, from Prop. 130 (4).

[Both propositions are also at once obvious since

$$PP'.p = DD'^2.]$$

Proposition 144.

[VII. 44, 45, 46.]

In a hyperbola,

(1) *if $AA' \not< p_a$, or*

(2) *if $AA' < p_a$, but $AA'^2 \not< \frac{1}{2}(AA' - p_a)^2$, then*

$$AA'^2 + p_a^2 < PP'^2 + p^2,$$

where PP' is any diameter, and $PP'^2 + p^2$ increases as P moves away from A;

(3) *if $AA''^2 < \frac{1}{2}(AA' \sim p_a)^2$, then there will be found on either side of the axis a diameter P_0P_0' such that $P_0P_0''^2 = \frac{1}{2}(P_0P_0' - p_0)^2$, and $P_0P_0''^2 + p_0^2$ will be less than $PP''^2 + p^2$, where PP' is any other diameter. Also $PP''^2 + p^2$ will be the smaller the nearer PP' is to P_0P_0'.*

(1) Let AA' be not less than p_a.

Then, if PP' be any other diameter, $p > p_a$, and p increases as P moves further from A [Prop. 137 (1)]; also $AA' < PP'$, which increases as P moves further from A;

$$\therefore AA''^2 + p_a^2 < PP''^2 + p^2,$$

and $PP''^2 + p^2$ increases continually as P moves further from A.

(2) Let AA' be less than p_a, but $AA''^2 \nless \frac{1}{2}(AA' \sim p_a)^2$.

Then, since $AA' : p_a = A'H : AH = AH' : A'H'$,

$$2AH''^2 \nless HH''^2,$$

and $$2MH' . AH' > HH''^2.$$

Adding $2AH . AH'$ to each side of the last inequality,

$$2(MH + AH')AH' > 2AH . AH' + HH''^2$$
$$> AH^2 + AH''^2;$$

$$\therefore 2(MH + AH')AM : 2(MH + AH')AH', \text{ or } AM : AH',$$
$$< 2(MH + AH')AM : AH^2 + AH''^2.$$

Therefore, *componendo*,

$$MH' : AH' < 2(MH + AH')AM + AH^2 + AH''^2 : AH^2 + AH''^2,$$

and $$MH^2 + MH''^2 = AH^2 + AH''^2 + 2AM(MH + AH'),$$

so that $$MH' : AH' < MH^2 + MH''^2 : AH^2 + AH''^2,$$

or $$A'H . MH' : MH^2 + MH''^2 < A'H . AH' : AH^2 + AH''^2;$$

$$\therefore AA''^2 : PP''^2 + p^2 < AA''^2 : AA''^2 + p_a^2. \text{ [Prop. 130 (5)]}$$

Thus $$AA''^2 + p_a^2 < PP''^2 + p^2.$$

Again, since $2MH''^2 > HH''^2$,

and (if $AM_1 > AM$) $2M_1H' . MH' > HH''^2$,

we prove in a similar manner, by substituting M for A and M_1 for M, that

$$PP''^2 + p^2 < P_1P_1''^2 + p_1^2.$$

(3) Let AA' be less than $\frac{1}{2}(AA' - p_a)^2$,

so that $$2AH'^2 < HH''^2.$$

Make $2M_0H''$ equal to HH''.

Now $$M_0H' : M_0H = P_0P_0' : p_0,$$ [Prop. 127]

so that $$P_0P_0''^2 = \frac{1}{2}(P_0P_0' - p_0)^2.$$

Next, if P be between A and P_0,

$$2M_0H''^2 = HH''^2,$$

and $$2M_0H' . MH' < HH''^2.$$

Adding $2MH . MH'$ to each side,

$$2(M_0H + MH') MH' < MH^2 + MH''^2,$$

and, exactly in the same way as before, we prove that

$$P_0P_0'^2 + p_0^2 < PP'^2 + p^2.$$

Again, if P_1 be between A and P,

$$2MH' . M_1H' < HH''^2,$$

whence (adding $2M_1H . M_1H'$)

$$2(MH + M_1H') M_1H' < M_1H^2 + M_1H''^2,$$

and, in the same way,

$$PP'^2 + p^2 < P_1P_1'^2 + p_1^2.$$

Similarly $$P_1P_1''^2 + p_1^2 < AA''^2 + p_a^2.$$

Lastly, if $$AM > AM_0,$$

$$2MH' . M_0H' > HH''^2,$$

and, if $$AM_1 > AM,$$

$$2M_1H' . MH' > HH''^2;$$

whence we derive in like manner that

$$PP'^2 + p^2 > P_0P_0'^2 + p_0^2,$$

$$P_1P_1''^2 + p_1^2 > PP''^2 + p^2,$$

and so on.

Proposition 145.

[VII. 47, 48.]

In an ellipse,

(1) *if* $AA''^2 \not> \frac{1}{2}(AA' + p_a)^2$, *then* $AA''^2 + p_a^2 < PP''^2 + p^2$, *and the latter increases as P moves away from A, reaching a* **maximum** *when P coincides with B;*

(2) *if* $AA''^2 > \frac{1}{2}(AA' + p_a)^2$, *then there will be on each side of the axis a diameter* P_0P_0' *such that* $P_0P_0''^2 = \frac{1}{2}(P_0P_0' + p_0)^2$, *and* $P_0P_0''^2 + p_0^2$ *will then be less than* $PP''^2 + p^2$ *in the same quadrant, while this latter increases as P moves from* P_0 *on either side.*

(1) Suppose $$AA''^2 \not> \frac{1}{2}(AA' + p_a)^2.$$

Now $$A'H.AH' : A'H^2 + A'H''^2 = AA''^2 : AA''^2 + p_a^2.$$

Also $$AA''^2 : BB''^2 = p_b : BB' = AA' : p_a = A'H : A'H'$$
$$= A'H.A'H' : A'H'^2,$$

and $$BB''^2 : (BB''^2 + p_b^2) = A'H'^2 : A'H^2 + A'H'^2;$$

hence, *ex aequali*,

$$AA''^2 : (BB''^2 + p_b^2) = A'H.A'H' : A'H^2 + A'H''^2,$$

and, as above,

$$AA''^2 : (AA''^2 + p_a^2) = A'H.AH' : A'H^2 + A'H''^2.$$

Again, $$AA''^2 \not> \frac{1}{2}(AA' + p_a)^2,$$

$$\therefore 2A'H.AH' \not> HH''^2,$$

whence $$2A'H.MH' < HH''^2.$$

Subtracting $2MH.MH'$, we have

$$2A'M.MH' < MH^2 + MH''^2 \ldots\ldots\ldots\ldots\ldots\ldots(1),$$

$$\therefore 2A'M.AM : 2A'M.MH', \text{ or } AM : MH',$$
$$> 2A'M.AM : MH^2 + MH''^2,$$

and, since $2A'M.AM + MH^2 + MH''^2 = A'H^2 + A'H''^2$,

we have, *componendo*,

$$AH' : MH' > A'H^2 + A'H'^2 : MH^2 + MH'^2,$$

$$\therefore A'H . AH' : A'H^2 + A'H'^2 > A'H . MH' : MH^2 + MH'^2,$$

whence $AA'^2 : (AA'^2 + p_a^2) > AA'^2 : (PP'^2 + p^2)$,

[Prop. 130 (5)]

or $$AA'^2 + p_a^2 < PP'^2 + p^2.$$

Again, either $MH < M_1H'$, or $MH \nless M_1H'$.

(*a*) Let $$MH < M_1H'.$$

Then $$MH^2 + MH'^2 > M_1H^2 + M_1H'^2,$$

and $$M_1H^2 + M_1H'^2 > M_1H' . 2(M_1H' - MH)^*;$$

$$\therefore MM_1 . 2(M_1H' - MH) : M_1H' . 2(M_1H' - MH), \text{ or } MM_1 : M_1H',$$
$$> MM_1 . 2(M_1H' - MH) : M_1H^2 + M_1H'^2.$$

But $MH^2 + MH'^2 - (M_1H^2 + M_1H'^2) = 2(CM^2 - CM_1^2)$;

$$\therefore MM_1 . 2(M_1H' - MH) + M_1H^2 + M_1H'^2 = MH^2 + MH'^2;$$

thus, *componendo*, we have

$$MH' : M_1H' > MH^2 + MH'^2 : M_1H^2 + M_1H'^2;$$

therefore, alternately,

$$A'H . MH' : MH^2 + MH'^2 > A'H . M_1H' : M_1H^2 + M_1H'^2,$$

and $AA'^2 : PP'^2 + p^2 > AA'^2 : P_1P_1'^2 + p_1^2$, [Prop. 130 (5)]

so that $$PP'^2 + p^2 < P_1P_1'^2 + p_1^2.$$

(*b*) If $$MH \nless M_1H',$$

$$MH^2 + MH'^2 \ngtr M_1H^2 + M_1H'^2,$$

and it results, in the same way as before, that

$$A'H . MH' : MH^2 + MH'^2 > A'H . M_1H' : M_1H^2 + M_1H'^2,$$

and $$PP'^2 + p^2 < P_1P_1'^2 + p_1^2.$$

Lastly, since

$$A'H . A'H' : A'H^2 + A'H'^2 = AA'^2 : BB'^2 + p_b^2,$$

and $A'H . M_1H' : M_1H^2 + M_1H'^2 = AA'^2 : P_1P_1'^2 + p_1^2$,

* As in (1) above,

$$M_1H^2 + M_1H'^2 > 2A'M_1 . M_1H'$$
$$> M_1H' . 2(M_1H' - A'H')$$
$$> M_1H' . 2(M_1H' - MH), \textit{ a fortiori.}$$

it is shown in the same manner that

$$P_1P_1'^2 + p_1^2 < BB'^2 + p_b^2.$$

(2) Suppose $AA'^2 > \frac{1}{4}(AA' + p_a)^2$,

so that $2AH'^2 > HH'^2$.

Make $2M_0H'^2$ equal to HH'^2, so that

$$M_0H'^2 = \tfrac{1}{2}HH'^2 = HH' . CH';$$

$$\therefore HH' : M_0H' = M_0H' : CH'$$

$$= HH' \sim M_0H' : M_0H' \sim CH',$$

whence $M_0H : CM_0 = HH' : M_0H'$,

and $HH' . CM_0 = M_0H . M_0H'$.

If then (a) $AM < AM_0$,

$$4CM_0 . CH' > 2MH . M_0H'.$$

Adding $2MM_0 . M_0H'$ to each side,

$$4CM_0 . CH' + 2MM_0 . M_0H' > 2M_0H . M_0H',$$

and again, adding $4CM_0^2$,

$$2(CM + CM_0)\, M_0H' > (M_0H^2 + M_0H'^2).$$

It follows that

$$2(CM + CM_0)\, MM_0 : 2(CM + CM_0)\, M_0H', \text{ or } MM_0 : M_0H',$$

$$< 2(CM + CM_0)\, MM_0 : (M_0H^2 + M_0H'^2).$$

Now $2(CM + CM_0)\, MM_0 + M_0H^2 + M_0H'^2$

$$= MH^2 + MH'^2,$$

so that, *componendo*,

$$MH' : M_0H' < MH^2 + MH'^2 : M_0H^2 + M_0H'^2,$$

and

$$A'H . MH' : MH^2 + MH'^2 < A'H . M_0H' : M_0H^2 + M_0H'^2,$$

whence $P_0P_0'^2 + p_0^2 < PP'^2 + p^2$.

Similarly, if $AM_1 < AM$,

$$2HH' . CM > 2M_1H . MH',$$

and we prove, in the same manner as above,

$$PP'^2 + p^2 < P_1P_1'^2 + p_1^2.$$

And, since $2HH'.CM_1 > 2AH.M_1H'$,
in like manner

$$P_1P_1'^2 + p_1^2 < AA'^2 + p_a^2.$$

Lastly (*b*), if $AM > AM_0$, the same method of proof gives

$$P_0P_0'^2 + p_0^2 < PP'^2 + p^2,$$

etc.

Proposition 146.

[VII. 49, 50.]

In a hyperbola,

(1) *if* $AA' > p_a$, *then*

$AA'^2 \sim p_a^2 < PP'^2 \sim p^2$, *where PP' is any diameter, and* $PP'^2 \sim p^2$ *increases as P moves further from A;*

also $PP'^2 \sim p^2 > AA'^2 \sim p_a . AA'$ *but* $< 2(AA'^2 \sim p_a . AA')$:

(2) *if* $AA' < p_a$, *then*

$AA'^2 \sim p_a^2 > PP'^2 \sim p^2$, *which diminishes as P moves away from A;*

also $PP'^2 \sim p^2 > 2(AA'^2 \sim p_a . AA')$.

(1) As usual, $A'H : AH = AH' : A'H' = AA' : p_a$;

$$\therefore A'H.AH' : AH'^2 \sim AH^2 = AA'^2 : AA'^2 \sim p_a^2.$$

Now $MH' : AH' < MH : AH$;

$$\therefore MH' : AH' < MH' + MH : AH' + AH$$

$$< (MH' + MH)HH' : (AH' + AH)HH',$$

i.e. $$< MH'^2 \sim MH^2 : AH'^2 \sim AH^2.$$

Hence

$$A'H.MH' : MH'^2 \sim MH^2 < A'H.AH' : AH'^2 \sim AH^2;$$

$$\therefore AA'^2 : PP'^2 \sim p^2 < AA'^2 : AA'^2 \sim p_a^2, \text{ [Prop. 130 (5)]}$$

or $$AA'^2 \sim p_a^2 < PP'^2 \sim p^2.$$

Again, if $AM_1 > AM$,

$$M_1H' : MH' < M_1H : MH;$$

$$\therefore M_1H' : MH' < M_1H' + M_1H : MH' + MH,$$

and, proceeding as before, we find

$$PP'^2 \sim p^2 < P_1P_1'^2 \sim p_1^2,$$

and so on.

Now, if PO be measured along PP' equal to p,

$$PP'^2 \sim p^2 = 2PO \,.\, OP' + OP'^2;$$

$$\therefore\ PP'^2 \sim p^2 > PP' \,.\, OP' \text{ but } < 2PP' \,.\, OP'.$$

But $$PP' \,.\, OP' = PP'^2 - PP' \,.\, PO$$
$$= PP'^2 - p \,.\, PP'$$
$$= AA'^2 - p_a \,.\, AA'; \qquad \text{[Prop. 129]}$$

$$\therefore\ PP'^2 \sim p^2 > AA'^2 \sim p_a \,.\, AA' \text{ but } < 2\,(AA'^2 \sim p_a \,.\, AA').$$

(2) If $AA' < p_a$,

$$MH' : AH' > MH : AH;$$

$$\therefore\ MH' : AH' > MH' + MH : AH' + AH,$$

and

$$A'H \,.\, MH' : A'H \,.\, AH' > (MH' + MH)\,HH' : (AH' + AH)\,HH',$$

i.e. $$> MH'^2 \sim MH^2 : AH'^2 \sim AH^2.$$

Therefore, proceeding as above, we find in this case

$$PP'^2 \sim p^2 < AA'^2 \sim p_a^2.$$

Similarly

$$P_1P_1'^2 \sim p_1^2 < PP'^2 \sim p^2,$$

and so on.

Lastly, if PP' be produced to O so that $PO = p$,

$$AA'^2 \sim p_a \,.\, AA' = PP'^2 \sim p \,.\, PP' \qquad \text{[Prop. 129]}$$
$$= PP' \,.\, OP'.$$

And $$PP'^2 \sim p^2 = PP'^2 \sim PO^2$$
$$= 2PP' \,.\, P'O + P'O^2$$
$$> 2PP' \,.\, OP'$$

or $$> 2\,(AA'^2 \sim p_a \,.\, AA').$$

Proposition 147.

[VII. 51.]

In an ellipse,

(1) *if PP' be any diameter such that $PP' > p$,*

$$AA'^2 \sim p_a^2 > PP'^2 \sim p^2,$$

and $PP'^2 \sim p^2$ diminishes as P moves further from A;

(2) *if PP' be any diameter such that $PP' < p$,*

$$BB'^2 \sim p_b^2 > PP'^2 \sim p^2,$$

and $PP'^2 \sim p^2$ diminishes as P moves further from B.

(1) In this case (using the figure of Prop. 141)

$$AH' : MH' < AC : CM$$

$$\therefore A'H \cdot AH' : A'H \cdot MH' < 2HH' \cdot AC : 2HH' \cdot CM$$

i.e. $$< AH'^2 \sim AH^2 : MH'^2 \sim MH^2.$$

Therefore, alternately,

$$A'H \cdot AH' : AH'^2 \sim AH^2 < A'H \cdot MH' : MH'^2 \sim MH^2.$$

Hence

$$AA'^2 : AA'^2 \sim p_a^2 < AA'^2 : PP'^2 \sim p^2, \text{ [Prop. 130 (5)]}$$

and $$AA'^2 \sim p_a^2 > PP'^2 \sim p^2.$$

Also, if $AM_1 > AM$, we shall have in the same way

$$A'H \cdot MH' : A'H \cdot M_1H' < MH'^2 \sim MH^2 : M_1H'^2 \sim M_1H^2,$$

and therefore $PP'^2 \sim p^2 > P_1P_1'^2 \sim p_1^2$, and so on.

(2) P must in this case lie between B and the extremity of either of the equal conjugate diameters, and M will lie between C and A' if P is on the quadrant AB.

Then, if M_1 corresponds to another point P_1, and $AM_1 > AM$, we have

$$MH' > M_1H', \text{ and } CM < CM_1;$$

$$\therefore A'H.MH' : A'H.M_1H' > CM : CM_1$$

$$> 2CM.HH' : 2CM_1.HH',$$

i.e. $$> MH^2 \sim MH''^2 : M_1H^2 \sim M_1H''^2,$$

whence, in the same manner, we prove

$$PP''^2 \sim p^2 < P_1P_1''^2 \sim p_1^2;$$

and $PP''^2 \sim p^2$ increases as P moves nearer to B, being a *maximum* when P coincides with B.

Printed in Great Britain
by Amazon